The Urban Book Series

Igor Vojnovic, Department of Geography, Michigan State University, East Lansing, MI, USA

Claudia van der Laag, Oslo, Norway

Qunshan Zhao, School of Social and Political Sciences, University of Glasgow, Glasgow, UK

The Urban Book Series is a resource for urban studies and geography research worldwide. It provides a unique and innovative resource for the latest developments in the field, nurturing a comprehensive and encompassing publication venue for urban studies, urban geography, planning and regional development.

The series publishes peer-reviewed volumes related to urbanization, sustainability, urban environments, sustainable urbanism, governance, globalization, urban and sustainable development, spatial and area studies, urban management, transport systems, urban infrastructure, urban dynamics, green cities and urban landscapes. It also invites research which documents urbanization processes and urban dynamics on a national, regional and local level, welcoming case studies, as well as comparative and applied research.

The series will appeal to urbanists, geographers, planners, engineers, architects, policy makers, and to all of those interested in a wide-ranging overview of contemporary urban studies and innovations in the field. It accepts monographs, edited volumes and textbooks.

Indexed by Scopus.

Enrique Navarro-Jurado ·
Remedios Larrubia Vargas ·
Fernando Almeida-García · Juan José Natera Rivas
Editors

Urban Dynamics in the Post-pandemic Period

Tourist Spaces and Urban Centres

Springer

Editors
Enrique Navarro-Jurado
Department of Geography
University of Malaga
Malaga, Spain

Remedios Larrubia Vargas
Department of Geography
University of Malaga
Malaga, Spain

Fernando Almeida-García
Department of Geography
University of Malaga
Malaga, Spain

Juan José Natera Rivas
Department of Geography
University of Malaga
Malaga, Spain

ISSN 2365-757X ISSN 2365-7588 (electronic)
The Urban Book Series
ISBN 978-3-031-36019-0 ISBN 978-3-031-36017-6 (eBook)
https://doi.org/10.1007/978-3-031-36017-6

This Springer imprint is published by the registered company Springer Nature Switzerland AG
The registered company address is: Gewerbestrasse 11, 6330 Cham, Switzerland

About This Book

This book analyses the transformations of urban centres and the changes in urban-tourist spaces on the coast and in cities in this post-pandemic period.

In the twenty-first century, the city and the urban space are the main protagonists of social, economic and environmental changes. One of these changes is tourism development which, on the one hand, is beginning to transform social relations, generating other dynamics in the urban centres of cities, and on the other, it is creating new urbanised spaces in coastal areas, without creating a city. At the same time, urban centres are the most emblematic, symbolic and distinctive places in cities, and where urban transformations are concentrated with processes of gentrification, new commercial typologies, transformations of the housing stock, changes in the functionality of public spaces, among others.

The pandemic generated by COVID-19 has accelerated the trends that have been investigated for years in urban space and tourism. This context of uncertainty does not start with COVID-19 in 2020. The 2010s have seen the consolidation of phenomena and processes on a planetary scale. The euphoria of the tourism sector, with record numbers of tourists, is concentrated in cities and coastal areas, with a growth in accommodation where, in addition to hotels, tourist accommodation has exploded, generating conflicts in the urban centres of cities. At the same time, real estate reactivation has developed after the 2008 crisis, and in some countries, such as the USA, Spain and Germany, difficulties in access to housing have been diagnosed due to the change in accommodation consumption habits on the part of tourists, associated with the platform economy. At the same time, there is an increase in new agents developing urban projects with the internationalisation of the real estate market and irrigation capital companies. To this are added urban gentrification strategies, increasing inequalities, public investment policies for the rehabilitation and restructuring of cities, and social conflicts that have turned into social outbursts expressing the non-conformity of citizens, which we could define as the "city of protest".

The chapters of the book are grouped into 3 blocks: tourist dynamics in urban centres; urban-tourist spaces on the coast with new urbanised spaces; and new urban dynamics, conflicts and urban challenges. The case studies focus mainly on Spain, together with cities in Europe and Latin America. As a postscript, it reflects on the social and territorial phenomena at intra-urban borders in Latin American border cities.

Contents

About the Editors

Enrique Navarro-Jurado is Ph.D. in Geography, Director of the University Institute for Research in Tourism Innovation at the University of Malaga. The focus of his research is on tourist and urban spaces, planning and management of tourist destinations, limits to growth and tourist carrying capacity, sustainability indicators and technologies applied to sustainability, tourism, urban territory and climate change. He is Principal Investigator of several R+D+i projects and has participated in more than 30 research projects (Brazil, Dominican Republic, Cuba, Mexico, Argentina...) and in technology transfer contracts. With more than 40 publications in journals of international impact, he is Member of several boards of trustees, scientific associations, journal editor and Advisor to the Blue Plan (UNEP-UN) and various regional and local plans.

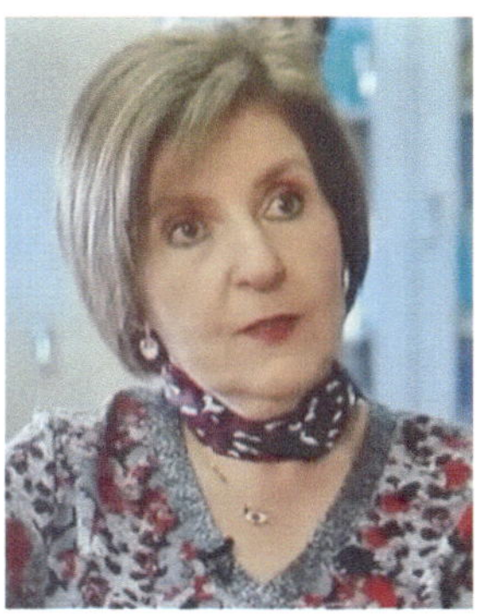

Remedios Larrubia Vargas is Ph.D. in Geography from the University of Malaga (2011). She is Professor of Human Geography at the University of Malaga, where she teaches in the subjects of Fundamentals of Human Geography and develops research topics related to agricultural marketing, recent transformations of rural space (rural development, labour market, social transfers), organic farming, cutting-edge crops, productive innovations; Welfare State Services (social services to the population, social policies and welfare state, tourism); Population-immigration.

Fernando Almeida-García is full Professor at the University of Malaga, Spain. He teaches at the Faculty of Tourism and the Faculty of Social Sciences at the University of Malaga. His research fields are tourism policy, management destination, residents' attitudes to tourism and tourism and East Asia. He has published articles in main international tourism journals and collaborated in international projects with European, Latin American and East Asia universities. He was Senior Researcher of the research group "Land and Tourism". He is currently Director of the Master in Tourism Planning at the University of Malaga, and he is studying the attitudes of residents in mass destinations, in historical destinations and the resident's attachment to places.

Juan José Natera Rivas is Senior Lecturer of Human Geography at the University of Malaga working on residential segregation in middle-sized cities in Spain and Argentina, and on the impact of migration on rural populations.

Part I
Tourism Dynamics in Urban City Centres

Chapter 1
Towards Resilient Urbanism in Tourist Cities: Post-pandemic Challenges

Irene N-Franco and Concepción Foronda-Robles

Abstract The objective of this study is to explore urban trends after the COVID-19 period in terms of safety, proximity, green spaces and/or socialisation, taking the cities of Paris, Milan and Barcelona as case studies. The methodology employed is based on the use of qualitative and quantitative techniques where, firstly, the urban planning actions developed during the pandemic context and aimed at developing a more resilient urban environment have been analysed. Subsequently, the relationships between these actions were analysed using UciNet 6.743 Social Network Analysis Software (SNA) to determine the connections and networks established between them. The results obtained demonstrate not only the links between the different areas of action (public space, mobility, facilities and services, housing), but also that COVID-19 has accelerated the trend of urban planning towards resilient urbanism to create more sustainable and healthy cities to address post-pandemic challenges such as adaptation to climate change or the emergence of future diseases.

Keywords Resilient urbanism · Chronourbanisme · Public space · Mobility · Facilities and services · Housing

1.1 Background

COVID-19 spread at breakneck speed, in late 2019, resulting in a global pandemic. Cities, being in continuous urban sprawl, axes of political and economic power, hubs of commercial exchange and people flow, as well as hyperconnected global networks and centres of tourist attraction, have been placed at the epicentre of this disease.

In this perspective, the dilemma of what future cities and lifestyles should be like, which are directly affected by the current context derived from the pandemic, is once

I. N-Franco (✉) · C. Foronda-Robles
Department of Human Geography, University of Seville, Seville, Spain
e-mail: infranco@us.es

C. Foronda-Robles
e-mail: foronda@us.es

E. Navarro-Jurado et al. (eds.), *Urban Dynamics in the Post-pandemic Period*,
The Urban Book Series,
https://doi.org/10.1007/978-3-031-36017-6_1

again posed (Wang and Tang 2020). Such a situation forces a reconsideration of long-held beliefs about the good shape of the city and the purpose of urban planning, reopening the debate on compact development versus sprawl (Badger 2020). In this sense, social distancing has been shown to help prevent the transmission of COVID-19, although it is also difficult to apply to compact urban spaces, so in order to achieve more open public spaces, it is necessary to change the design and use of the urban fabric, allowing for flexibility and variability of use (Abusaada and Elshater 2020).

Emphasis has also been placed on the need to have essential basic services close to residential areas. A compact city model based on proximity translates into easier access to these services through shorter trips and by sustainable means (Sisson 2020). This implies configuring a city on a human scale through neighbourhoods connected by quality public transport and pedestrian and cycling infrastructure. In this way, the aim is to break the urban paradigms that have traditionally predominated, where the city has been functionally fragmented (shops, industry, residence and leisure) and its morphology adapted to private motorised vehicles (Paquot 2021).

Consequently, a number of cities are implementing urban models based on chronourbanism (Boucher 2020), which refers to urban planning that prioritises the spatio-temporal relationship and underlines territorial inequality at the individual level (López 2015). Examples are: vital neighbourhoods (Bogotá), superblocks or superilles (Barcelona), 15-minute neighbourhoods (Paris, Milan) or 20-minute neighbourhoods (Buenos Aires, Melbourne, Portland, Ottawa). In this sense, the aim is to create an urban future that is more resilient and resistant to climate change, health and economic crises, boosting proximity and reducing displacement while helping to mitigate pre-existing inequalities with respect to the essential elements of the neighbourhood (Gower and Grodach 2022).

In the case of Europe, public administrations have been forced to face this new challenge and to rethink the city model implemented and inherited from the different stages of this historical context (Kunzmann 2020). In the past, the Industrial Revolution changed the shape of medieval cities, enlarging them and endowing them with large and heterogeneous populations that brought about demographic changes that resulted in overcrowding and migration, thus favouring the transmission of infectious diseases (Hall 1997). In this context, urban planning emerged as a discipline to deal with such diseases, as one of the main means to prevent their spread (Lynch 1981).

Today, as in previous centuries, urban planning has to be considered as one of the key axes in dealing with possible resurgences of COVID-19 or the emergence of other future diseases. Guiding urban transformation through the resilient approach would, among other objectives, aim to achieve sustainable urban planning and foster a methodological-normative framework that contributes to better self-management and governance of the city. This would translate into greater responsiveness to change (Jabareen and Eizenberg 2021). Indeed, the UN-Habitat (2021) report "Cities and Pandemics: Towards a More Just, Green and Healthy Future", based on the documentation of 1,700 cities and a comprehensive analysis of their policies, demonstrates how cities have the capacity to reduce the impact of future pandemics and to become more equitable, healthy and sustainable.

Against this backdrop, it should be remembered that tourism is subsumed and embedded in the postmodern city, and while it may be a dominant element in those localities that actively promote their tourist attractions, it is an integrated aspect of the city's morphology and functionality (Taylor and Hochuli 2015). In this way, tourism activity complicates policy decisions on issues of urban space, having to deal with conflicts related to the privatisation of public space or its influence on housing and the geography of labour and workers, as well as the distribution and functioning of local businesses (Alraouf 2021). Furthermore, although in some cases easily recognisable, the tourist city is not necessarily spatially distinct, as there are spatially dispersed consumption experiences that are often grouped into districts and zones, which have their own symbols and a wide range of icons that differentiate them from other urban spaces (Freytag and Bauder 2018).

In recent decades, managing sustainable urban tourism has been a challenge for local governments, residents and tourists. Consequently, the rise of urban tourism has brought with it an unprecedented concentration of tourist activity, where urban spaces have undergone complex transformations resulting from overtourism or gentrification, leading to their functional transformation and touristification (De la Calle-Vaquero 2019).

The pandemic has given rise to an urban crisis, a time of unprecedented change and uncertainty in which tourism and urban space are in the spotlight and where urban planning is fundamental (Alraouf 2021). In this sense, the aim of the chapter is to explore urban trends after COVID-19 in terms of security, proximity, green spaces and/or socialisation. To this end, two specific objectives are set out: (1) to analyse the urban planning discourse and establish relationships between the main areas of action (public space, mobility, facilities and services, and housing); and (2) to analyse as case studies major tourist cities, Paris, Barcelona and Milan.

1.2 Methodology

Firstly, an analysis has been carried out of the urban planning actions developed during the COVID-19 pandemic and aimed at creating an urban environment that is more resilient to the emergence of future diseases. Many of the actions had been initiated or planned prior to the outbreak of the disease but have been accentuated after its emergence. In the case of both Paris and Milan, the actions proposed under the 15-minute city approach are taken into consideration. Regarding the former, the measures contemplated in the Local Urban Plan, projected in the programmatic document Planning and Sustainable Development Project (City of Paris 2018), have been considered. As for the second, the Milan 2020 Adaptation Strategy (Comune di Milano 2020) has been considered. In the city of Barcelona, the actions revolving around the "Superilles Programme", included in the Sustainable Urban Mobility Plan (SUMP) 2013–2018 and in the SUMP 2019–2024, are analysed. The Paris and Barcelona plans were previously proposed to COVID-19 and have undergone modifications to streamline procedures and adapt to the pandemic situation. In the case of Milan, the approaches were designed during the pandemic period.

The planning instruments have been classified into four main areas of action: (1) Public space (PS), dealing with all those actions of transformation, improvement, adaptation and regeneration of public urban space; (2) Mobility (M), in relation to the different forms of movement and transit in cities; (3) Facilities and services (Fs), which refers to the uses and functionalities to satisfy needs (shops, education, health, etc.); and (4) Housing (H), which encompasses specific actions of the urban residential aspect.

Twenty-one actions were analysed for Paris, 22 for Milan and 25 for Barcelona, all of which are classified, coded and detailed in Annex. It should be noted that, although they are presented together, each of these actions addresses different issues and has no connection between them, even though they share the same code.

In order to consider the relationships between the urban planning actions and the proposed classification areas of Paris, Milan and Barcelona, the relationships between the different nodes previously generated in the matrix analysis are analysed, using the UciNet 6 version 6.742 Social Network Analysis (SNA) software (Borgatti et al. 2002).

1.3 Results

1.3.1 Paris and Milan, 15-Minute Cities

In Paris, the central idea of public space (PS) revolves around its reorganisation and reuse through tactical planning. The aim is to increase space for pedestrians and cyclists, as well as to make the use of space more natural, safe and flexible, allowing for issues such as social distancing and the creation of healthier urban environments.

Furthermore, it is committed to "ecomobility" (M) centred on the intermodal and fluid use of walking, cycling and public transport. The actions related to this correspond to the prioritisation of collective public transport over private vehicles and the creation of pedestrian and cyclable spaces ("coronapistes"), as well as measures to regulate their flow and limit motorised traffic.

The actions of facilities and services (Fs) are aimed at satisfying the basic needs of citizens through proximity and the slowing down of the economy. Issues such as the promotion of teleworking to reduce commuting, the prioritisation of land use for security, health or local commerce, as well as the creation of green spaces for meeting and socialising, allow the practice of sports or children's leisure activities.

Housing (H) is the least addressed action during the pandemic. However, it is interesting to note the transformation of underused offices and business districts into residential spaces due to the increase in teleworking. Prior to the pandemic, the need for more social housing was already being considered. In the same way, the creation of socialisation spaces in the construction of new housing that would allow for greater social integration in the face of future confinements is also a priority.

Figure 1.1 shows the relationships between the different developments. There is a degree of centrality around pedestrianisation (PS4). This node marks the distribution of the rest of the actions, establishing dyadic relationships that form a network with an egocentric tendency. It can be affirmed that the transformation of public space in Paris is largely influenced by the processes of pedestrianisation, conditioning the rest of the actions under this measure. Likewise, there is a certain centrality with the meeting and socialisation spaces (Fs3). Although there is a distance between both central nodes, they are indirectly linked to public spaces (PS5) and facilities and services (Fs2). It is possible to see the close relationship between these two areas, so it can be inferred that the degree of centralisation is high given that the structure of the network is concentrated around these nodes.

In the case of mobility, the distribution is more dispersed and transversal, with proximity to public space and facilities/services, acting as a connecting axis for links with other actions (M2-Fs2-EQ1, M5-PS2-PS3-PS4). In the same way, the coronapistes (M3) and the regulation of pedestrian flows (M5) are integrated into this centrality.

Finally, urban housing actions are scarce and poorly connected. In fact, the transformation of offices and business districts (H1) and the increase in social housing (H2) are only linked to the network through the promotion of teleworking (Fs1) which, in turn, is linked to the reduction of individual commuting (M2) and the reservation and prioritisation of land use for security, health and local commerce (Fs2). In the case of the provision of socialisation spaces in the construction of new housing (H3), it is more integrated, establishing links with equipment and service

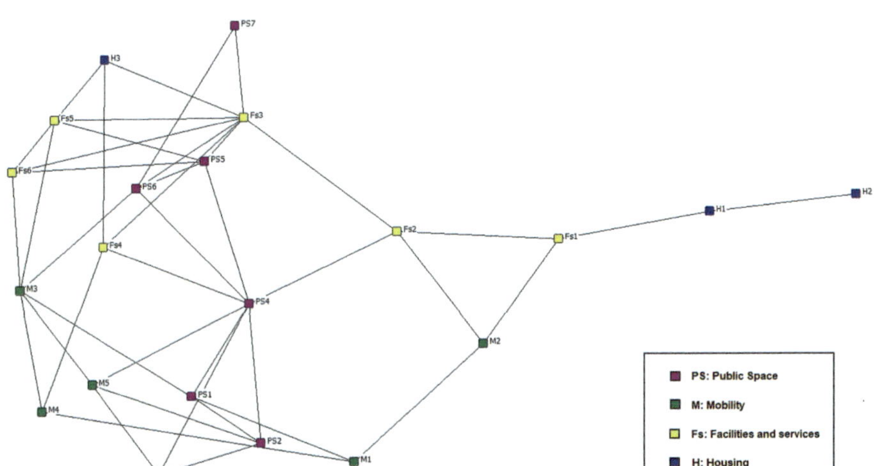

Fig. 1.1 Network analysis of urban planning actions in Paris in the face of the pandemic challenge. *Source* Own elaboration

actions that encourage sociability and interactions between citizens (Fs3, Fs4 and Fs5).

In Milan, PS measures aim to promote proximity through active and sustainable mobility, to the detriment of the use of private vehicles. They also advocate the extension of pedestrianisation processes and the improvement and adaptation of space (green and safe areas) to pedestrian and cycling traffic through car restrictions such as lane reduction, the reuse of car parks or the creation of a network of pedestrian and cycle paths.

In mobility, Paris is committed to sustainability with the implementation of zones where it is not possible to drive faster than 30 km/h, as well as the diversification of public transport and intermodality where active means of transport such as walking and cycling are combined, adding the use of scooters. Milan also aims to promote the use of VTC and taxis for essential services and certain groups such as the elderly who have greater difficulties in using intermodal transport.

The Fs measures focus on promoting the proximity of basic services by encouraging teleworking and adding, in this case, an increase in co-working spaces. They also focus on the desynchronisation of timetables and the spatial redistribution of public services, shops and productive activities. Likewise, the measures focus on meeting the needs of the elderly and children, through the redesign and adaptation of services and the creation of play and leisure spaces, especially in the summer months.

Other proposals are to provide a more equitable housing market (H) through the activation of empty properties and their prioritisation for social renting, as well as the climate-proofing of newly built housing and the use of renewable energies, as well as their capacity to include teleworking.

Figure 1.2 shows the urban relationships in Milan. At first glance, it is similar in shape to the Paris network, although a closer look reveals that the number of relationships is much higher. In general terms, there is no central node or nodes that condition the distribution of the rest of the relationships, so the degree of internal cohesion is quite high and transversality in the network is more present than centrality. In relation to the latter, it is worth highlighting the high degree of interrelation between the variables of the PS, M and Ps spheres, especially about the actions for the transformation of space towards pedestrianisation and cycling (PS1, PS2, PS3, PS4), sustainable mobility measures (M1, M2, M4, M5) and redistribution (Fs3) and the redesign of public, commercial and cultural services (Fs5).

Despite the high number of links with bidirectional flows present in the network, there are some nodes that present a lower number of relations, and which correspond to the reorganisation (Fs2) and reuse (Fs6) of public facilities and services, aimed at promoting their proximity to the resident population. The same applies to the promotion of teleworking and co-working (Fs1) which, as well as being related to the measures that follow this line, is also linked to the aforementioned mobility actions. Furthermore, it acts as a link with the only cluster in the network, which corresponds to the four measures in area V related to encouraging social housing and adapting residential buildings both to climate change and to the new working formulas that have been encouraged after COVID-19.

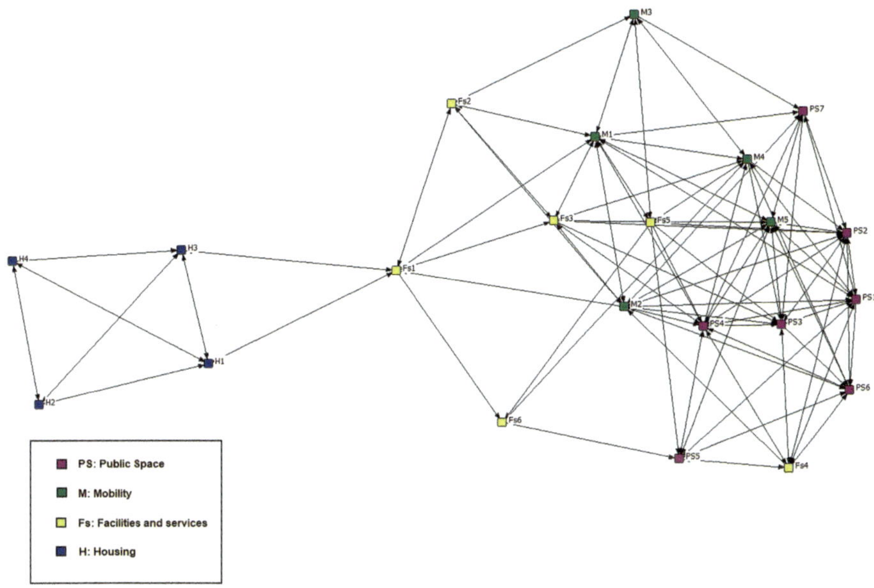

Fig. 1.2 Network analysis of Milan's urban planning actions in the face of the pandemic challenge. *Source* Own elaboration

1.3.2 Barcelona and the Superilles

The Superilles programme aims to transform the public space (PS) into a safer, more ecological, closer, healthier and more social space. The actions are framed by the pedestrianisation of large spaces (superblocks), providing them with new urban furniture and creating new green axes and squares to promote urban sociability and develop more ecological and safer environments for pedestrian and cyclist traffic.

Mobility (M) is focused on reducing car use and promoting public transport combined with individual transport through cycling and walking. In this sense, it is proposed to limit motorised speed, create cycle lanes and pedestrian paths and extend tram, bus and metro services.

The facilities and services (Fs) focus on proximity, sociability and the enhancement of urban culture, and the creation of spaces that enhance the value of Barcelona's architectural heritage, which forms part of its cultural identity. Similarly, in order to reduce individual journeys and meet the needs of residents, local commerce is encouraged, allowing suppliers to load and unload and limiting the use of private vehicles for other purposes.

The actions for housing are based on criteria of health, sustainability, and sociability, and are more numerous than in Paris and Milan. In this sense, solutions are proposed to adapt housing to the effects of climate change and to establish the regulation of the intermediate space between housing and the street for community use, reception of goods or collective storage, as well as the introduction of community and flexible spaces within residential buildings that contribute to weaving neighbourhood networks.

In Fig. 1.3, Barcelona presents clear differences with the network of the other tourist cities. Firstly, there is a greater propensity to group nodes into clearly segregated clusters, which are related to each other with some intermediate connection. There is a degree of centrality around the public space cluster, mainly regarding the pedestrianisation of large blocks (PS4) which, as in Paris, is the central axis from which most of the actions start, creating a network of dyadic links starting from this area (PS1, PS2, PS3, PS6 and PS7).

Likewise, PS4 is the linking node between public space and mobility, through the reduction of speed to give priority to pedestrians (M1), and also, with facilities and services through the creation of meeting and recreational streets (Fs2). In relation to this last area, it should be noted that its nodes are very close and linked to those corresponding to the public space actions. In the same way, these nodes (Fs1, Fs2 and Fs4) are connected and linked to the network of housing actions to promote sociability (H5 and H6).

As for mobility, this area generates its own cluster through nodes that only establish links between them, in a more or less balanced way. The integration of the mobility actions in the network is established through the aforementioned M1 and the permission of loading and unloading for local commerce (Fs3) which in turn is

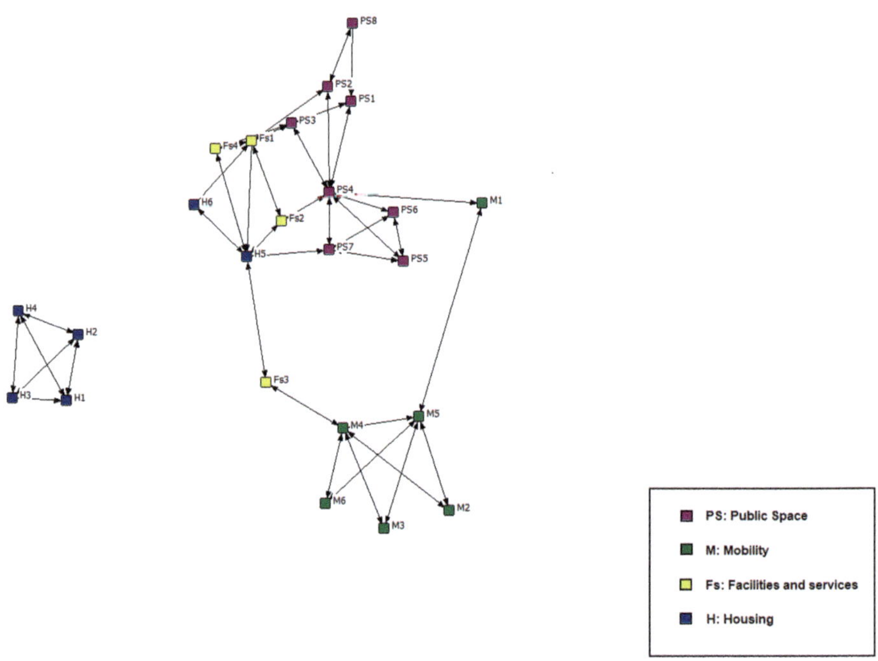

Fig. 1.3 Network analysis of Barcelona's urban planning actions in the face of the pandemic challenge. *Source* Own elaboration

related to the regulation of the intermediate space with housing (H5), which acts as a connector to the network.

Finally, very similar to Milan, it is remarkable the presence of an independent cluster (H1, H2, H3 and H4). This is because such rehabilitation, transformation and endowment measures are exclusive to residential buildings, so although they are intended to contribute to a healthier and more sustainable urban environment, they are not directly related to issues of mobility, public space and facilities and services.

1.4 Conclusions

The pandemic has accelerated the change in the urban planning paradigm, especially in terms of proximity and sustainable mobility. Both the city of Paris and Barcelona are references in urban planning issues and their urban models are the result of important processes developed in previous centuries (Hausmanisation and Plan Cerdà, respectively) and which were replicated in other cities (Jordan 2015; Martí 1999). In the case of the first, it is betting on a fairer and more sustainable development model that revolves around the "15-minute city", an ambitious renovation based on a political, theoretical and technical framework that reverses the traditional central model with a strong ecological focus, which is materialised in the progressive decrease in car use to give way to walking and cycling (Moreno et al. 2021). In the case of the second, it has accelerated the implementation of a new urban model based on the "Superblocks" programme, which proposes the transformation of all the city's streets to recover and provide other uses for the space given over to the car. In both cases, the actions analysed were, for the most part, planned in the context prior to the pandemic and revolved around the adaptation of urban spaces to climate change. As the study demonstrates, COVID-19 has further emphasised the need to carry them out and implement them in the short term, acting as an accelerator of these emerging practices. Milan takes Paris as a reference point and after the pandemic context starts to develop the 15-minute city model, putting urban planning back in the spotlight as a tool to tackle the effects of diseases as it did in previous centuries.

The predominant areas of action in the three cities are changes and transformations in public space and mobility, aimed at reducing car use, increasing pedestrianisation and promoting walking and cycling. The aim is to create more sustainable, healthy and resilient urban environments under the criteria of proximity through urban planning models based on chronourbanism. As explained above, this is a global trend, however, after analysing the planning documents in this study, it should be noted that this is not an operationalised issue in terms of urban planning concepts.

As Gower and Grodach (2022) argue, the effective implementation of a chronourban model, be it 15, 20 or 30 minutes, requires a set of measurable and quantifiable benchmarks that can be adapted to any neighbourhood and city, whatever its characteristics. Although the research highlights that the three cities follow very similar trends based on proximity and sustainable mobility, the city models associated with chronourbanism must provide more precise metrics so that the measures implemented can be put into practice since, although there is a certain degree of coincidence, there is no definitive consensus, not even in the case of Paris and Milan, which follow the same model.

With regard to the analysis of networks, there are notable differences in terms of the number and distribution of connections, centrality and the greater or lesser presence of clusters. The morphology of the networks of Paris and Milan are similar, although the latter shows greater cohesion and transversality. In the case of Barcelona, it is the least integrated with a high tendency towards the creation of clusters. This analysis shows that, to a greater or lesser extent, these areas are related to each other and generate connections that reaffirm the need to address these issues from a cross-cutting and integrated perspective. Whatever the nature of the plan, the relationships between public spaces, mobility, as well as facilities and services to meet the needs of residents and housing, have to be taken into account, the latter being more isolated in all three cases.

Likewise, in relation to the need to provide urban planning with greater transversality and integrity, it should be noted that these new urban models of proximity are not taking urban tourism into account in their initial theoretical constructs, even though it is an activity that is fully integrated into urban life, coexisting directly with the resident (Freytag and Bauder 2018). This fact can make such approaches difficult, since, in the case of highly touristified spaces, this activity affects all the urban planning areas analysed and conflicts arise in issues such as the use of public space (overcrowding), increase in residential rental (tourist gentrification) or loss of traditional commerce (commercial gentrification).

Another aspect to take into account is that most of the actions that have been analysed are developed through tactical and ecosystemic urbanism, which are also setting the trend in urban planning. However, and especially in the case of the former, although during the pandemic it has been shown to be one of the most effective techniques for making the urban environment more flexible (Jabareen and Eizenberg 2021), it would be complicated to implement in heritage urban spaces that have higher levels of protection, which, moreover, in the case of European cities, usually correspond to historic centres which, in turn, are also usually the epicentres of urban tourism.

Finally, although the present study is exploratory, it could be carried out in other European urban spaces that are also developing resilient urbanism that started in the pre-pandemic period, but whose areas of action have accelerated in the current context, such as Berlin or London (Abusaada and Elshater 2020; Law et al. 2021; Jabareen and Eizenberg 2021).

Acknowledgements This chapter is part of the research project US-1381628 US/JUNTA/FEDER, UE.

Annex: Urban Developments in Paris, Milan and Barcelona for the COVID-19 Context

Scope	Actions	Paris	Barcelona	Milan
Public Space	PS1	Reduction of private vehicle lanes	New street furniture	Reduction of private vehicle lanes
	PS2	Widening of pavement (2.8 m)	Creation of green axes	Pavement widening and adaptation to fragile populations
	PS3	Security barriers	Creation of 2000 m2 squares	Barriers and night-time security measures
	PS4	Pedestrianisation (streets)	Pedestrianisation (blocks)	Pedestrianisation and creation of residential streets
	PS5	Renaturalisation	Single platform pavement	Increasing green public spaces
	PS6	Re-use of car parks	Use of panot and ecological granite	Re-use of car parks
	PS7	Extension of terraces in establishments	Widening of pavements (4 m)	Creation of a network of pedestrian and cycle paths
	PS8	–	Green spaces (14% on average)	–
Mobility	M1	Increase of bus, taxi lanes and priority vehicles	Speed reduction 10 km/h	Improving and diversifying public transport and intermodality
	M2	Reduced individual travel, increased collective travel	Creation of bus, taxi and bicycle lanes	Promotion of active transport (walking, cycling, scooters)
	M3	Coronapistes (cycle path)	Extension of tramway lines	Increase use of VTC and taxi for essential services
	M4	Creation of limited traffic areas	Construction of exchanger	Increasing delivery stops for goods and emergencies
	M5	Regulation of pedestrian flows	Improving air quality	Implementation of 30 km/h zones
	M6	–	Orthogonal bus systems	–
Facilities and services	Fs1	Promoting telework	Provision of green axes or squares (-200 m)	Promoting teleworking and co-working spaces

(continued)

(continued)

Scope	Actions	Paris	Barcelona	Milan
	Fs2	Land use prioritisation	Creation of meeting and recreational streets	Desynchronisation of timetables of essential services
	Fs3	Meeting and socialising spaces	Loading and unloading permit	Redistribution of public, commercial and cultural services
	Fs4	School streets	Urban integration and architectural heritage	Play streets
	EQ5	Healthy streets	–	Redesigning summer services (seniors)
	EQ6	Green micro-spaces	–	Re-use of schools in summer
Housing	H1	Residential office use	Building with double skin for climate improvement	Activation of empty flats and reuse of office space
	H2	Increase in social housing	Biomatic ventilation and sunshine	Prioritisation of renting at subsidised prices
	H3	Socialisation spaces in housing	Renewable energy sources	Climate adaptation of housing
	H4	–	Refurbishment energy efficiency	Adaptation of housing to telework
	H5	–	Regulation of the buffer space	–
	H6	–	Community and flexible spaces	–

References

Abusaada H, Elshater A (2020) COVID-19's challenges to urbanism: social distancing and the phenomenon of boredom in urban spaces. J Urban Int Res Placemaking Urban Sustain 15(2):258–260. https://doi.org/10.1080/17549175.2020.1842484

Alraouf AA (2021) The new normal or the forgotten normal: contesting COVID-19 impact on contemporary architecture and urbanism. Archnet-IJAR 15(1):167–188. https://doi.org/10.1108/ARCH-10-2020-0249

Ayuntamiento de Barcelona (2015) Programa Superilles de Barcelona. Recuperado de https://ajuntament.barcelona.cat/superilles/sites/default/files/Presentacio_SUPERILLA_BARCELONA.pdf

Ayuntamiento de París (2018) Proyecto de Ordenación y Desarrollo Sostenible (PADD). Recuperado de http://pluenligne.paris.fr/plu/sites-plu/site_statique_50/documents/1064_Plan_Local_d_Urbanisme_de_/1068_Projet_d_amenagement_et_de/C_PADD-V03.pdf

Badger, E. (24 de marzo de 2020). Density is normally good for us. That will be true after coronavirus, too. The New York Times. Retrieved from https://www.nytimes.com/2020/03/24/upshot/coronavirus-urban-density-risks.html

Borgatti SP, Everett MG, Freeman LC (2002) UCINET for Windows: software for social network analysis. Analytic Technologies, Harvard, MA

Boucher D (2020) Local living, rise of 20 minute cities post-Covid [online]. Urban Dev. Available from https://theurbandeveloper.com/articles/local-living-rise-of-20-minute-cities-post-covid

Comune di Milano (2020) Estrategia de adaptación Milano 2020

De la Calle-Vaquero M (2019) Turistificación de centros urbanos: clarificando el debate. Boletín de Asociación de Geógrafos Españoles (83), 2829:1–40. https://doi.org/10.21138/bage.282

Freytag T, Bauder M (2018) Bottom-up touristification and urban transformations in Paris. Tour Geogr 20(3):443–460. https://doi.org/10.1080/14616688.2018.1454504

Gower A, Grodach C (2022) Planning innovation or city branding? Exploring how cities operationalise the 20-minute neighbourhood concept. Urban Policy Res 40(1):36–52. https://doi.org/10.1080/08111146.2021.2019701

Hall T (1997) Planning Europe's capital cities: aspects of nineteenth-century urban development. Taylor & Francis e-Library.

Jabareen Y, Eizenberg E (2021) The failure of urban forms under the COVID-19 epidemic: towards a more just urbanism. Town Planning Rev 92(1):57–64. https://doi.org/10.3828/tpr.2020.42

Jordan DP (2015) Paris: Haussman and after. J Urban Hist 41(3):541–549. https://doi.org/10.1177/0096144215571567

Kunzmann K (2020) Smart cities after COVID-19: ten narratives. DisP-the Planning Rev 56(2):20–31. https://doi.org/10.1080/02513625.2020.1794120

Law L, Azzali S, Conejos S (2021) Planning for the temporary urbanism and public space in a time of Covid-19. Town Planning Rev 92(1):65–74. https://doi.org/10.3828/tpr.2020.48

López L (2015) Diccionario de Geografía aplicada y profesional. Universidad de León, España

Lynch K (1981) A theory of good city form. MIT Press, Cambridge, MA

Martí C (1999) Cerdà: Un puente entre dos civilizaciones. Ciudad y Territorio Estudios Territoriale XXXI:119–120. Recuperado de https://recyt.fecyt.es/index.php/CyTET/article/view/85561/62430

Moreno C, Allam Z, Chabaud D, Gall C, Pratlong F (2021) Introducing the "15-Minute City" sustainability, resilience and place identity in future post-pandemic cities. Smart Cities 4(1):93–111. https://doi.org/10.3390/smartcities4010006

ONU-Habitat (2021) Cities and pandemics: towards a more just, green and healthy future. United Nations Human Settlements Programme (UN-Habitat). Recuperado de https://unhabitat.org/sites/default/files/2021/03/cities_and_pandemicstowards_a_more_just_green_and_healthy_future_un-habitat_2021.pdf

Paquot T (2021) La ville du quart d'heure. Sprit 4:22–24. Recuperado de https://www.cairn.info/revue-esprit-2021-4-page-22.htm

Sisson P (2020) Mayors tout the "15-minute city" as Covid recovery—Bloomberg [online]. Bloom. CityLab. Available from https://www.bloomberg.com/news/articles/2020-07-15/mayors-tout-the-15-minute-city-as-covid-recovery?srnd=citylab

Taylor L, Hochuli DF (2015) Creating better cities: how biodiversity and ecosystem functioning enhance urban residents' wellbeing. Urban Ecosyst 18:747–762. https://doi.org/10.1007/s11252-014-0427-3

Wang Z, Tang K (2020) Combating COVID-19: health equity matters. Nat Med 26(4):458–464. https://doi.org/10.1038/s41591-020-0823-6

Chapter 2
The Pain of Being a Resident in Granada. Analysis of the Accommodation Offer and Residents' Perception

Francisco Antonio Navarro-Valverde, Alberto Capote-Lama, María Barrero-Rescalvo, and Ibán Díaz-Parra

Abstract The process of touristification of cities has intensified in the last decade and the interruption of the pandemic has not improved it. Many destinations perceived overtourism, understood as the negative impact of tourism in a destination where it is perceived as saturated and has an excessive and negative influence on the quality of life of citizens and/or on the quality of the visitor experience. This congestion is not only due to the number of visitors, but also to the capacity to manage them. The seasonality and concentration in certain areas exacerbate this perception. It is understood that this congestion is not just a tourism problem and that technological solutions alone do not solve the problem. In Granada, a historical city in Europe, residents reveal that, prior to the onset of the pandemic, there was a perception of tourism-phobia. However, research shows that the pandemic has not improved the management of the destination, but rather increased the perception of overcrowding. Residents see the tourism tax or the limitation of tourist homes marketed on platforms. The debate is of enormous complexity, and the solution in the immediate future is unclear.

Keywords Touristification · Overtourism · Airbnb · Sustainable tourism in urban areas · Granada

F. A. Navarro-Valverde (✉) · A. Capote-Lama
Department of Human Geography, University of Granada, 18071 Granada, Spain
e-mail: favalver@ugr.es

A. Capote-Lama
e-mail: alama@ugr.es

M. Barrero-Rescalvo · I. Díaz-Parra
Department of Human Geography, University of Sevilla, 41004 Seville, Spain
e-mail: mbrescalvo@us.es

I. Díaz-Parra
e-mail: ibandiaz@us.es

E. Navarro-Jurado et al. (eds.), *Urban Dynamics in the Post-pandemic Period*,
The Urban Book Series,
https://doi.org/10.1007/978-3-031-36017-6_2

2.1 Introduction

In 2018, the World Tourism Organization (WTO), after the observation of the serious problem in some historic centres of cities caused by the tourism activity, defined the problem of *overtourism*, and understanding this concept as "the impact of tourism on a destination or part of it, which is perceived, excessively and negatively, by the affected residents; and also, the citizens' quality of life, and/or the quality of visitors" (WTO 2018, 4). Also, the document pointed that tourism congestion is due to the number of visitors and the capacity to manage them; that it is aggravated by seasonality; the disproportionately negative impact of visitors; their concentration in specific spaces; and finally, that intelligent and technological solutions do not fully solve the problem.

The *overtourism* phenomenon has been studied in big Spanish tourist cities such as Barcelona (González-Reverté 2021; Pirillo-Ramos and Mundet 2021; among others), or Madrid (Ruano de la Fuente et al. 2019; among others). However, the scientific and mediatic attention in other medium cities, such as Granada, has been less analysed, despite for this city, this is becoming an increasing problem. This research seeks to show the evidence of *overtourism* in the city of Granada, noted by quantitative and qualitative data, for official statistics and key informants and residents, detecting concrete proofs, moments and places where this phenomenon is highly relevant. Therefore, is the *overtourism* a problem in Granada in the same magnitude as in other cities? Our hypothesis is that this process is evident and significant in this city, but it is focused on the historical centre.

2.2 Theoretical Framework

A growing number of researchers are working on *touristification*, *overtourism* and *tourism-phobia*. *Touristification* is understood as a complex process in which various stakeholders interfere, transforming a territory through tourist activity (Ojeda and Kieffer 2020), and having its highest level and perception through *overtourism*, a phenomenon linked with the development of low-cost flights, AirBnB (Butler and Dodds 2022). Thus, around 2016, Colomb and Novy (2016) showed in their work the growing position taken by residents and decision-makers in the face of a phenomenon of excess tourism in certain cities, such as Venice and Prague, and which causes an excess of negative impacts.

The following year, Goodwin (2017, 1) defined this process as one that "describes destinations where residents or guests perceive that there is an excess of tourists and that the quality of life in the area or the quality of the tourist experience, or both, have been unacceptably deteriorated". Milano (2017), on the other hand, conceptualized it that same year as a set of unsustainable mass tourism practices, leaving the activity out of control by the administrative authorities, and with a total absence of planning and management, being the maximum expression of the phenomenon the case of Venice.

Bobic and Akhavan (2022), for the cases of several Mediterranean cities, especially in Venetia, highlighted the necessity of establishing concrete bottom-up initiatives to tackle tourism gentrification. In the Venetian case, for example, was limited the arrival of cruisers to the city.

For Budapest, suffering the same process of *overtourism*, Pérez-Garrido et al. (2022) pointed out the high number and concentration of Airbnb apartments in the centre of the city, which is one of the leading causes, and the necessity of marketing policy to promote a sustainable tourism model.

In the case of Berlin, Wrede (2022) shows that short-term rentals are removing to long-term and permanent rentals in the Centre of the city, being the lists and offer of Airbnb flats excessively high in this part of German capital.

Other similar research for Poland (Rozmiarek et al. 2022), in the city of Poznan, reveals that the activity in the city started to show signals of *overtourism* in the local habitants: unfamiliar streets and traditional lifestyle lost in the centre of the city.

Almeida-Garcia et al. (2022), compared *tourism-phobia* and *gentrification* in Dansk (Poland) and Málaga (Spain), studied the impact of some tourism problems that affect residents (noise, dirt, occupation of public spaces, etc.), analysing the spatial distribution of tourism-phobia. Both historic centres are affected by the processes of increased tourist flows and the growth of new forms of tourism accommodation. The resident's annoyance caused by tourism gentrification is more intense than *tourism-phobia* and the inadequate management of these problems by public stakeholders.

For the Spanish situation, Ruano de la Fuente (2019, 31) specifies that "there is no precise diagnosis of the dimension of the problem, nor sufficient and systematic inspection or control activities: there are no studies that use indicators of the transformation of the social and business fabric (…); and there is no effective collaboration between Administrations to provide a joint response to the problem", so the activity is outside the control and regulation of the competent public administrations. López et al. (2019) noted in the case of Santiago de Compostela, several signals of this phenomenon: saturation in the streets; the loss of services and shops for residents; increasingly problematic access to housing; and *gentrification* and loss of identity of these areas.

One more step in this excessive tourist pressure, which shows the feeling of rejection and protest on the part of the residents, is that of *tourism-phobia*. Almeida-Garcia et al. (2019), Álvarez-Sousa (2018), Blanco-Romero et al. (2019), Escudero-Gómez (2019), González-Pérez (2019), Milano (2017), defined it as the social opposition towards tourist activity and a critical attitude towards its management. Blanco-Romero et al. (2019) also relate it to perceptions of *urban-philia*.

Almeida-Garcia et al. (2019) and González-Pérez (2019) highlighted the main causes of this rejection: the increase in real estate prices and speculation, even reaching evictions, the privatization of public spaces, a decrease in the purchasing power of residents, precarious employment in the tourism sector, the loss of traditional businesses compared to others oriented to foreign customers, the *gourmetization* of restaurants and food stores, an imbalance between the number of visitors and residents, excessive noise, an increase in the offer of accommodation via online platforms, hotels and tourist apartments and the symbolic and commercial transformation of the city.

The final consequence is the gentrification resulting from the displacement of the population with less purchasing power (Ardura-Urquiaga et al. 2019). Smith (2012) pointed out this phenomenon in the cases of Edinburgh or Harlem in New York. In New Orleans, Gotham (2005) observed a displacement of the original population due to the proliferation of entertainment and tourist places, with the resident population being replaced by a non-population. And Wachsmuth and Weisler (2018) showed short-term rentals, which significantly increased the rentals' prices.

In the case of Lisbon, Cocola-Gant and Gago (2021) highlighted that platforms such as Airbnb, instead of promoting the collaborative economy, are active for landlords and big companies, which can obtain massive benefits using short stays tenants.

Janoschka (2018), for the Spanish case, considered that hotel expansion, international residential demand, and the impact derived from the growth of tourist houses on platforms such as Airbnb, as highly relevant in tourism gentrification. For Seville, Díaz and Jover (2020) pointed out that gentrification and international tourism are linked and complementary phenomena, used to generate surplus value, deriving in a loss for residents of daily life in the historic centre, alienation of those spaces, diminishing the authenticity of the city, even for the residents. Urban centres become "thematic parks" (Cócola-Gant 2011), generating urban scenes where tourist elements predominate and generating landscapes of globalization, in its most banal, aesthetic and deteriorated vision (De la Calle-Vaquero 2019).

González-Reverte (2021), having Barcelona as a case study, pointed out that resident's opinion towards tourism is shaped by emotional response factors, identifying three types of residents: tourism supporters, who show a tourism-acceptance response; tourism-opposed residents, who show tourism-rejecting and tourism-phobic attitudes; and neutral residents, who show emotional indifference in their response. Crowding is the key factor explaining the residents' negative emotional responses towards tourism. For Pirillo-Ramos and Mundet (2021), in Barcelona, the negative impacts of tourism are creating a general malaise in the City, with social movements protesting and calling for a decrease in tourism numbers; and even the municipal government established forms of control over the sector, becoming the city a laboratory for pioneering measures to create a new model of tourist city.

For Granada, Smith (1996 [2012]) highlighted the phenomenon of tourist gentrification in the city; De la Calle-Vaquero (2002), already pointed to tourist specialization in the historic neighbourhoods. And Cerezo-Medina et al. (2021) pointed to the concentration of the tourist rental offer in the historical centre, similar to the spatial distribution of tourist accommodation and the high involvement of short-stay tourist apartment rentals. And finally, in 2022, Navarro Valverde et al. (2022) showed, according to the testimonies of different key actors, a clear overtourism process and vicious circle in the city's historical neighbourhoods.

The debate on sustainability, post-growth, the saturation of destinations, overtourism and the parenthesis caused by the COVID-19 virus could be an opportunity to rethink tourism activity and emerge from the crisis promoting a new tourism model that overcomes the difficulties and problems acquired by the previous uncontrolled growth (Palomo et al. 2020).

2.3 Methodology and Study Area

This chapter has as a key reference the contribution "Overtourism in Granada. Distribution and perception of the tourism activity", presented for the XVI Urban Geography Conference (Navarro Valverde et al. 2022). As the first step, to introduce the concrete topic, quantitative data extracted from the Register of Tourism for Andalusian Region, the number of tourist accommodations in Granada. The typologies of the accommodation considered in this study are houses for tourist use, hotels and tourist apartments. The concept of the first one is: houses for tourist purposes are those properties whose location is on buildings for residential use and are offered as accommodation services for tourist purposes regularly and for a price. It has been elaborated a unique and homogeneous database for houses of tourist use, hotels and hostels, being all the registers georeferenced inside a shapefile to represent it cartographically. These tourist houses are promoted through tourist offer channels such as travel agencies, mediation companies, organizers of tourist services and other media (Regional Government).

The spatial distribution of bed places are studied, thanks to specific software that takes the data from different web pages, such as Airbnb and Booking. This information was represented using different maps, analysing the distribution of the accommodation offer of the city. Additionally, the Moran Index has been used to analyse the spatial auto-correlation; and the tourist rate, considering bed places and residents in each neighbourhood.

For the next step, qualitative data were extracted from in-depth interviews with key informants: managers of housing for tourist use (6), residents of tourist neighbourhoods (6) and activists (3), in total, 15. Two agents' profiles related to the tourist rental subsector have been considered: medium-scale management companies and individual owners. This profile selection was based on obtaining the representation of tourist entrepreneurs and social and local networks working in the three tourist neighbourhoods of Granada. These interviews were conducted between the years 2020 and 2021. Informants were selected using the snowball technique. The script of the interviews has questions about the touristification process of the last ten years in the city of Granada: the presence of tourist accommodation in historic neighbourhoods and the impact on the local economy and businesses. The informants lived in the three historic areas most visited: Albaicín, Realejo and Centro-Sagrario.

Granada is one of the most tourist cities in Spain. It is a relevant destination that experienced an increase in the number of visits until 2020, when the pandemic began. The city has a relevant cultural and historical heritage and a university recognized internationally. The tourist area coincides with the historic centre: Albaicín-Sacromonte, Centro-Sagrario and San Matías-Realejo. This area coincides with the city's tourist area (Fig. 2.1). In these neighbourhoods are located the more emblematic monuments, some of these declared as World Heritage by UNESCO: the castle and palaces of Alhambra-Generalife, the Albaicín neighbourhood, the Cathedral and the Troglodytsm of Sacromonte; where the most significant number of hotels and apartments for tourist use are placed. Considering the number of tourist places, it is the second for Andalusian cities, behind Seville (Granada City Council, 2019 data).

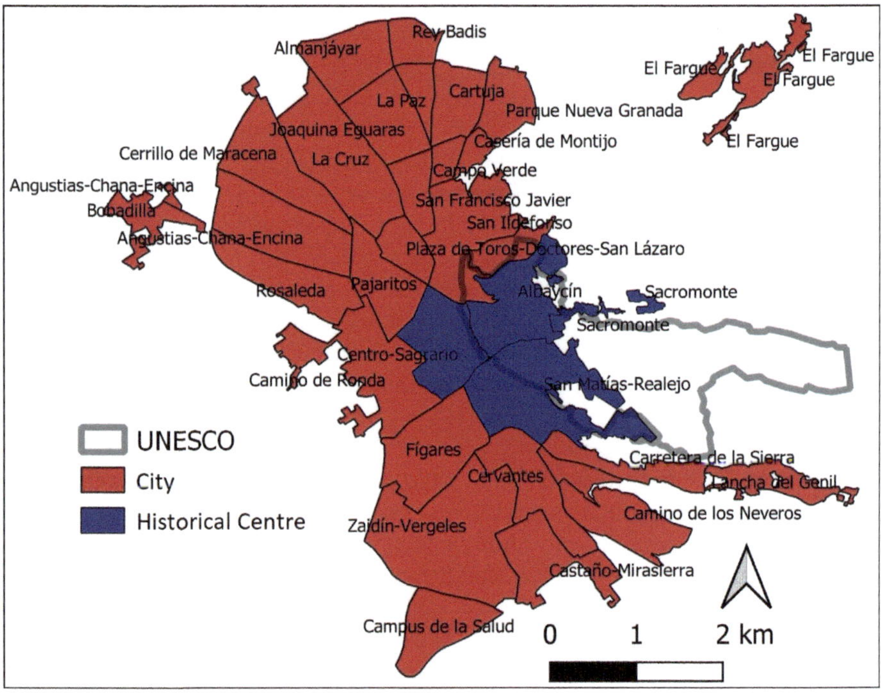

Fig. 2.1 Study area. *Source* Navarro Valverde et al. (2022, 197)

2.4 Results

2.4.1 Where and When

The territorial distribution of the accommodation in the city of Granada shows a clear concentration around the more relevant touristic places and monuments. The higher number of tourism accommodations is located mainly in Centro-Sagrario, San Matías-Realejo and Albaicín; and concretely, in Albaicín Bajo, Centro-Sagrario and San Matías. On the other hand, the rest of the city has a low number of touristic offers. The distribution of bed places obeys four main rules and locations: main streets, entrances to the city, inside of the tourist neighbourhoods and around the more attractive monuments (Fig. 2.2).

Regarding the intensity of the concentration of the number of beds places, in the case of hotels and hostels, the Moran Index shows a "random" index, 0,040, but in the case of houses for tourist use, the result is 0,131, "clustered" (Fig. 2.3).

This consequence is clearly represented by the maps of bed places for hotel establishments, which show a higher dispersion across the city—main monuments and neighbourhoods and relevant streets and entrances to the city—and conversely, the map of tourist houses reflects a higher concentration in touristic neighbourhoods and

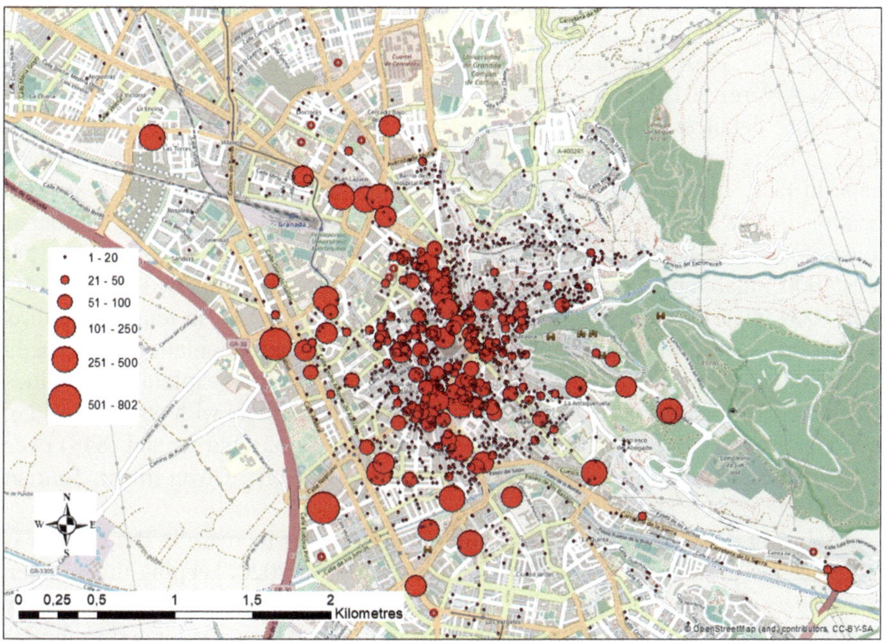

Fig. 2.2 Distribution of bed places in the city of Granada in 2020. *Source* Navarro Valverde et al. (2022, 198)

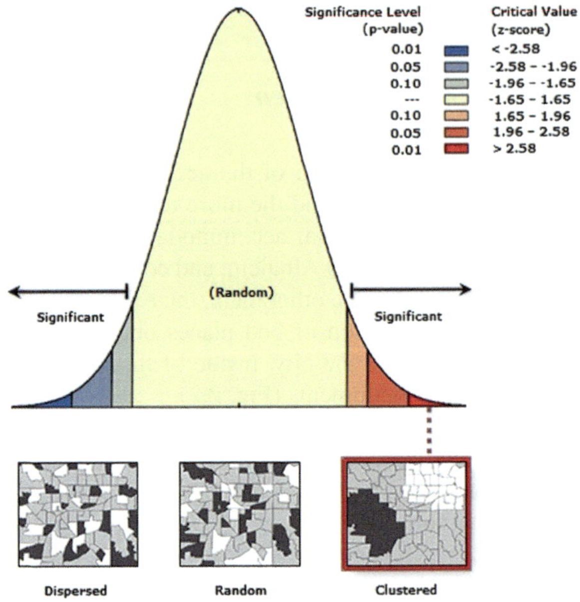

Fig. 2.3 Moran Index of bed places of houses for tourist use in the city of Granada in 2020. *Source* Navarro Valverde et al. (2022, 198)

the most attractive monuments (Figs. 2.4 and 2.5); being located their main concentration in the historical centre and around the monuments; mainly in Elvira Street, Albaicín Bajo, Plaza Nueva and around the streets of Gran Vía and Reyes Católicos (García-Álvarez 2019).

A more detailed analysis shows that is in Realejo-San Matias neighbourhood, where a more significant tourist rate is detected, 595 bed places per 1000 inhabitants. In fact, this part of the city has lost 42% of its inhabitants in 12 years. The location in this sector of a higher number of buildings with a greater number of houses for tourist use would explain this fact. In second place, Centro-Sagrario (545 beds/1000 inhabitants, losing 20% of the population). And finally, Albaicín (468 beds/1000 inhabitants, losing 15% of inhabitants) (Table 2.1). In total, these three neighbourhoods suffered a decrease of more than 1/4 of their population (28%) in the last 12 years. Gentrification and touristification would help to explain these values, which are extremely serious. In fact, compared these numbers with other cities, i.e., in the case of Seville, only the areas of Santa Cruz (1,065) and Arenal (848) overcome the neighbourhoods of Granada (Andalusian Regional Government, Tourism Council).

Analysing now the historical evolution, a continuous increase in the number of bed places is observed, from 218 in 1910 to 30,188 in 2020, keeping the numbers in the years of economic crisis and increasing for the rest of the years. The informants share the opinion that the touristification process was accelerated in the 80s of the last century, beginning the gentrification phenomenon at the start of the new millennium, and the feeling of *overtourism* by residents in 2016–2017, with the rise of platforms to rent houses for tourist use. The accommodations were of hotels

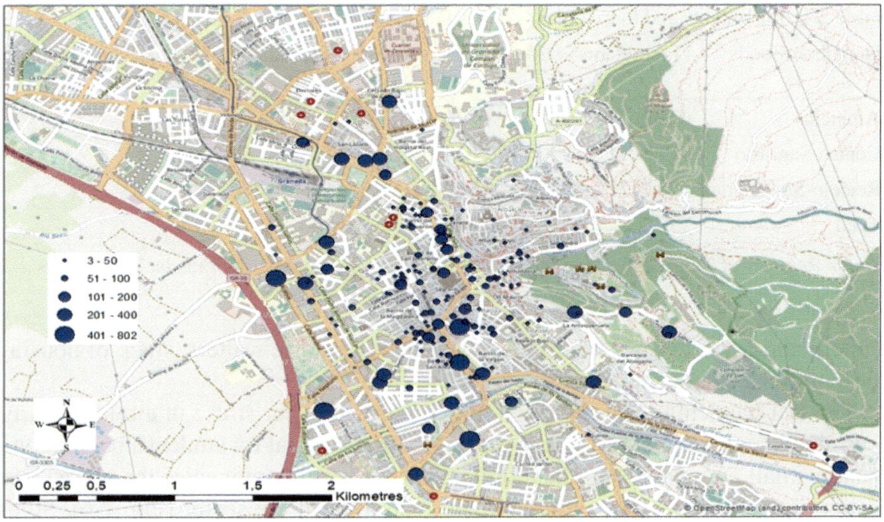

Fig. 2.4 Distribution of bed places for hotels and hostels in the city of Granada in 2020. 2022 *Source* Navarro Valverde et al. (, 199)

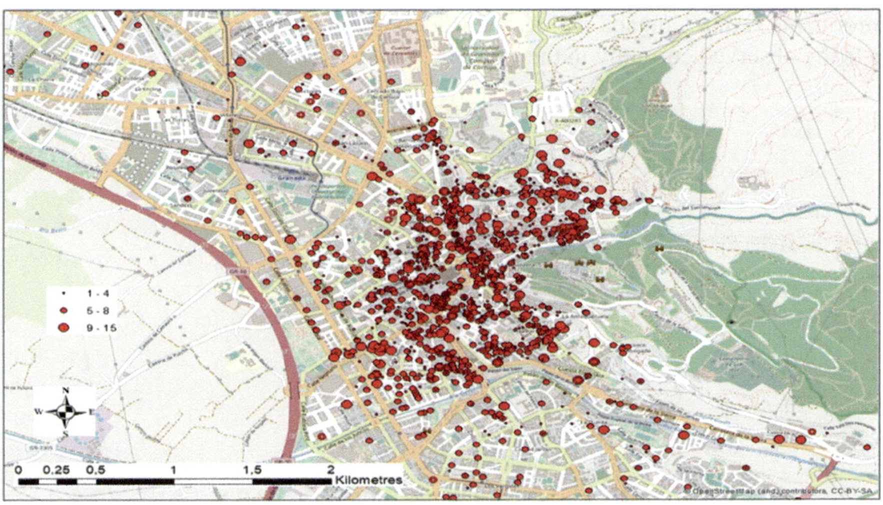

Fig. 2.5 Distribution of bed places for tourist houses in the city of Granada in 2020. *Source* Navarro Valverde et al. (2022, 199)

and hostels until 2000. The first houses for tourist use were created in the city in 2016, having a massive increase between 2017 and 2018. In fact, the more relevant processes of gentrification and touristification happened over the last ten years, specifically between 2016 and 2019. At the same time, a process of precariousness and *McDonaldization* is dominating the activity: homogenization, rationality, speed

Table 2.1 Inhabitants and tourist rate of touristic neighbourhoods of Granada

Neighbourhoods	Inhabitants 2020	Bed places 2020	Tourist rate (bed places/inhabitants) 2020
Albaicín	7,206	3,320	468
Centro-Sagrario	12,803	7,014	545
Realejo-San Matías	10.012	5,854	595
Total	30,021	16,188	543

Source Register of Tourism for Andalusian Region and Municipality of Granada. The authors

and efficiency (Ritzer & Miles 2019), within tour guides, waiters, stock of housing for tourist use not declared, etc.

The last pandemic years affected these houses for tourist use in a relevant way, returning a high number of these to the conventional rental for students, but predominantly small companies and landlords, keeping the big companies the tourist use, favouring the professionalization and concentration of the sector (Fig. 2.6). Some informants said that during the pandemic years, a new type of tourist appeared in the city, the virtual and networker, who enjoyed and lived for several months in the city. During these months, residents recovered good feelings about the city, and for them, it was a pleasure to walk for the city's historic centre. The shared opinion of these informants is that Granada will recover the tourist numbers of pre-COVID-19, as it is being observed currently in the year 2023.

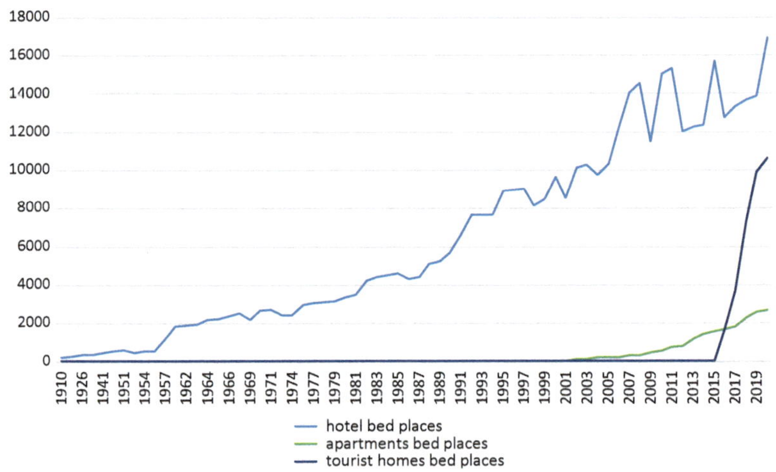

Fig. 2.6 Tourist bed places in the city of Granada between 1910 and 2020. *Source* Elaborated by the authors

2.4.2 *The Perception of the* Overtourism

There are a high number of signals and circumstances, revealing the clear tourist gentrification process in Granada and being corroborated by the key informants in the interviews. In the case of the Albaicín neighbourhood, according to all these key actors, exists *overtourism*, the "local identity is totally lost", and "it's a neighbourhood more oriented to tourism than to the local population itself". As in other Mediterranean cities (Bobic and Akhavan 2022), the UNESCO declaration of a Global Heritage in 1984 was when this part of the City began to change and leave its traditional costumes and peasantry. Gradually, the area closed small food shops and stores; foreigners bought traditional houses called Cármenes, fostering residential tourism and modifying the interior of these traditional houses, removing the "patios"; houses for tourist use were increasing; even pedestrian mobility was affected by tourist groups, etc. The process has been increasing in the last ten years. Even the morphology of housing has begun to change.

> For about ten years, one part of the neighbourhood has now been transformed by speculation. Companies building new apartaments and selling them. This was later modified, and the dimensions of the dwellings were reduced […]. This has caused a reduction in the inhabitant, in those who were permanent inhabitants.

The Albayzin experience has created controversy with the current project to declare another neighbourhood, Sacromonte, as a World Heritage Site by UNESCO. However, practically all the informants affirm that it can benefit the neighbourhood if it is for the benefit of the local population. An old popular neighbourhood with a relevant gipsy population, it is now characterized by flamenco shows for tourists in its typical houses, the cave houses.

> Everything that is named Patrimony by UNESCO worries me. The question is: why what? What is behind this initiative? I think the main interest is the tourism promotion and not so much protection. There are already caves where conflict has been generated.

The case of the Realejo neighbourhood also presents its peculiarity due to several aspects. Geographically, it is located in the City's historic centre and the transit area to the Alhambra Castle/Palace. Socially, the increased presence of university centres, is the prelude, on many occasions, of gentrification (Revington et al. 2021). The location of two faculties, the Faculty of Architecture and, above all, the Institute of Modern Languages, has contributed to the mix and combination of a profile of students, mostly foreigners but also tourists, with high purchasing power. The university's presence has not been the determining factor for tourist speculation, but it has contributed to changing the neighbourhood's image as one of the most attractive areas for a certain social profile. The university presence has not been the determining factor for tourist speculation, but it has contributed to changing the image of the neighbourhood as one of the most attractive areas for a certain social profile: young people looking for a more cosmopolitan lifestyle. It is currently the City's neighbourhood with the largest number of tourist apartments.

> People from relevant universities have come, who are people who have money, and it's a reality. These people have a bit of a budget, even if they are students, but they have a different budget. So, although I do believe that the percentage is tiny, it does influence a little.

And Centro-Sagrario has also transformed due to its orientation towards tourism. As a result, several emblematic spaces in the City have been transformed. For example, the squares, such as Birrambla, restaurants and terraces occupy a high part of the public space. Another emblematic case is the Market of San Agustín, which has suffered a "gourmetization" process, but with poor success. In fact, the residents of the City try to avoid this part of the City because of the mobility problems, not only by car but also by walking.

Therefore, one of the characteristics shared by these neighbourhoods is the increase in apartments for tourist use: gentrification has given way to housing tourism. It is one of the most critical points for the associative network. Its impact has gone in several directions: the increase in housing prices, the displacement, direct or indirect, of the local population and problems of coexistence between neighbours. As one person interviewed said: "Airbnb is the beginning of the end; the link between host and tourist is removed". Therefore, all the informants share the necessity to control the houses for tourist use, even the managers and landlords: the number in each neighbourhood, in each building and the number of hosts in each house.

2.5 Discussion and Conclusions

Key informants of this research and other experts (García-Hernández et al. 2019) share the opinion that to tackle *overtourism* is necessary more than "technological" solutions. Additionally, the destinations need participative and governance processes promoted by the municipalities and public–private collaborations and partnerships. In the case of the city of Granada, informants noted the low interest of the municipality in promoting the participative process to debate and face *overtourism* phenomenon in the City. As noted in other cities and studies (Ruano de la Fuente et al. 2019), residents feel outside of local political decision-making to tackle *overtourism*, by the lack of a participative culture and the non-possibility to control the capitalist process of appropriation of public spaces. Even these key actors suggested the creation of neighbourhood councils (Navarro Valverde et al. 2022). Other measures could be: the control and regulation (prices, number) of houses for tourist use (Horn and Merante 2017; Ardura-Urquiaga et al. 2019; Mirashi 2017; Donaire-Benito et al. 2019); the limitation of bed places in these overcrowded neighbourhoods; and the instauration of a tourism tax; among others.

Unfortunately, the parenthesis of the COVID-19 pandemic in the arrival of tourists has been not used to rethink the *overtourism* problems in the city and improve the quality and sustainability of this destination (Palomo et al. 2020). The negative opinion of these informants for the tourism activity also shows a feeling of *urban-philia*, as a natural answer from residents to *overtourism* phenomenon (Blanco-Romero et al. 2019).

In these touristic neighbourhoods, the excessive presence of hotels and houses for tourist use, and their growth in the last five years, has increased the perception of tourism gentrification. This process has been exacerbated by the activity's pauperization and macdonalization; the tendency to increase the vicious circle of depopulation, at least the abandonment of permanent residents; and the loss of local identity. However, it has to be mentioned the qualitative differences of this phenomenon inside of the city. Realejo maintains its identity, probably by the type of tourism, less aggressive, long-term stays, mixing academic and tourism aims, for travellers and students; and its recognition as a fashion, hipster and cool area. In the Albaicín, conversely, the residents move to other parts of the city or other municipalities, suffering an aggressive touristification process and huge mobility problems. And Centro-Sagrario, sharing, as well, this abandonment, increasing as shopping and leisure area, with squares dominated by restaurants.

Definitively, *overtourism* is an existing phenomenon in the city of Granada, such as it is occurring in other big cities, focused on historic and centric neighbourhoods, with a high relevance of AirBnB tourists, the abandonment of residents of these areas and the speed of the process. *Overtourism* and airbnbfication are new forms of gentrification (Mermet 2022), being the owners and entrepreneurs of these apartments, promoters and victims, at the same time, of these processes, such it happens in Granada.

In the short-term, and for the city of Granada, some of the possible proposals to face these problems lead to the elaboration of specific diagnoses in each city (Ruano de la Fuente 2019), the study of the number of visitors to the monuments, the application of tourist taxes (opened and current debate) and the control of the number and concentration of Airbnb apartments (Pérez-Garrido et al. 2022). In the long-term, some of these measures to reach a more controlled, responsible and sustainable tourism are linked to the search for more personalized demands, the reinforcement of planning by the public and local government, sustainable mobility, the review of tourist fees and taxes, the incorporation of peri-urban space into the offer and the reinforcement of governance processes and public–private partnerships. However, probably the necessity to overcome some post-COVID-19 tendencies could contribute to tackling *overtourism*: the lack of support of stakeholders, with the renouncement of these to apply the concept of tourism capacity; economic priorities before others; global tendencies and the increase of the number of medium classes raising the problem; and the prioritization of marketing before of management and planning (Butler and Dodds 2022).

Clearly, the limitations of this research are related to the analysis tools to size *overtourism* and tourism sustainability. For sure, the indicators must be extended. Also, the profiles of actors included, profiles such as own tourists or a broader range/profile of residents. The study of Navarro and Capote (2021) could be complementary research to this. However, this research could be an initial step and referent to evaluate this phenomenon for medium and small cities.

Acknowledgements Research Project: Desarrollo urbano e impactos socio-espaciales del sector turístico en grandes ciudades andaluzas. PAIDI 2020. Reference. P18-RT-2427. Key informants: residents, managers of houses for tourist use and activists.

References

Almeida-García F, Cortés-Macías R, Balbuena-Vázquez A (2019) Tourism-phobia in historic centres: the case of Málaga. Boletín de la Asociación de Geógrafos Españoles 83(2823):1–32. https://doi.org/10.21138/bage.2823

Almeida-García F, Cortés Macías R, Parzych K (2022) Tourism impacts, tourism-phobia and Gentrificacion in Historic centers: the cases of Málaga (Spain) and Gdansk (Poland). Sustainability 13(1):408. https://doi.org/10.3390/su13010408

Álvarez-Sousa A (2018) The problems of tourist sustainability in cultural cities: socio-political perceptions and interests management. Sustainability 10(2):503. https://doi.org/10.3390/su10020503

Ardura-Urquiaga A, Lorente-Riverola I, Mohino I, Ruiz Sánchez J (2019) "No estamos tan mal como Barcelona": análisis de la proliferación y regulación de las viviendas de uso turístico en Madrid y Barcelona. Boletín de la Asociación de Geógrafos Españoles 83(2828):1–47. https://doi.org/10.21138/bage.2828

Blanco-Romero A, Blázquez-Salom M, Morell M, Fletcher R (2019) Not tourism-phobia but urban-philia: understanding stakeholders' perceptions of urban touristification. Boletín de la Asociación de Geógrafos Españoles 83(2834):1–30. https://doi.org/10.21138/bage.2834

Bobic S, Akhavan M (2022) Tourism gentrification in Mediterranean heritage cities. The necessity for multidisciplinary planning. Cities 124.https://doi.org/10.1016/j.cities.2022.103616

Butler RW, Dodds R (2022) Overcoming overtourism: a review of failure. Tour Rev 77(1):35–53. https://doi.org/10.1108/TR-04-2021-0215

Cerezo-Medina A, Romero-Padilla Y, García-López A, Navarro-Jurado E, Sortino-Barrionuevo J, Guevara-Plaza A (2021) Comparative analysis of short-term rental homes and traditional accommodation in Andalusian tourist cities: intensity, density, and potential expansion areas. Curr Issues Tour 25(11):1782–1797. https://doi.org/10.1080/13683500.2021.1983522

Cócola-Gant A (2011) El Barrio Gótico de Barcelona. De símbolo nacional a parque temático. Scr Nova 15. https://doi.org/10.1344/sn2011.15.3393

Cocola-Gant A, Gago A (2021) Airbnb, buy-to-let investment and tourism-driven displacement: a case study in Lisbon. Environ Plann Econ Space 53(7). https://doi.org/10.1177/0308518X19869012

Colomb C, Novy J (2016) Protest and resistance in the tourist city. Routledge, London

De la Calle-Vaquero M (2002) La ciudad histórica como destino turístico. Ariel, Barcelona

De la Calle-Vaquero M (2019) Turistificación de centros urbanos: clarificando el debate. Boletín de la Asociación de Geógrafos Españoles 83(2829):1–40. https://doi.org/10.21138/bage.2829

Díaz I, Jover J (2020) Overtourism, place alienation and the right to the city: insights from the historic centre of Seville, Spain. J Sustain Tour 29(2–3):158–175. https://doi.org/10.1080/09669582.2020.1717504

Donaire-Benito JA, Zerva K, Palov i Rubio S, Blanco S (2019) Do not cross the line: planning the limits of tourism in Barcelona. Boletín de la Asociación de Geógrafos Españoles 83:1–26. https://doi.org/10.21138/bage.2835

Escudero-Gómez LA (2019) Residents' opinions and perceptions of tourism development in the historic city of Toledo, Spain. Sustainability 11:3854. https://doi.org/10.3390/su11143854

García-Álvarez D (2019) Análisis espacial del alojamiento ofertado a través de nuevas plataformas de economía colaborativa en la ciudad de Granada (España). XXVI Congreso de la Asociación

Española de Geografía. Crisis y espacios de oportunidad. Retos para la geografía. Asociación de Geógrafos Españoles, Valencia

García-Hernández M, Ivars-Baidal J, Mendoza de Miguel S (2019) Overtourism in urban destinations: the myth of smart solutions. Boletín de la Asociación de Geógrafos Españoles 83(2830):1–38. https://doi.org/10.21138/bage.2830

González-Pérez JM (2019) The dispute over tourist cities. Tourism gentrification in the historic Centre of Palma (Majorca, Spain). Tour Geogr 22(1):171–191. https://doi.org/10.1080/14616688.2019.1586986

González-Reverté F (2021) The perception of overtourism in urban destinations. Empirical evidence based on residents' emotional response. Tour Plann Dev. https://doi.org/10.1080/21568316.2021.1980094

Goodwin H (2017) Seasonality, cruises and overtourism: coping with success. https://news.wtm.com/seasonality-cruises-and-overtourism-coping-with-success/

Gotham KF (2005) Tourism gentrification: the case of New Orleans' Vieux Carre (French Quarter). Urban Stud 42(7):1099–1121. https://doi.org/10.1080/00420980500120881

Horn K, Merante M (2017) Is home sharing driving up rents? Evidence from Airbnb in Boston. J Hous Econ 38:14–24. https://doi.org/10.1016/j.jhe.2017.08.002

Janoschka M (2018) Gentrificación en España reloaded. Papers: Regió Metropolitana de Barcelona: Territori, estratègies, planejament 60:24–33.

López L, Pazos Otón M, Piñeiro Antelo Mª (2019) ¿Existe overtourism en Santiago de Compostela? Contribuciones para un debate ya iniciado. Boletín de la Asociación de Geógrafos Españoles 83(2825):1–48. https://doi.org/10.21138/bage.2825

Mermet AC (2022) Can gentrification theory learn from Airbnb? Airbnbfication and the asset economy in Reykjavík. Environ Plann A Econ Space 54(6):1147–1164. https://doi.org/10.1177/0308518X221094616

Milano C (2017) Overtourism y Turismofobia. Tendencias globales y contextos locales. Ostelea School of Tourism & Hospitality, Barcelona

Misrahi T (2017) Wish you weren't here: what can we do about over-tourism? World Economic Forum Weekly Update, 19 September. https://www.weforum.org/agenda/2017/09/what-can-we-do-aboutovertourism/

Navarro F, Capote A (2021) ¿Overtourism en la ciudad de Granada? Cuadernos Geográficos 60(1):35–53. https://doi.org/10.30827/cuadgeo.v60i1.13717

Navarro Valverde F, Barrero Rescalvo M, Díaz-Parra I, Capote Lama A, Carracedo Pandelet J (2022) Overtourism in Granada. Distribution and perception of the tourism activity. XVI Coloquio de Geografía Urbana. Asociación de Geógrafos Españoles y Universidad de Málaga, Málaga, pp 194–204

Ojeda A, Kieffer M (2020) Touristification. Empty concept or element of analysis in tourism geography? Geoforum 115:143–145. https://doi.org/10.1016/j.geoforum.2020.06.021

Palomo G, Jurado E, Medina A, Bernier E (2020) Turismo poscoronavirus, ¿una oportunidad para el poscrecimiento? In: Simancas Cruz M, Hernández Martín R, Padrón Fumero N (eds) Turismo pos-COVID-19: Reflexiones, retos y oportunidades. Cátedra de Turismo Caja Canarias-Ashotel de la Universidad de La Laguna, pp 161–173.

Pérez-Garrido B, Szilárd Sebrek S, Semenova V, Bal D, Michalko D (2022) Adressing the phenomenon of Overtourism in Budapest from multiple angles using unconventional methodologies and data. Sustainability 14(4):2268. https://doi.org/10.3390/su14042268

Pirillo-Ramos S, Mundet L (2021) Tourism-phobia in Barcelona: dismantling discursive strategies and power games in the construction of a sustainable tourist city. J Tour Cult Chang 19(1):113–131. https://doi.org/10.1080/14766825.2020.1752224

Revington N, Zwick A, Hartt M, Schlosser J (2021) Universities and urban social structure: gentrification, studentification and youthification in five United States legacy cities. Urban Geogr. https://doi.org/10.1080/02723638.2021.1985306

Ritzer G, Miles S (2019) The changing nature of consumption and the intensification of McDonaldization in the digital age. J Consum Cult 19(1):3–20. https://doi.org/10.1177/1469540518818628

Rozmiarek M, Malchrowicz-Mośko E, Kazimierczak M (2022) Overtourism and the impact of tourist traffic on the daily life of city residents: a case study of Poznan. J Tour Cult Chang.https://doi.org/10.1080/14766825.2022.2029463

Ruano de la Fuente JM, Iglesias Jiménez E, Polo Villar C (2019) El Madrid vivido: los problemas urbanos desde la perspectiva de la ciudadanía en el contexto del turismo de masas. Boletín de la Asociación de Geógrafos Españoles 83(2826):1–38. https://doi.org/10.21138/bage.2826

Smith N (2012 [1996]) La nueva frontera urbana. Ciudad revanchista y gentrificación. Traficantes de sueños, Madrid

Wachsmuth D, Weisler A (2018) Airbnb and the rent gap: gentrification through the sharing economy. Environ Plann A Econ Space 50(6):1147–1170. https://doi.org/10.1177/0308518X18778038

World Tourism Organization (2018) "Overtourism"? Understanding and managing urban tourism growth beyond perceptions. UNWTO, Madrid. https://doi.org/10.18111/978928449999

Wrede M (2022) How short-term rentals are changing the neighborhood. Int Reg Sci Rev 45(4):417–443. https://doi.org/10.1177/0160017621105998

Chapter 3
Recent Socio-Spatial Transformations in the San Pedro Neighbourhood (Santiago de Compostela)

Miguel Pazos-Otón, Lucrezia Lopez, and María de los Ángeles Piñeiro Antelo

Abstract Santiago de Compostela has undergone many important transformations in recent decades, mainly due to the increase in tourist activity. In fact, the revival of modern Compostela pilgrimages has had an impact on the local economy and on the urban space. For some years, these changes have been spreading from the old walled enclosure to the different neighbourhoods of the city. Among them, the Barrio de San Pedro stands out, through which the French, Northern and Primitive Ways access the city. Formerly a former working-class neighbourhood of Compostela, this part of town retains a traditional identity that has favoured the appearance of incipient gentrification processes and is affected by the growing touristification of the city, which is leading to the proliferation of tourist flats and new commercial activities. For all these reasons, the main objective of this work is to recognise and analyse the recent functional and landscape transformations of the neighbourhood. To carry out this research we have used a methodology that combines several phases of analysis in order to obtain, from different perspectives, approaches to complex problems such as the one we are analysing, in particular: a review of planning documents and participant observation. The results point to a transformation that implies a renovation of businesses to adapt to new audiences, and in parallel the closure of a significant number of stores due to lack of generational replacement or the effects of the crises of 2008 and later of the COVID-19 pandemic.

Keywords Neighbourhoods in Transition · Gentrification · Touristification · Barrio San Pedro · Historic Cities

M. Pazos-Otón (✉) · L. Lopez · M. de los Ángeles Piñeiro Antelo
Department of Geography, University of Santiago de Compostela, Santiago, Spain
e-mail: miguel.pazos.oton@usc.es

L. Lopez
e-mail: lucrezia.lopez@usc.es

M. de los Ángeles Piñeiro Antelo
e-mail: manxeles.pineiro@usc.es

E. Navarro-Jurado et al. (eds.), *Urban Dynamics in the Post-pandemic Period*, The Urban Book Series,
https://doi.org/10.1007/978-3-031-36017-6_3

3.1 Introduction

Before and after the pandemic, several European World Heritage-listed cities experienced a large increase in tourist flows, which have threatened their sustainable development transforming the living conditions of their inhabitants (Gaztelumendi 2009; Lopez et al. 2019; Troitiño and Troitiño 2009). These flows have their epicentre in historic city centres and from there they spread to nearby streets. The most affected areas have been some neighbourhoods, which have experienced changes in residential, commercial and tourism patterns, and which, above all, have seen their uniqueness in the urban complex threatened.

This proposal analyses the case of the Santiago de Compostela San Pedro neighbourhood (Barrio de San Pedro), and just like the historic centre, is one of the most emblematic and traditional in the city. Located outside the city walls, that is, outside the historic-walled city area known as the "almond", it has been a place of residence for craftsmen, merchants and the urban working class since the Middle Ages (Alsina 1988). The main artery and transport route, which gives the neighbourhood its name and extends over it, is San Pedro Street (which coincides entirely with the last metres of the French Way, before reaching the Santiago de Compostela Cathedral). Since the beginning of the 2000s, the neighbourhood started to be known as the "SoHo compostelano" (Montero 2014), as it has progressively attracted the attention of artists, urban creative classes and developed a very active associative citizen civic movement. The attraction that the neighbourhood has exerted on bohemian-bourgeois social sectors has caused an incipient process of gentrification, which has affected the vast majority of streets in the neighbourhood, and which has considerably increased the price of housing as well as commercial sales and rentals (Lopez et al. 2019).

The aim of this research is to reconstruct and analyse the main reasons that have motivated the transformations in recent years. From a methodological point of view, we combine different techniques to ensure different perspectives and approaches to complex problems: a review of planning documents and participant observation. Through our fieldwork, we will reconstruct the urban and social dynamics that have characterised this unique neighbourhood of Santiago de Compostela, which has also been the focus of attention of municipal policies. The proposal will compare neighbourhood and political views that, despite acting in this central space, do not move in the same direction.

Among the results, we will highlight the emergence of new dynamics, since, in addition to gentrification, there is now talk of "touristification". The success of the Way of St. James and the pilgrimages in recent years has triggered the appeal of the neighbourhood, located in the final stages of the French, Northern and Primitive Ways. The proliferation of holiday rentals and the number of businesses targeting tourists, and pilgrims, has, in particular, been increasing (Ramil et al. 2018). This situation has generated discontent among a part of the neighbours and traders partly due to recent public works to improve the landscape to integrate the French Way upon its entry into Santiago de Compostela that in their view overlook the interests of the neighbours (Villa Rey 2019).

3.2 Gentrification, Touristification and Other Associated Phenomena

The notion of gentrification was coined in the 1960s (Glass 1964) to describe the innumerable processes of change in a neighbourhood (Atkinson 2008). This social process traditionally took place in inner city areas where the arrival of the middle class in working-class neighbourhoods increased housing prices and resulted in the displacement of low-income residents (Newman and Wyly 2006; Smith 2002; Watt 2008).

According to Davidson and Lees (2005), gentrification has four defining characteristics: reinvestment of capital, social improvement of the neighbourhood, changes in the landscape and perspective and the direct or indirect displacement of low-income groups. Early studies of gentrification have observed how the arrival of middle-class residents in working-class neighbourhoods contributed to the displacement of their original inhabitants (Glass 1964). However, more recent studies suggest that gentrification could occur without widespread displacement being a major concern (Boddy 2007; Butler 2007).

The driving cause behind gentrification is the commodification of consumable authenticity. Authenticity is often staged and has been identified as a criterion for measuring quality and adding value. In addition, authenticity is associated with reality, history, traditions, locality and otherness (Özdemir and Seyitoğlu 2017). When it comes to gentrification, authenticity has a transformative power because it explains and justifies the combination of products, the design of new businesses and the intangible atmosphere of restaurants, boutiques and gourmet shops. The most prominent pioneers and "authenticity seekers" who contribute to gentrification are the members of a culturally prosperous new white middle class (Butler 1997, 2003).

This new middle class assures its social status through its cultural capital, which is expressed in the willingness to appreciate the authentic and uses it as a vehicle of distinction from other social groups. It is this group that has increasingly settled in low-income neighbourhoods as residents and consumers in recent decades (Butler 1997).

While retaining basic similarities such as landscape change, capital reinvestment, population inflow and class-based displacement, gentrification is now expressed in different ways. Among them, the present research pays attention to commercial gentrification, understood as the gentrification of streets or commercial areas (Lees et al. 2008). Commercial gentrification is closely related to residential gentrification, as commercial gentrification emerges to meet the tastes and preferences of newcomers (Bridge and Dowling 2001; Ley 1996). As the impact of commercial gentrification has become widespread, the social conflict surrounding displacement intensifies, as do social and spatial transformations. Neighbourhoods that cater to local needs become hotspots for shopping, dining and art. Among the various commercial activities, dining out plays a decisive role in the expansion of commercial gentrification processes (Hubbard 2017).

There are two other social phenomena associated with gentrification, namely the concepts of tourist gentrification and touristification. Both emphasise the interaction between tourism and urban change, focusing on the transformation and improvement of middle-class neighbourhoods in tourist areas marked by a proliferation of tourist and entertainment venues, cultural events, museumification processes and spectacularisation of urban space (Gotham 2005).

In these urban centres, gentrification is exemplified in different processes and changes, such as *foodification*. According to Loda et al. (2020), the concept of *foodification* is understood as the process through which the urban economic landscape vanishes to be progressively absorbed by a single type of commercial activity related to the sale and consumption of food.

In 1999, Joanne Finkelstein (1999) coined the neologism *foodatainment* to draw attention to how food—and in particular the construction of an aesthetic, cultural and urban imaginary around food, drink and going out to eat can constitute a touristic experience. In other words, *foodification* can be understood as one of the many consequences of urban overtourism, similar to the "airbnbisation" neologism to express changes and tensions in local real estate markets (Nofre and Sequera 2019). Thus, the same social process of touristification can manifest in the form of foodification or airbnbisation, among others, since both refer to the growing presence of tourists, who sometimes gentrify the city, increasing the segregation between leisure spaces and residential neighbourhoods.

3.3 Methodology

To carry out this research, we have used a methodology that combines several phases of analysis in order to obtain, from different perspectives, approaches to complex problems such as the one we are analysing, in particular: a review of planning documents and participant observation.

First, we have consulted territorial planning documents, such as the General Municipal Planning Plan (PGOM) and, above all, the 1997 Special Plan for the Protection and Rehabilitation of the Historic City of Santiago de Compostela (Fig. 3.1), revised in 2005 and 2019, just before the outbreak of the COVID-19 pandemic.

Following the 2005 revision, the Special Plan provides a very useful cartography with complete differentiation of the main uses of the premises, namely: residential, tertiary, industrial, garage, facilities and unused. Likewise, within the category of

Fig. 3.1 PE-1-walled area (in orange), to which the Special Plan for the Protection and Rehabilitation of the Historic City of Santiago de Compostela applies. *Source* City Council of Santiago de Compostela

tertiary use, the following subcategories were included: commercial, offices, hotels and hospitality (Fig. 3.2).

Taking into account that San Pedro Street, which gives its name to the homonymous neighbourhood and constitutes its main axis, is included in the PE-1, from the data included in the Special Plan, we have made a comparison with the uses corresponding to the years 2005, 2019 and 2022, in order to analyse the functional evolution of the case study. To examine the changes undergone by this street, in addition to the uses, we have considered the type of businesses and activities established at street level. For the year 2019, we have gathered information from Google Street View, with photographs corresponding to June of that year. As for the present year 2022, we have carried out field work visiting and photographing the premises located in all the ground floors of San Pedro Street.

As a result, the comparative analysis of changes in uses and activities between the years 2005, 2019 and 2022 allows us to analyse the evolution of the neighbourhood, especially focusing on commercial uses.

Fig. 3.2 Mapping of uses of San Pedro Street corresponding to the 2005 Special Plan Review. *Source* City Council of Santiago de Compostela

In order to analyse the evolution of changes in the use of commercial ground floors in the Barrio de San Pedro, we believe that it is necessary to differentiate several typologies, based on their evolution throughout the period studied. To this end, in the results section we elaborated a table (Table 3.2), in which we advance a categorisation of commercial activities. In this way, we intend to highlight changes or permanence, and more in detail, issues that allow us to point to a process of gentrification and/or touristification.

3.4 The Area of Study: Santiago De Compostela

Santiago de Compostela, declared a World Heritage Site by UNESCO in 1985, was designated in 1986 as the capital of the Autonomous Community of Galicia (Northwestern Spain). Is one of the Spanish cities with one of the most recognised tourism characters today (Calle Vaquero 2019; Ramil 2018; Santos and Lopez 2019). Over the last few decades, the rise of urban and cultural tourism in Santiago de Compostela has joined the phenomenon of contemporary pilgrimages, undoubtedly the main distinctive feature of the city (Moscarelli et al. 2020).

To understand the importance of the tourism boom in Santiago de Compostela in recent years, it is necessary to consider the great urban transformation that the city experienced in the 1980s and 1990s, prompted by its declaration as a World Heritage City, as the capital of the Galicia region and as a pilgrimage destination. From the hand of the mayor Xerardo Estévez, Compostela was endowed with a PGOU (General Plan for Urban Planning) at the height of its heritage, cultural importance and its new role as the capital of the autonomous Galicia. Signed by the architects Juan Luis Dalda and Ánxel Viña, the general plan was committed to consolidating the enhancement of heritage elements and increasing the quantitative and qualitative importance of green spaces. Even despite the limitation of the residential function, this plan defined a very attractive city as an urban setting, complemented by ambitious cultural policies. This laid the foundations for later tourist attractiveness (Estévez 2001; Dalda 2007).

Although there were attempts to revitalise medieval pilgrimages and reinterpret them from the contemporary point of view in the second half of the twentieth century, the turning point was the 1993 Xacobeo Holy Year when the arrival of pilgrims in Santiago began to grow significantly (Hernández Soriano 2016; Santos 1999, 2006). The fact that Santiago de Compostela is the seat of the regional administration must also be taken into account when explaining the prioritisation of many investments focused on enhancing heritage, cultural identity or rehabilitation and conservation of buildings in the historic centre. In this context, the reassessment of Santiago de Compostela as a destination of urban and cultural tourism experienced a very clear trend with the change of millennium (Martínez Roget et al. 2009). At the beginning of the twenty-first century, the revolution of low-cost airlines and the growing trend towards tourism *short-breaks* increased the tourist nature of the city. To this, we must add the marketing and promotion work carried out by official institutions, where considerable resources and efforts were invested (Pérez and Lois 2018; Ramil 2018).

This increase in the importance of tourism within the city's economy has led, as expected, to a number of changes in various aspects. Firstly, the increase in tourism pressure results in tensions in the housing market. In a city of around 100,000 inhabitants, with a university population of 25,000 students, the traditional student rental flat has had a lot of weight, which in recent years has been replaced by the promotion of holiday rentals. The progressive expansion of holiday rentals or homes for tourism use within the urban fabric has contributed to rising tensions (Santos and Lopez 2019; Lopez et al. 2019). Secondly, the commercial sector of the city has been changing to adapt to the demands of the increasing number of pilgrims and tourists in general. Throughout these years, we have witnessed (and not only in Santiago de Compostela) the replacement of traditional businesses with others with a clearly modern imprint and designed for a foreign audience (Lestegás 2019). Thirdly, the increase in tourist flows is posing serious crowd management problems, since the urban space of a city with a historic plan such as Santiago is clearly small, with an area of 169.9 hectares between the old intramural old city and the historic neighbourhoods. In previous works, we have already investigated the existence of the phenomenon of overtourism in Santiago, with a whole series of negative consequences, especially for the local population (Lopez et al. 2019; Piñeiro Antelo et al. 2019).

Considering all this background information, this text analyses the "San Pedro neighbourhood", a historic neighbourhood with a strong identity, and in which the working classes that lived outside the walls resided. This neighbourhood is the object of this research due to the changes that have occurred in recent years, related to the fact that three of the pilgrimage routes to access the city of Santiago de Compostela run through it, which means that approximately 60% of the pilgrims enter the city via San Pedro Street, its main artery. According to data from the Office of the Pilgrim, in 2019, it is estimated that at least 224,671 pilgrims crossed this street on their way to the Cathedral.

In addition, the peculiarities and idiosyncrasies of the neighbourhood have made it a place cherished and valued by certain sectors of the population, who have been empowered and entrenched (from a symbolic and identity point of view) in this urban area. The most evident signs of this sense of identity are manifested in the existence of an active neighbourhood and cultural movement, in the importance of premises linked to artistic collectives, in the widespread use of the Galician language or in the identification of the neighbourhood as gay-friendly, to highlight the most relevant examples (ScqLab 2017). The San Pedro neighbourhood has been chosen by young professionals, students, creative sectors and the world of culture and the arts, so that today it has a strong identity based not only on the sense of place, but above all on the wealth of neighbourhood and associative relationships (Liñeira 2014).

3.5 Discussion and Results

In this research, we propose the analysis of the businesses located on San Pedro Street, in order to study the different commercial evolution behaviours recorded from 2005 to the present day. The response of the different businesses has also been different, depending on a number of complex factors. Business ownership, trade regime, location, economic capacity of the business owners to deal with crises, type and orientation of the business, target audience and many other factors constitute the basis for explaining the evolution of commercial activity in the San Pedro neighbourhood. In any case, the changes in commercial uses, in the activity of the businesses, in their typology and in their main economic, cultural and social traits help to understand the transformation of the neighbourhood in the last two decades (see Table 3.1).

In 2005, students were still residing in the San Pedro neighbourhood. However, the rise in rental prices due to the increase in urban-cultural tourism and pilgrimages was a determining factor in the displacement of students to other parts of the city, and even to towns in the urban periphery of Santiago, outside of the compact city. In the San Pedro neighbourhood, the offer of accommodation and catering businesses has become increasingly oriented towards tourism (Piñeiro Antelo et al. 2019). This trend peaked before the COVID-19 pandemic.

Secondly, the San Pedro neighbourhood has become increasingly attractive to certain segments of the young or young-adult population with more or less bohemian profiles. The emergence of the neighbourhood movement, together with the popularisation of San Pedro as a fashionable area linked to culture and the arts, has also resulted in a proliferation of businesses targeting this new proto-gentrified class (Liñeira 2014). We can use the word hipsterisation or indicate that these places are part of what we can call *trendy*, in the sense of "*fashionable*" or "*very modern*".

Thirdly, many of the neighbourhood's traditional businesses came to the end of a cycle and of a historical journey as their traditional owners retired. This process was already visible before the turn of the century and remains so today. In many cases, the traditional businesses that have closed down were managed by families who own

Table 3.1 Evolution of selected businesses in San Pedro Street

Property number	2005	2019	2022
9	*Barbería* Milenio (name in Spanish)	Milenio *Barbershop* (translation into English, perceived as more trendy)	Milenio barbershop
23	Velocípedo bicycleshop	Velocípedo bicycleshop	Unoccupied due to retirement
26	San Clodio grill	San Clodio grill	Restaurante DeLito (Traditional grill restaurant with modern decotation and presentation)
47	Bar O Mosquito	Unoccupied due to retirement	Unoccupied due to retirement
52	A Rúa hardware shop	A Rúa hardware shop	*Os mundos de Carlota* children bookshop
112	Sareliña fashion	Sareliña fashion	Unoccupied due to retirement

Source Own work from photographs (2005 and 2019) and direct observation (2022)

Table 3.2 Typology of trade changes in San Pedro Street (2005–2022)

Type	Examples of premises
(1) Unchanged iconic traditional establishments	María Luisa haberdashery (No. 53); Suso shoeshop(No. 19); O Dezaseis restaurant (No. 16), Sande bar (No. 35); Bar O Cruceiro (No. 119); San Pedro churro shop (No. 121); Espazo Aberto theatre and dance school (No. 48)
(2) Traditional establishments that have closed	Velocipedo bicycleshop (No. 23); O Mosquito bar (No. 47); Sareliña fashion (No. 112); Prisma academy (No. 110); A Rúa hardware shop (No. 52)
(3) Trendy establishments that have emerged since 2005	Cadrado Dóce Coffee Shop (No. 17); Mononoke (No. 33); Dr Mortis tattoo (No 62); A Maceta restaurante (No. 120); TS A Casa vegetarian restaurant (No. 113); A Moa restaurant (No. 32); WAX! Cultura DJ record shop (No. 69)
(4) Establishments that have been *trendified*	Milenio barbershop (No. 9); DeLito restaurant (No. 26); Café de Clem (No. 118); Carmiña foodstore (No. 54)

Source Own work from photographs (2005 and 2019) and direct observation (2022)

the premises in which they developed their professional activity over a long time, and who chose to sell or rent the premises in the absence of generational replacement. There arc also cases in which the owners have decided to keep the premises closed and empty, as a social dormant asset, since there is no urgency to obtain an economic return from it.

Finally, the neighbourhood, just like the city of Santiago and almost the entire world, has been affected by two major crises: the economic crisis that broke out in 2008 and the pandemic that began in 2020. The occurrence of these two crises in

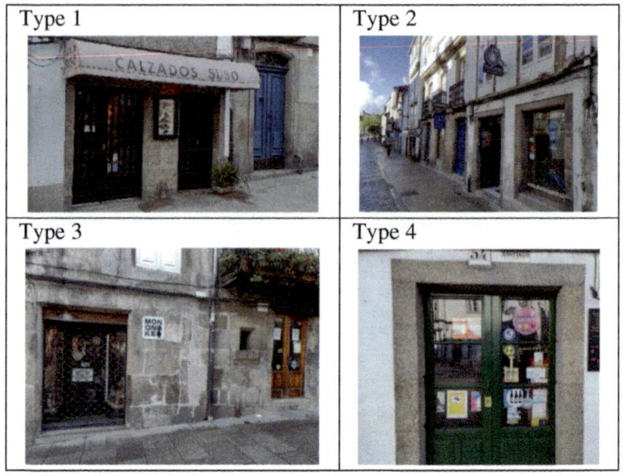

Fig. 3.3 Typology of trade changes in San Pedro Street (2022). *Source* Authors

such a short period of time is fundamental to understanding why many establishments have not survived from 2005 to the present day. Some had to close in the first and others in the second crisis.

Taking this into account, and in light of the field work carried out, we can differentiate the following types of business in relation to their evolution in recent decades (Table 3.2, Fig. 3.3).

The first typology of establishments is composed of traditional businesses which have not changed over the years. Here, we can differentiate two subgroups.

First of all, the traditional establishments and businesses in which the main customers are the local, mainly elderly neighbours, who have been residing in this area for decades. They are part of a Santiago trade prior to the appearance of the peripheral shopping centres and will stay open as long as their owners are still active. It is based on very close social ties, consolidated over many years, and constitutes the paradigm of what we could call "local neighbourhood trade". The traditional spirit of the neighbourhood, which will then be claimed and reinterpreted by the proto-gentry and the youth of the neighbourhood movement, lives in this type of distinguished premises.

Second, the existence of businesses that also have a long tradition and that have continued open throughout this time is worth highlighting, although their clientele could be qualified as mixed as it integrates traditional residents with new residents, visitors and tourists. Although it is common to see the elderly population of the neighbourhood in the winter months, in medium and high tourist season it is usually more common to see pilgrims and even groups of tourists. The local guides lead tourists through this area, which they show as an example of the authentic life of the San Pedro neighbourhood.

A second type of premises is establishments that have closed in this period. As we have explained above, the main reason has been the retirement of the business owner, which is almost always the owner of the premises. The lack of generational replacement explains the decision to close the business, since in many cases the owners usually have a secure financial situation and can live off the income generated.

The third and fourth types of establishments show the great changes that have occurred in the neighbourhood in recent years and its evolution towards the aforementioned concept of "SoHo compostelano". The third group of establishments that we have classified are those that we can describe as *trendy* and that are oriented to this new urban class that has been identified and territorially empowered in the neighbourhood. In all these cases, we are dealing with modern businesses, renewed with all the attributes of postmodernity. The fourth and final type corresponds to those traditional establishments that have adapted their aesthetics, their business model, their decoration and the type of marketing over these years to the new type of inhabitant and customer.

In this way, there is a minority of establishments that have not noticeably changed over the last few years. They offer highly valued and well-established products or services, and the reason for their business survival is based on their traditional character and orientation to a majority audience. Among them, we can mention shops such as haberdasheries or shoe stores for the older residents of the neighbourhood, which continue to be their loyal clientele. These long-lived businesses currently have the greatest problems, due to the absence of generational replacement at their owner's time of retirement. This circumstance favours the existence of a large number of closed commercial premises, which become part of the real estate market with properties for sale, to rent or simply destined to close. When these establishments are rented or sold, the new owners of the businesses usually adapt them to the demands of the new public: new residents of the neighbourhood, young people and tourists. Thus, the transformation that certain places have undergone is particularly relevant, adapting to the new social environment and contributing to the transformation of the commercial and cultural landscape of the San Pedro neighbourhood.

The following map (Fig. 3.4) shows the results of the above-mentioned changes. We have used the viewfinder of the online Land registry (Catastro) records, which provide a cartographic representation of the area of study with the numbering and characteristics of each of the properties.

3.6 Conclusions

Since its first conceptualisation and definition, the process of gentrification has displayed different social and economic features that combine differently depending on the context of reference, and no less relevant than its territorial scope. Hence, the plurality of definitions of gentrification, some of which have been stated above. However, it is true that the end result of gentrification is a transformation of the urban landscape (from a material and immaterial point of view).

In this research, we wanted to pay attention to the socio-spatial dynamics in the San Pedro neighbourhood, which because of its dynamic transformation, could be defined as a "neighbourhood in transformation". Here, the commercial sector

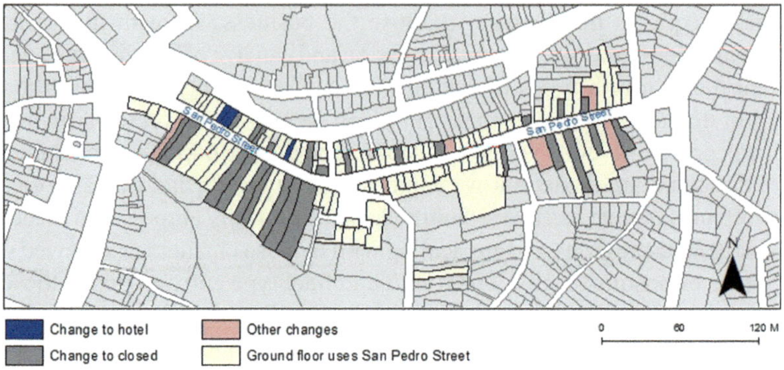

Fig. 3.4 Changes of use in San Pedro Street (2005–2022). *Source* Own elaboration based on Cadastral Electronic Site and direct observation (2022)

very clearly reflects the amount and magnitude of these changes, which greatly affect, and are related to, its identity. This dynamic of continuous change implies that many businesses have closed down, almost always due to a lack of generational replacement. Others remain true to their traditional essence with little change. Finally, there are new businesses that open to adapt to new types of audiences or as a result of the transformation and modernising pre-existing businesses.

At the same time, and in parallel with this gentrification process, the Barrio de San Pedro has known, during the study period, a growing importance of tourism and its repercussions in the space itself. It is not just that three paths cross the neighbourhood (The French Way, The Northern Way, and The Primitive Way), but the very expansion of tourist activity in the city has introduced important changes in economic functionality and in urban landscapes. The combination of gentrification and tourism has given rise to phenomena such as the aforementioned foodification.

The management of a neighbourhood as emblematic as the San Pedro neighbourhood becomes increasingly complex precisely because of the inter-action of new contemporary urban dynamics. Therefore, new forms of governance should be designed; they should be less oriented by economic criteria, and more concerned with keeping the centres of the cities inhabited, still desirable for their citizens (Koens et al. 2018; Milano 2018).

Ultimately, this research aims to open a new line of work, supported by new sources and methods which we will expand upon in the near future, as we recognise possible advances in research with the use of semi-structured interviews with citizens, associations and local agents of interest. Our intention is to continue working with evolutionary analyses that allow us to assess not only the changes that have occurred in the neighbourhood but also the understanding of the logics that explain these transformations, in accordance with demographic, social and economic evolution, as well as the perception of the neighbourhood both by the residents and by visitors and tourists.

References

Alsina FL (1988) La ciudad de Santiago de Compostela en la alta Edad Media. Ayuntamiento de Santiago de Compostela

Atkinson R (2008) Commentary: gentrification, segregation and the vocabulary of affluent residential choice. Urban Stud 45(12):2626–2636. https://doi.org/10.1177/0042098008097110

Boddy M (2007) Designer neighbourhoods: new-build residential development in nonmetropolitan UK cities—the case of Bristol. Environ Plan A 39:86–105. https://doi.org/10.1068/a39144

Bridge G, Dowling R (2001) Microgeographies of retailing and gentrification. Aust Geogr 32:93–108. https://doi.org/10.1080/00049180020036259

Butler T (1997) Gentrification and the middle classes. Hants

Butler T (2003) Living in the bubble: gentrification and its 'others' in North London. Urban Stud 40:12.https://doi.org/10.1080/0042098032000136165

Butler T (2007) For gentrification? Environ Plan A 39(1):162–181. https://doi.org/10.1068/a38472

Calle Vaquero M de la (2019) Turistificación de centros urbanos: clarificando el debate. Boletín de la Asociación de Geógrafos Españoles 83(2829):1–40. https://doi.org/10.21138/bage.2829

Dalda JL (2007) Planes y políticas urbanas: la experiencia urbanística de Santiago de Compostela desde 1988. Urban 12:102–125

Davidson M, Lees L (2005) New-build 'gentrification' and London's riverside renaissance. Environ Plan A 37(7):1165–1190. https://doi.org/10.1068/a3739

Estévez X (2001) Santiago de Compostela, conservación y transformación, Arbor: Ciencia, pensamiento y cultura 671–672:473–488. https://doi.org/10.3989/arbor.2001.i671-672.941

Finkelstein J (1999) Foodatainment. Perform Res 4(1):130–136. https://doi.org/10.1080/13528165.1999.10871653

Gaztelumendi I (2009) Nuevas tendencias del turismo en ciudades Patrimonio de la Humanidad. El proyecto COMPOSTUR, en Turismo cultural en ciudades Patrimonio de la Humanidad, Ediciones de la Universidad de Castilla-La Mancha, pp 53–62

Glass R (1964) London: aspects of change. MacGibbon & Kee, London

Gotham KF (2005) Tourism gentrification: the case of New Orleans' Vieux Carre (French Quarter). Urban Stud 42(7):1099–1121. https://doi.org/10.1080/00420980500120881

Hernández Soriano RN (2016) Santiago de Compostela, historia y progreso el Xacobeo como instrumento de transformación urbana

Hubbard P (2017) The battle for the high street: retail gentrification, class and disgust. Palgrave Macmillan, London

Koens K, Postma A, Papp B (2018) Is overtourism overused? Understanding the impact of tourism in a city context. Sustainability 10:4384. https://doi.org/10.3390/su10124384

Lees L, Slater T, Wyly EK (2008) Gentrification. Routledge, London

Ley D (1996) The new middle class and the remaking of the central city. Oxford University Press, Oxford

Lestegás I (2019) Lisbon after the crisis: from credit-fuelled suburbanization to tourist-driven gentrification. Int J Urban Reg Res 43(4):705–723

Liñeira M (2014) Santiago de Compostela: fact and fetish. A companion to galician culture. Suffolk, UK, pp 53–71

Loda M, Bonati S, Puttilli M (2020) History to eat. The foodification of the historic centre of Florence. Cities 103. https://doi.org/10.1016/j.cities.2020.102746

Lopez L, Pazos M, Piñeiro MA (2019). ¿Existe overtourism en Santiago de Compostela? Contribuciones para un debate ya iniciado. Boletín de la Asociación de Geógrafos Españoles (83)

Martínez Roget FM, Novello S, Fernández MPM (2009) Análise da lealdade dos turistas que visitan a cidade de Santiago de Compostela. Revista Galega de Economía: Publicación Interdisciplinar da Facultade de Ciencias Económicas e Empresariais 18(2):5–20

Milano C (2018) Overtourism, malestar social y turismofobia. Un debate controvertido. PASOS: Revista de Turismo y Patrimonio Cultural 16(3):551–564. https://doi.org/10.25145/j.pasos.2018.16.041

Montero T (2014) El Barrio de San Pedro, el Soho compostelano. *La* Voz de Galicia. Available at https://www.lavozdegalicia.es/noticia/santiago/2014/05/17/soho-compostelano-san-pedro/0003_201405SY17P20991.htm. Accessed 2 Sept 2022

Moscarelli R, Lopez L, Lois González RC (2020) Who is interested in developing the way of Saint James? The pilgrimage from faith to tourism. Religions 11:24. https://doi.org/10.3390/rel11010024

Newman K, Wyly EK (2006) The right to stay put, revisited: gentrification and resistance to displacement in New York City. Urban Stud 43(1):23–57. https://doi.org/10.1080/00420980500388710

Nofre J, Sequera J (2019) Dinámicas urbanas y turistificación en el barrio de Alfama (Lisboa): una aproximación a los condicionantes históricos y políticos en el proceso de Airbnbización. Estudios Geográficos 80(287):e018. https://doi.org/10.3989/estgeogr.201935.015

Özdemir B, Seyitoğlu F (2017) A conceptual study of gastronomical quests of tourists: authenticity or safety and comfort? Tour Manage Perspect 23:1–17. https://doi.org/10.1016/j.tmp.2017.03.010

Pérez Guilarte Y, Lois González RC (2018) Sustainability and visitor management in tourist historic cities: the case of Santiago de Compostela, Spain. J Herit Tour 13(6):489–505

Piñeiro Antelo MA, Lopez L, Pazos Otón M (2019) Gentrificación y overtourism en destinos urbanos. El caso del Barrio de San Pedro en Santiago de Compostela (Galicia, España). In: Márquez Domínguez JA, Llamas Chávez JL (eds) Hélices y anclas para el desarrollo local. Consorcio Universidades Huelva-Cartagena de Indias, pp 1231–1241

Ramil F (2018) Consolidación de Santiago de Compostela como destino turístico global. Estudios Turísticos 216:77–105

Ramil F, Casado C, Fernández J, Neira C (2018) As vivendas de uso turístico na cidade histórica. Ayuntamiento de Santiago de Compostela. Disponible online en https://www.santiagoturismo.com/files/2018/03/Estudovivendasusotursticofinal.pdf. Recuperado el 8 de marzo de 2022.

Santos XM (1999) Mitos y realidades del Xacobeo. Boletín de la Asociación de Geógrafos Españoles 28:103–118

Santos XM (2006) El Camino de Santiago: turistas y peregrinos hacia Compostela. Cuadernos de Turismo 18:135–150

Santos X, Lopez L (2019) Gestión Turística en Ciudades Históricas. Santiago de Compostela. Estudios Turísticos: 217–218

ScqLab (2017) Mapa emocional de San Pedro. https://www.scqlab.info/mapa-emocional-de-san-pedro/

Smith DP (2002) Extending the temporal and spatial limits of gentrification: a research agenda for population geographers. Int J Popul Geogr 8(6):385–394. https://doi.org/10.1002/ijpg.267

Troitiño Vinuesa MÁ, Troitiño Torralba L (2009) Turismo y patrimonio en Castilla y León: las ciudades Patrimonio de la Humanidad (Ávila, Salamanca y Segovia) como destinos turísticos de referencia. Polígonos: revista de geografía 19:145–178

Villa Rey M (2019) Rehabilitación y humanización de rúa dos Concheiros (Santiago de Compostela). University of A Coruña, Degree

Watt P (2008) The only class in town? Gentrification and the middle-class colonization of the city and the urban imagination. Int J Urban Reg Res 32(1):206–211

Chapter 4
The Touristification of Historic Centres Through Commercial Gentrification in Times of COVID-19

Fátima Santos-Izquierdo, Mario Blanco-Vílchez, Yolanda Romero-Padilla, and Enrique Navarro-Jurado

Abstract COVID-19 has questioned the model of production and consumption, especially that of economies overly dependent on tourism. This has dramatically impacted city centres with a commercial fabric excessively adapted to global tourist demand. After the 2008 financial crisis outbreak, the growth of tourism activity emerged as one of the main mechanisms of economic recovery, leading to increased tourism specialization in southern European cities. This led to a new intensification of tourism in urban centres through the proliferation of tourist accommodations and commercial typologies aimed at visitors and tourists. At the same time, neighbourhood protests arose in tourist cities over conflicts in the commodification of housing, the disappearance of traditional commerce, and the privatization of public space. In this context, the touristification of many historical centres is associated with commercial gentrification. The aim is to study the urban transformations in a European historic centre, specifically in the capital of the Costa del Sol, Malaga. The research focuses on the changes in the commercial fabric caused by tourism in the context of the pandemic. The objective is to show how the most tourist areas are becoming less diverse, concentrating only on specific uses, and turning more vulnerable. The results show how the collapse of international tourist flows and the vulnerability of the study area have not led to a change in urban strategy, but an intensification of the tourist function in Malaga through commercial gentrification.

F. Santos-Izquierdo (✉) · M. Blanco-Vílchez · E. Navarro-Jurado
Department of Geography, University of Malaga, C. de León Tolstoi, 4, 29010 Malaga, Spain
e-mail: fatimasantos@uma.es

M. Blanco-Vílchez
e-mail: mb@uma.es

E. Navarro-Jurado
e-mail: enavarro@uma.es

Y. Romero-Padilla
Department of Business Organisation, Universidad Complutense de Madrid, Av. de Filipinas, 3, 28003 Madrid, Spain
e-mail: yromerop@ucm.es

E. Navarro-Jurado et al. (eds.), *Urban Dynamics in the Post-pandemic Period*,
The Urban Book Series,
https://doi.org/10.1007/978-3-031-36017-6_4

Keywords Touristification · Gentrification · Urban transformations · Commercial fabric · COVID-19

4.1 Introduction. The Transformations of Historic Centres in the Context of the Pandemic

Since the 1970s, historic centres have begun urban regeneration processes in response to globalization and economic restructuring (Soja 2008). These processes have affected all Western cities, especially in Europe, to a greater or lesser extent.

This commitment to the revaluation of degraded areas in Western cities allowed the implementation of planned urban rehabilitation and regeneration processes. They increased these areas' residential attractiveness, attracting tourism in the process (Jover and Díaz-Parra 2020; de la Calle 2019). This context gives rise to the so-called tourist restructuring of the city (Blanco-Romero et al. 2019), where tourism is one of the main space-producing forces (Lefebvre 1974). The result is the prioritization of the tourist over the resident's life, in a process known as touristification (Barata-Salgueiro and Guimarães 2020; Campos and Sequeira 2020; Gil and Sequera 2020).

One of the 2008 crisis's consequences is the intensification of cities' tourist specialization as a formula for recovery, where tourist rental platforms have been the protagonists (Celata and Romano 2020). Although an economic recovery was generated, it aggravated some previous problems and developed new ones, especially with residents, transforming their residential daily life. Also, the vulnerability of highly tourism-dependent destinations was manifested when tourists flow decreased during the COVID-19 pandemic. In this sense, Duro et al. (2021) developed a methodology to assess how vulnerable Spanish destinations are to COVID-19, accounting for factors like density of lodging, market structure, and seasonality. Their results reached peaks in Mediterranean and Island destinations, highly dependent of tourism.

In this context, housing becomes the new boundary of tourist commodification, and the social content of cities, the new object of appropriation (Martínez-Caldentey et al. 2020). This has caused a progressive loss in the residential function of cities, together with the appearance of new investors willing to speculate with housing and tourist rentals (Ardura Urquiaga et al. 2020; García-Hernández et al. 2020). To illustrate this, historic centres of European capitals like Venice have halved their population since the 1960s (Marin-Cots et al. 2017). Toledo shows a similar trend, reducing its population to a third of its figures in the 1950s (Escudero et al. 2019).

Commercial activities are also transforming, and some processes that arose with the 2008 crisis are being accelerated: the accumulation of capital in fewer companies, logistical reliance, the growing virtualization of commerce and its consequent destruction of jobs, etc. (Frago 2021; Nanda et al. 2021).

This new wave of touristification occurred alongside the deployment of a whole series of liberalizing measures that promoted tourism investments and produced a notable increase of visitors to major European cities (Capocchi et al. 2020; Oklevik et al. 2019). Progressively, the terms touristification and gentrification are linked because tourist development generates dynamics of displacement of residents. Those residents, with lower purchasing power, are replaced by other residents (or users) with higher purchasing power (substitution of residential uses and increase in prices of conventional rent). Similarly, shops and commerce change (loss of functionality and substitution of traditional uses), and so does the public space's residential and tourist uses (privatization of public space, transformation of the social/identity fabric) (Cócola-Gant 2015; Gil 2020).

In this framework, the objective is to study the transformations in a European historic centre, specifically in the capital of the Costa del Sol, Malaga. The research focuses on the changes in the commercial fabric caused by tourism in the context of the pandemic and is related to other data (tourist flows, tourist housing, residents' perceptions…) to show how the most tourist areas are becoming less diverse, concentrate only specific uses, and are more vulnerable.

4.2 Gentrification and Touristification in Historic City Centres

The latest studies on the relationship between gentrification and touristification are being approached from the impacts of the growth of short-term rentals, especially in Spanish cities such as Madrid (Ardura Urquiaga et al. 2020; Gil and Sequera 2018), San Sebastian (Aguado-Moralejo and Del Campo-Echevarría 2020; Etxezarreta-Etxarri 2020), Seville (Jover-Baez and Barraquero-Díaz 2020), and Santa Cruz de Tenerife (García-Hernández et al. 2020).

Initially, the term gentrification was used to refer to the physical improvement of an environment linked to the replacement of the working class by middle-class groups. In central London, this phenomenon was studied by Ruth Glass (1964) in the 1950s and 1960s. The classic components of gentrification involve the reinvestment of capital, the upgrading of the locality by high-income groups, a change in the landscape, and a direct or indirect displacement of low-income groups (Davidson and Lees 2005).

Gotham (2005) proposed the concept of tourist gentrification as a "heuristic device that explains the transformations of middle-class neighbourhoods into relatively affluent and exclusive enclaves marked by a proliferation of corporate entertainment and tourism areas" (p. 1102). Tourist gentrification indirectly displaces businesses and residents with a symbolic substitution of uses and users (Cócola-Gant 2015). The agents of change are not only middle-class residents but a cosmopolitan middle class that demands and reproduces similar urban environments wherever it goes (Judd 2003).

Gentrification is part of an overall urban strategy within a neoliberal city management paradigm, identified with corporate urbanism. One immediate effect is how gentrification capitalizes on culture as an axis of economic development (Scott

1997, Castro-Noblejas et al. 2022): the introduction of cultural services increases the city's tourist and commercial attractiveness. It also enhances a city branding designed to attract tourists and outside investment whilst giving confidence to residents (Bramwell and Rawding 1996) and modulating the discourse of local corporations. It only highlights the perceived positive effects, in the form of economic development, hiding the appropriation of city centres for the upper-middle classes and capital (Lees 2008).

In Spain, the study of gentrification has been heterogeneous. In general, the Spanish-speaking world has political, economic, social, and administrative structures that influence the implementation of a gentrifying neoliberal urbanism (Janoschka et al. 2014), so the relevance of the phenomenon for academia is growing. Thus, Janoschka et al. (2014) managed to identify up to 6 current approaches or debates on gentrification in Spain: this research is framed within what the author identifies as the gentrification of public influence (or tourist gentrification), symbolic gentrification, and commercial gentrification. Their common point lies in the effort of the local corporation to provide favourable conditions for gentrification processes.

In short, we understand commercial gentrification as a process in which the commercial fabric is transformed, changing to meet the needs of a higher-income population. This causes displacement of shopkeepers, uses, and users (Salinas 2015; Cócola-Gant 2015), and in tourist spaces, commercial gentrification is understood as a change of uses for reasons related to leisure and tourism (Cócola-Gant 2015).

4.3 Commercial Gentrification in Tourist Areas

Early work on commercial gentrification already spoke about its symbolic dimension. Thus, from the "commercial landscapes of gentrification" approach, Bridge and Dowling (2001) studied the consumption practices associated with the gentrifying middle class and their relationship with identity. Other noteworthy works advance from the analysis of the commercial structure, such as Anguera-Torrell and Cerdan (2021) in the case of Barcelona and Barata-Salgueiro and Guimarães (2020) in Lisbon. According to Blázquez-Salom et al. (2019), tourists play a central role in the construction of current commercial landscapes of gentrification in tourist spaces. They consume types of food and clothing that symbolize "lifestyles and self-confirmation of identity".

The studies that address this issue do so with various methodological techniques. However, they revolve around three critical criteria: evolution (changes) of the commercial fabric, commercial diversity, and concentration of types of activity.

In terms of changes in the commercial fabric, Blázquez-Salom et al. (2019) analyse the tourist gentrification of the commercial landscape through the evolution of shop fronts in the centre of Palma de Mallorca, applying Visual Grounded Theory. They highlight the theming of high street shops in a new landscape, where franchises and brands of major retail chains proliferate, as well as souvenir, food, clothing, and accessories shops. In their analysis of commercial gentrification in Garosu-Gil (South Korea), Park et al. (2020) propose four factors to identify and evaluate changes in establishment types: (1) housing changing to retail; (2) shops changing their typology; (3) general businesses changing to franchises; (4) increases in the rate of business closures.

From the commercial gentrification approach, Yoon and Park (2018) analyse diversity by applying an entropy index to measure the complexity of commercial uses in different streets in Seoul (South Korea). Also, Celata and Romano (2020) determine the "tourist city" based on the concentration of Airbnb ads using a fractal methodology to compare the impact of tourist rentals in Italian cities.

Recent research analyses transformations in retail structure in the context of the pandemic. Nanda et al. (2021) analyse the UK high street land-scape, paying particular attention to the process of digitalization that the pandemic has accelerated, as well as changes in the retail real estate sector. Frago (2021) compares the proportion of shops closed before and after the pandemic in Barcelona's shopping areas. However, there is not as much literature on commercial gentrification in medium-sized, touristified cities.

4.4 Study Area

The case study is in Malaga (Spain), southern Europe (Fig. 4.1). It is the capital of Andalusia's leading tourist brand: the Costa del Sol. It is located at a strategic point on the Andalusian Mediterranean coast, in a position of geographical centrality. Malaga is a strategic entry point for Andalusia, especially by air: its airport received more than 19 million passengers in 2019 (AENA 2020), being the fourth airport in terms of arrivals that year. The port infrastructure also stands out, being the fifth cruise port in Spain, with 476,000 passengers in 2019 (Puerto de Malaga 2020).

The city made tourism and construction its main economic sectors since Spain's international isolation ended after the Civil War. From the 1960s onwards, its coastline became a global sun and beach destination: the new opportunities on the coast generated a rural exodus in search of better living conditions.

After two decades of growth and tourist real estate construction, land consumption multiplies exponentially (Castro-Noblejas et al. 2022), and urban developments on the periphery take over the centre as the primarily residential area. The centre was abandoned, residentially and functionally, in the 1970s and 1980s. Its physical and social degradation turned it into a place of insecurity, inhabited by an aging, low-income population, and isolated from the rest of the city.

By the 1990s, the city was physically and socially degraded. Earlier, in 1983, the General Urban Development Plan of Malaga (PGOU) was developed. It left behind expansionist urban planning and paid particular attention to the centre and its functions. Its status as an Historic-Artistic Site, protected as a Cultural Heritage Site (in Spanish, *Bien de Interés Cultural)*, led to the drafting of a protection plan for this area, which materialized in 1990 with the first Special Plan for the Protection

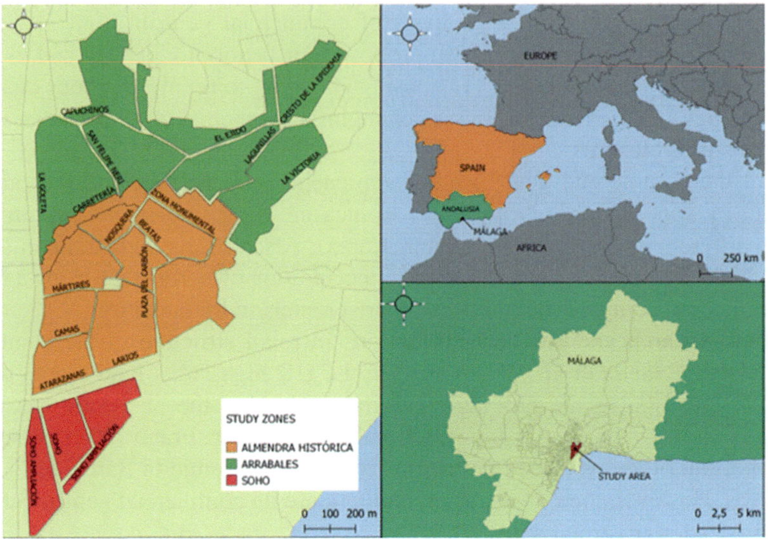

Fig. 4.1 Location and study area. *Source* Own elaboration

and Interior Reform of the Centre of Malaga (PEPRI). However, the real boost came in 1995 with the Urban regional policies of the European Union, which promoted the rehabilitation of historical quarters to improve the quality of life of residents and visitors, with funds to do so. With the first recovery interventions, these spaces, and the city, began to be seen as attractive products for private investment (Castro-Noblejas et al. 2022; Ozomek 2017).

Urban interventions included pedestrianization, renovation of buildings and squares, expropriations, demolition of some streets' surroundings for urban sponging, and the creation of green areas and car parks, amongst other actions. These interventions were aligned with cultural strategies around the historic centre, such as *Malaga Picassiana.* The use of culture as an economic engine continued in the following decade, favouring the opening of international museum branches, along with other public and private initiatives linked to culture. Today, the city boasts more than 35 museums that constitute a tourist asset (Romero-Padilla et al. 2019).

This chapter analyses the transformations of the commercial fabric, related to the tourist development of the historic centre of Malaga. We differentiate three zones (Fig. 4.1): *Almendra Histórica* (intramural city: orange), *Arrabales* (green), and *Soho* (red).

4.5 Methodology

In this research, different techniques have been used for data collection, processing, and analysis.

First, 5,810 single data on ground floor premises were obtained from a previous study carried out in the study area in 2019 (OMAU 2019). This data has been updated through fieldwork between January and December 2021 and documented with photographs and direct observation. The capture was complemented with data from Google Street View, Google My Business, and Bing Maps StreetSide.

The uses of the premises were recategorized into 24 Groups of Activities (AG) and 84 Detailed Activities (DA) based on: (1) direct observation and visual documentation during fieldwork; (2) classifications made by other studies on commercial fabric structure (Anguera-Torrell and Cerdan 2021; Blázquez-Salom et al. 2019; Bridge and Dowling 2001). They were subsequently characterized based on the classification of tourism activities from the Tourism Satellite Account of Spain (2019) as follows: (1) tourism businesses, premises offering products characteristic of tourism that would cease to exist in the absence of tourism, as well as products that visitors consume in significant quantities (gourmet food shops, souvenir shops, currency exchange offices, and luggage locker shops); (2) non-businesses, plots of land, closed commercial premises, residential dwellings, and other non-commercial use properties were included; (3) non-tourism businesses, the remaining businesses were included here.

Subsequently, the data was added to a Geographic Information System to analyse changes in commercial uses and to delve into the diversity and concentration of uses. Statistical and spatial analyses were carried out with QGIS 3.16 as detailed below:

- Changes. When interpreting the evolution of the commercial landscape (Park et al. 2020; Blázquez-Salom et al. 2019), we analyse changes in typology (tourist business, non-tourist business, and non-business). Any change implies a substitution and two interpretations: a business opens in 2021, for another one that was closed in 2019, and vice versa.
- Diversity. Following Yoon and Park (2018), we assess the diversity of Activity Groups (AG) using an entropy index: we use the Shannon–Weaver Index, already applied by Ontiveros and Castro (2014), to measure commercial diversity in different areas of the metropolitan area of San Luis Potosí (Mexico). A grid of homogeneous areas of 2,500 square meters is used (OMAU 2019). The index is calculated as follows:

  $$H' = -\sum_{i=1}^{u} pi \times \log^2 pi$$

  Where u is the total number of AGs and pi is the relative frequency of i AGs. The index measures the diversity of AGs in different zones of the study area. The value of the index increases as there is a greater number of AGs in an area, and/or the different premises are distributed homogeneously amongst the different groups. The higher the index, the greater the diversity of a particular area. The value "0" represents the minimum diversity when only one AG exists in an area.
- Concentration. Following studies that analyse concentration patterns in contexts of touristification (Martínez-Caldentey et al. 2020; Celata and Romano 2020), the

concentration zone of non-business, tourism business, and non-tourism businesses are identified through a hot–cold spot analysis. This analysis identifies statistically significant concentration zones based on the Getis-Ord Gi* statistic, indicating where entities with higher (hot spots) and lower (cold spots) values are spatially concentrated. This way, tourist, commercial, and residential areas are identified to compare the behaviour of their commercial fabric.

4.6 Results

Regarding the changes of use of commercial premises between 2019 and 2021 (Table 4.1), 45% of the changes correspond to new tourism businesses. Specifically, almost half of them (40%) are new tourist accommodations, whilst one out of three corresponds to restaurant. Most of these changes come from closed premises or premises dedicated to fashion, food, and specialized products. The most notable changes occurred in tourist accommodation (the most significant increase) and fashion (the largest decrease).

Regarding accommodation (hotel, pension, tourist apartment), they grew by 81% in 2021, with 106 new premises, almost double the figure in 2019. However, this should be put into context because opening new tourist accommodations (mainly hotels and tourist apartment buildings) implies a reduction in the premises available for other uses. An example: the building of *La Mundial* was demolished to build the so-called "*Hotel Moneo*". This not only meant the change in the use of the premises of that building and adjacent ones (8 in total): they are gone, which means a loss.

Other growing activities are those related to buying, selling, and pawning. Although the relative number is high (71.4%) compared to the absolute figure (10 new premises), this activity can be related to contexts of crisis and difficult access to bank loans.

In absolute terms, the two groups of activities with the highest number of new premises are those related to food and restaurants. Restaurant is the type of business with the highest number of premises: 722 in 2019 and 734 in 2021. They remain stable in 2021, with a weight of 46% with respect to total businesses. Meanwhile, the food group (take-away, gourmet, supermarkets, food shops, bakery/pastry shops), similar in terms of the needs it satisfies, increased its presence by 16% with 23 more units. In this case, the trend can be explained by the increase of Short-Term Rentals (STRs) before the pandemic: according to Cerezo-Medina et al. (2021), the cumulative annual variation rate of bed-places was 46.7% for the 2016–2019 period. However, it may have increased during the different confinement periods of the pandemic, causing an increased demand for takeaway food and home consumption.

Regarding the businesses experiencing a loss of activity, the fashion group stands out, falling by 21% (66 fewer premises). Businesses dedicated to ancillary tourist services (car hire, souvenirs, luggage storage, and leisure) also reduced their units, losing more than 18%. However, in absolute terms their impact is minor (9 premises).

Table 4.1 Activity groups 2019–2021

Groups of activities	2019 Activities		2021 Activities		% Change period
	Freq	Prop. (%)	Freq	Prop. (%)	
Tourist accommodation	131	2,30	237	4,10	80,90
Buying, selling, and pawning	14	0,20	24	0,40	71,40
Real estate	34	0,60	40	0,70	17,70
Food	145	2,50	168	2,90	15,90
Tourist attraction	95	1,60	105	1,80	10,50
Aesthetics	93	1,60	101	1,70	8,60
Stationery, graphic arts, and photography	36	0,60	39	0,70	8,30
Watertight	17	0,30	18	0,30	5,90
Residential	616	10,60	645	11,10	4,80
Other businesses	264	4,50	274	4,70	3,80
Jewellery	42	0,70	43	0,70	2,40
Restaurant	722	12,40	734	12,60	1,70
Bookshop	21	0,40	21	0,40	0
Other non-business	254	4,40	251	4,30	−1,20
Garage	277	4,80	272	4,70	−1,20
Gateway/Plot access	1362	23,40	1314	22,60	−3,50
Decoration and tableware	54	0,90	51	0,90	−5,60
Closed	740	12,70	698	12	−5,70
Solar	220	3,80	207	3,60	−6
Pharmacy	28	0,50	26	0,40	−7,10
Electronics and Hardware	35	0,60	32	0,60	−8,60
Specialized products	247	4,30	222	3,80	−10,10
Ancillary tourist services	48	0,80	39	0,70	−18,80
Fashion	315	5,40	249	4,30	−21
Total	5810	100	5810	100	-

Source Own elaboration

Specialized products (crafts, beauty, dressmaking, sports, flamenco, toys, music, opticians, etc.) decreased by around 10% (−25 premises).

Concerning the analysis of concentration, based on the count by typology of premises, three different zones have been identified (Fig. 4.2): an area of concentration of tourist uses (tourist accommodation, restaurants, food, travel agencies), another of commercial uses (supermarket, fashion, telephony, pharmacy, etc.), and another one with a predominance of non-business uses (housing, plots of land, museums, churches, theatres, garages, closed premises) which has been called the residential zone.

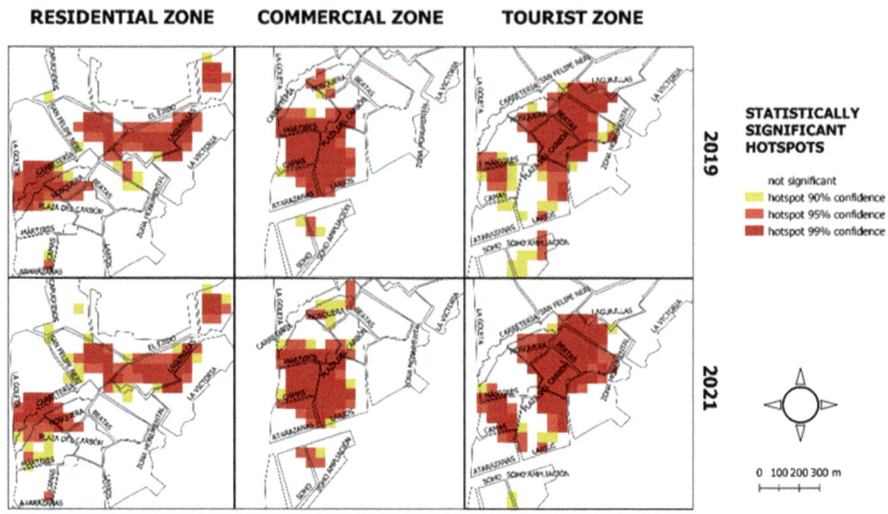

Fig. 4.2 Statistically significant concentration zones of non-business (residential), non-tourist business (commercial), and tourist business. The intensity of the colour indicates the statistical confidence level of the concentration of each type of business. Years 2019–2021. *Source* Own elaboration

The highest number of businesses (both tourist and non-tourist businesses) falls in the area where tourist businesses are concentrated, overlapping the area classified as a tourist zone. Consequently, the highest number of changes in the use of premises is also located in this area, signalling its volatility. Closures of tourist businesses were balanced out by new openings of the same type. In contrast, non-tourist businesses have lost presence. The number of non-tourist businesses in the remaining areas has also decreased, whilst tourist businesses have increased somewhat (Fig. 4.3).

Lastly, regarding diversity, the calculation of the Shannon–Weaver Index for AGs (excluding non-business, Fig. 4.4) shows that the tourist zone is not very diverse in

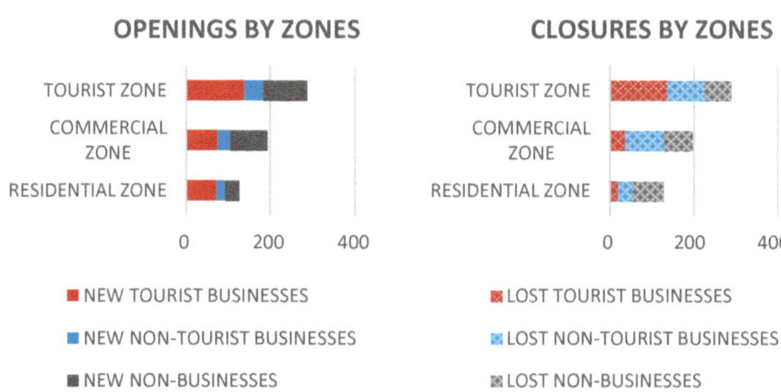

Fig. 4.3 Business openings and closures by area. *Source* Own elaboration

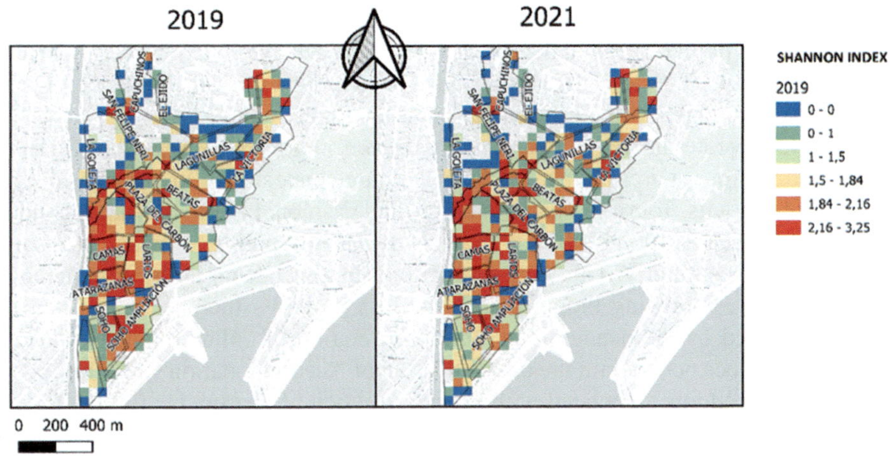

Fig. 4.4 Shannon Diversity Index for the years 2019–2021. *Source* Own elaboration

both years, despite being the zone with the most businesses. More than 40% of the premises correspond to a single type of activity (44% fits restaurant). The commercial area has a greater diversity of uses, but this decreases the closer we get to the tourist area, an indicator of its influence.

4.7 Discussion and Conclusions

This study made it possible to measure two moments in terms of touristification and commercial gentrification, with revealing results: tourist areas are less diverse. In addition, tourist uses penetrate commercial and residential areas. It was also found that the entry of tourist uses does not result in less immediate diversity: residential areas still remain diverse. However, tourist uses are likely phagocytizing expansion processes that will gradually reduce it.

The recovery policies implemented in the centre since the 1990s provided the conditions to achieve Malaga's current tourist development. These policies are in line with Harvey's (2002) entrepreneurial urbanism. Added to this is the cultural strategy aligned with the urban policy. They enabled more than 35 museums as new elements of tourist leisure that, without denying their social function, implant new symbolism in the roles, uses, and users of the centre (Gotham 2005; Cócola-Gant 2015). The refuge in tourism after the 2008 crisis was a lever for economic recovery that made destinations more dependent (Celata and Romano 2020; Martínez Caldentey 2020; Ardura Urquiaga et al. 2020; García-Hernández et al. 2020).

Considering the economic effects of the pandemic, a reorientation towards a more diversified model might be expected, but the results point to an expanding intensification of tourism. Many academics, as well as certain tourist agents, agree that the pandemic was an opportunity to reconsider our tourism model towards a sustainable one; some even point directly at degrowth as a way to achieve it (Palomo et al. 2020; Romagosa 2020). This study, like others developed around the same issue, confirms that the main trend is still the intensification of tourist activities.

The diversity rates in commercial uses are uneven, although less varied the further we go into the tourist areas, where 60% of uses are concentrated in premises dedicated to Restaurants, Tourist Accommodation, and Fashion. In turn, the main change observed between 2019 and 2021 is the loss of Fashion premises, whilst Accommodation grows similarly. This inverse relationship also occurs in the commercial area of the city. Following Frago (2021), the loss in the Fashion sector may have been accelerated by the pandemic and the proliferation of online retail. However, the pandemic does not explain the intensification of Accommodation since the operations and procedures associated with the construction of apartments (or any other lodging type) are planned urban actions that take a long time (Cerezo-Medina et al. 2021). Tourist dwellings, on their part, do not have this barrier (Cerezo-Medina et al. 2021). However, the administrative simplification measures implemented during the pandemic may have accelerated these actions (Junta de Andalucía 2021). In this regard, Boto-García et al. (2022) showed that a simplification in the regulation of short-term vacation rentals multiplied the number of units in various northern Spain destinations by 5 and 26 units.

Tourist uses are intensifying and expanding in areas subject to urban interventions (Rein Lorenzale 2013; Barreiro Cavestany 2013), and in those where actions are already planned: tourist uses grow in commercial areas by means of new accommodation. The same occurs in residential areas where tourist accommodation is the second largest use. Therefore, it is observed that areas with early urban renewal interventions have shown lower diversity. They concentrate significant tourist uses linked to accommodation and restaurants, too, which makes it relevant to observe the behaviour of future interventions. The results of this work are aligned with those of Castro-Noblejas et al. (2022): in their work, they show how the original residents of the historic centre are displaced by a new profile of inhabitant. They also verify that the housing market is adapting towards smaller residential units tagged with higher prices. The changes in the commercial fabric happen concurrently, accounting for that transformation. In a similar note, Jover and Díaz-Parra (2020) concluded that the historic centre of Seville had undergone a specialization process towards restaurants and lodging: those two uses are the largest in our study zone for the former, and the highest growing for the latter.

Amongst the study's limitations, it is worth mentioning the complexity in justifying what a tourist business is; therefore, it is essential to consider the relationship with the types of customers (Cócola-Gant 2015). This analysis finds the Tourism Satellite Account as a functional categorization but insufficient to cover the tourism reality.

The very methodology was a limiting challenge itself: the data was gathered during lock-down periods, with curfews for both businesses and people. Confirming that the main trend is still the continuous growth of tourist activities, there is a need to further develop this research line in the future. Finally, in the future, it will be interesting to study how sectors identified with the tourism industry co-agglomerate, developing indices that compare the co-agglomeration of different business typologies. Other possible studies may involve the commercial landscape generated by touristification and franchises and chains in the city's commercial areas. Another line of research may address the limitation detected concerning the classification of tourism businesses.

Acknowledgements This work is part of the research project I+D+i FEDER Andalucía 2014–2020 "Territorial intelligence for the management of tourist destinations. Governance and TIT tools in the use of public spaces" (Reference: P20_00982).

The research leading to this study is part of the research project entitled "Tourist saturation in Spanish coastal destinations. Strategies for tourism degrowth. An approach from the social dimension" (RTI2018-094844-B-C33) of the Ministry of Science, Innovation, and Universities (National R&D&I Plan).

This work has been funded by the University of Malaga through grant A.2. Predoctoral Contracts of the I Plan Propio de Investigación, Transferencia e Innovación.

References

AENA (2020) Tráfico de pasajeros, operaciones y carga en los aeropuertos. Aeropuertos Españoles y Navegación Aérea, Madrid

Aguado-Moralejo I, Del Campo-Echeverria J (2020) Airbnb in Donostia-San Sebastián: a sharing economy model? Ciudad y Territorio Estudios Territoriales 52(206):841–860. https://doi.org/10.37230/CyTET.2020.206.08

Anguera-Torrell O, Cerdan A (2021) Which commercial sectors coagglomerate with the accommodation industry? Evidence from Barcelona. Cities 112(December 2020). https://doi.org/10.1016/j.cities.2021.103112

Ardura Urquiaga A, Lorente-Riverola I, Ruiz Sanchez J (2020) Platform-mediated short-term rentals and gentrification in Madrid. Urban Stud 57(15):3095–3115. https://doi.org/10.1177/0042098020918154

Barata-Salgueiro T, Guimarães P (2020) Public policy for sustainability and retail resilience in Lisbon city centre. Sustainability (Switzerland) 12(22):1–22. https://doi.org/10.3390/su12229433

Barreiro Cavestany F (2013) Evaluación de los impactos del proceso de recuperación y regeneración urbana integral del centro histórico de málaga. 1994–2013. OMAU. Málaga. Available in https://static.omau-malaga.com/omau/subidas/archivos/0/8/8580/evaluacion-de-los-impactos-del-proceso-de-recuperacion-y-regeneracion-urbana-integral-del-centro-historico-de-malaga.-1994-2013.pdf

Blanco-Romero A, Blázquez-Salom M, Morell M, Fletcher R (2019) Not tourism-phobia but urban-philia: understanding stakeholders' perceptions of urban touristification. Boletin de La Asociacion de Geografos Espanoles 83:1–30. https://doi.org/10.21138/bage.2834

Blázquez-Salom M, Blanco-Romero A, Gual Carbonell J, Murray I (2019) Tourist gentrification of retail shops in Palma (Majorca). In: Milano C, Cheer JM, Novelli M (eds) Overtourism: excesses, discontents and measures in travel and tourism. CABI, p 244

Boto-García D, Baños-Pino JF, Del Valle E, Sustacha I (2022) Vacation rental market regulation and accommodation supply growth. Tour Econ. https://doi.org/10.1177/13548166221110530

Bramwell B, Rawding L (1996) Tourism marketing images of industrial cities. Ann Tour Res 23(1):201–221. https://doi.org/10.1016/0160-7383(95)00061-5

Bridge G, Dowling R (2001) Microgeographies of retailing and gentrication. Austr Geogr 32(1):93–107. https://doi.org/10.1080/00049180020036259

Campos R, Sequeira Á (2020) Urban art touristification: the case of Lisbon. Tour Stud 20(2):182–202. https://doi.org/10.1177/1468797619873108

Capocchi A, Vallone C, Amaduzzi A, Pierotti M (2020) Is 'overtourism' a new issue in tourism development or just a new term for an already known phenomenon? Curr Issue Tour 23(18):2235–2239. https://doi.org/10.1080/13683500.2019.1638353

Castro-Noblejas H, Sortino-Barrionuevo JF, Reyes-Corredera S (2022) Proceso de gentrificación de un espacio degradado convertido en escaparate cultural. El caso del Centro Histórico de Málaga. Rev de Estud Andaluces 43:0–1. https://doi.org/10.12795/rea.2022.i43

Celata F, Romano A (2020) Overtourism and online short-term rental platforms in Italian cities. J Sustain Tour:1–20. https://doi.org/10.1080/09669582.2020.1788568

Cerezo-Medina A, Romero-Padilla Y, Navarro-Jurado E, Francisco Sortino J, Guevara-Plaza A (2021) Comparative analysis of short-term rental homes and traditional accommodation in Andalusian tourist cities: intensity, density, and potential expansion areas. Curr Issues Tour. https://doi.org/10.1080/13683500.2021.1983522

Cócola-Gant A (2015) Tourism and commercial gentrification. RC21 International Conference on "The Ideal City: between myth and reality. Representations, policies, contradictions and challenges for tomorrow's urban life". http://www.rc21.org/en/conferences/urbino2015/

Davidson M, Lees L (2005) New build "gentrification" and London's riverside renaissance. Environ Plan A 37(7):1165–1190. https://doi.org/10.1068/a3739

Decreto-ley 26/2021, de 14 de diciembre, por el que se adoptan medidas de simplificación administrativa y mejora de la calidad regulatoria para la reactivación económica en Andalucía. Boletín Oficial de la Junta de Andalucía. Sevilla, 17 de diciembre de 2021, núm. 241.

De la Calle Vaquero M (2019) Turistificación de centros urbanos: clarificando el debate. Boletín de La Asociación de Geógrafos Españoles 83:1–40. https://doi.org/10.21138/bage.2829

Duro JA, Perez-Laborda A, Turrion-Prats J, Fernández-Fernández M (2021) Covid-19 and tourism vulnerability. Tour Manag Perspect 38(August 2020). https://doi.org/10.1016/j.tmp.2021.100819

Escudero L, Ruiz-Apilánez B, Solís E (2019) Las dinámicas de los centros históricos europeos en el presente: Toledo. Boletin Geográfico 2(41):99–122

Etxezarreta-Etxarri A, Izagirre-Olaizola J, Morandeira-Arca J, Mo-zo Carollo I (2020) Urban touristification in Spanish cities: con-sequences for the rental-housing sector in San Sebastian. Economic research-Ekonomska Istrazivanja 33(1):1294–1310. https://doi.org/10.1080/1331677X.2020.1755878

Frago L (2021) Impact of covid-19 pandemic on retail structure in Barcelona: from tourism-phobia to the desertification of city centre. Sustainability (Switzerland) 13(15). https://doi.org/10.3390/su13158215

García-Hernández JS, Armas-Díaz A, del Carmen Díaz-Rodríguez M (2020) Home dispossession and touristification in Santa Cruz de Tenerife (Canary Islands-Spain): tenant evictions in the neighborhood of El Toscal. Boletin de La Asociacion de Geografos Espanoles 87. https://doi.org/10.21138/BAGE.2982

Gil García J (2020) Turistificación, rentas inmobiliarias y acumulación de capital a través de Airbnb. El caso de Valencia. Cuad Geográficos 60(1):95–117. https://doi.org/10.30827/cuadgeo.v60i1.13916

Gil J, Sequera J (2018) Expansión de la ciudad turística y nuevas resistencias. El caso de Airbnb en Madrid. Empiria. Revista de Metodología de Ciencias Sociales 41. https://doi.org/10.5944/empiria.41.2018.22602

Gil J, Sequera J (2020) The professionalization of Airbnb in Madrid: far from a collaborative economy. Curr Issues Tour:1–20. https://doi.org/10.1080/13683500.2020.1757628

Glass R (1964) London: aspects of change. MacGibbon & Kee, London

Gotham KF (2005) Tourism gentrification: the case of New Orleans' Vieux Carre (French Quarter). Urban Stud 42(7):1099–1121. https://doi.org/10.1080/00420980500120881

Harvey D (2002) The art of rent: globalization, monopoly, and the commodification of culture. Socialist Register 2002: A World of Contradictions 32:94–110. https://socialistregister.com/index.php/srv/article/view/5778/2674

Janoschka M, Sequera J, Salinas L (2014) Gentrification in Spain and Latin America—a critical dialogue. Int J Urban Reg Res 38(4):1234–1265. https://doi.org/10.1111/1468-2427.12030

Jover J, Díaz-Parra I (2020) Gentrification, transnational gentrification and touristification in Seville, Spain. Urban Stud 57(15):3044–3059. https://doi.org/10.1177/0042098019857585

Jover-Báez J, Berraquero-Díaz L (2020) Residents or visitors? The impact of vacational rental on the housing market in Seville. Ciudad y Territorio Estudios Territoriales 52(206):823–840. https://doi.org/10.37230/CyTET.2020.206.07

Judd DR (2003) Visitors and the spatial ecology of the city. In: Hoffman L, Fainstein S, Judd DR (eds) Cities and visitors: regulating people, markets, and city space, vol 1. Blackwell, Oxford, pp 23–38

Lees L (2008) Gentrification and social mixing: towards an inclusive urban renaissance? Urban Stud 45(12):2449–2470. https://doi.org/10.1177/0042098008097099

Lefebvre H (1974) La producción del espacio (2013th ed). Capitán Swing Libros.

Marin-Cots P, Guevara-Plaza A, Navarro-Jurado E (2017) Renovación urbana y masificación turística en la ciudad antigua: pérdida de población y conflictos sociales. Ciudad y territorio Cy TET No 193, 49(1):453–468

Martínez-Caldentey MA, Murray I, Blázquez-Salom M (2020) In the city of Madrid all roads lead to Airbnb. Investigaciones Turisticas 19:1–27. https://doi.org/10.14198/INTURI2020.19.01

Nanda A, Xu Y, Zhang F (2021) How would the COVID-19 pandemic reshape retail real estate and high streets through acceleration of E-commerce and digitalization? J Urban Manage 10(2):110–124. https://doi.org/10.1016/j.jum.2021.04.001

Observatorio de Medio Ambiente Urbano de Málaga (2019) Análisis de los usos en planta baja en el ámbito del PEPRI Centro. Available in https://static.omau-malaga.com/omau/subidas/archivos/6/7/8176/analisis-de-los-usos-en-planta-baja-en-el-ambito-del-pepri-centro.pdf

Oklevik O, Gössling S, Hall CM, Steen Jacobsen JK, Grøtte IP, McCabe S (2019) Overtourism, optimisation, and destination performance indicators: a case study of activities in Fjord Norway. J Sustain Tour 27(12):1804–1824. https://doi.org/10.1080/09669582.2018.1533020

Ontiveros AA, Castro MAB (2014) Medición de la diversidad comercial minorista en áreas urbanas a través del uso de los índices de Shannon-Weaver y de Ullman-Dacey. Estudios Geográficos 75(277):455–478. https://doi.org/10.3989/ESTGEOGR.201413

Ozomek I (2017) Centro histórico de Málaga 2001–2016. Evolución del comercio y la hostelería. Actualización de datos para 2016–2017. DondeNegocios, Málaga. Avaiable at https://revistaelobservador.com/images/stories/envios_17/junio/centro.pdf

Palomo G, Jurado E, Medina A, Bernier E (2020) Turismo poscoronavirus, ¿una oportunidad para el poscrecimiento? In: Simancas Cruz MR, Hernández Marín R, Padrón Fumero N (eds) Turismo pos-COVID-19. Reflexiones, retos y oportunidades. Universida, La Laguna, pp 161–173. https://doi.org/10.25145/b.Turismopos-COVID-19.2020

Park J, Kim J, Yang B (2020) Spatializing an artist-resident Community Area at a Building-Level: a case study of Garosu-Gil, South Korea

Puerto de Malaga (2020) Annual report 2019. Puerto de Málaga, Málaga

Rein Lorenzale M (2013) informe sobre intervenciones urbanísticas y transformaciones físicas en el ámbito del pepri centro (1994–2013). OMAU. Málaga. Available in https://static.omau-malaga.com/omau/subidas/archivos/9/7/8579/informe-sobre-intervenciones-urbanisticas-y-transformaciones-fisicas-en-el-ambito-del-pepri-centro-%281994-2013%29.pdf

Romagosa F (2020) The COVID-19 crisis: opportunities for sustainable and proximity tourism. Tour Geogr 22(3):690–694. https://doi.org/10.1080/14616688.2020.1763447

Romero Padilla Y, Cerezo Medina A, Navarro-Jurado E, Romero Martínez JM, Guevara-Plaza A (2019) Conflicts in the tourist city from the perspective of local social movements. Boletín de la Asociación de Geógrafos Españoles 83(2837):1–35. https://doi.org/10.21138/bage.2837

Salinas L (2015) Transformación de mercados municipales de Madrid: de espacio de consumo a espacio de esparcimiento. Revista INVI 31(November 2015):179–201

Scott AJ (1997) The cultural economy of cities. Int J Urban Reg Res 21(2):323–339. https://doi.org/10.1111/1468-2427.00075

Soja EW (2008) Postmetrópolis. Estudios críticos sobre ciudades y regiones (Traficante)

Tourism Satellite Account of Spain (2019) Revisión Estadística 2019. Notas Metodológicas. Instituto Nacional de Estadística, Madrid. Available in https://www.ine.es/prensa/cst_2019.pdf

Yoon Y, Park J (2018) Stage classification and characteristics analysis of commercial gentrification in Seoul. Sustainability 10:2440

Chapter 5
Citizen Resistance in Touristified Neighborhoods. A Post-pandemic Analysis

Montserrat Crespi-Vallbona and Cristina López Villanueva

Abstract Consolidated tourist destinations transform cities in terms of both their urban characteristics and territorial nature. So intense are some of these processes of touristification that they have a profound impact on cities' popular neighborhoods, which find themselves struggling to contain the effects and preserve their roots. These dynamics are accelerated by the exposure of the local real estate market to global investment processes and its tourist specialization, which trigger ongoing neighborhood protests. Against this backdrop, touristification has a number of repercussions: first, the expulsion of local residents due to the reduction in housing stock, as residential uses are replaced by tourist uses; second, the disappearance of local shops and services; third, the saturation of public space and fourth, the conflicts of coexistence between residents and tourists. This chapter examines the response of residents to the touristification of the city of Barcelona by combining a quantitative study—drawing on data taken from Barcelona City Council's Cens d'establiments d'allotjament turístic (Census of Tourist Accommodation), official Registres de la Propietat (Property Registers) and the Censo de Población y Viviendas (Census of Population and Housing) conducted by the Instituto Nacional de Estadística (INE) (National Institute of Statistics)—and a qualitative analysis—based on in-depth interviews with members of local government and leaders of the city's neighborhood associations. While the pandemic represented a temporary interruption in tourism activity and provided an opportunity to rethink the dominant model of tourism development, the resumption of activity in the sector has resurrected longstanding problems and old resistances in a new scenario.

Keywords Touristification · Collective action strategies · Sustainability · Barcelona

M. Crespi-Vallbona (✉)
Business Department, University of Barcelona, Barcelona, Spain
e-mail: mcrespi@ub.edu

C. L. Villanueva
Sociology Department, University of Barcelona, Barcelona, Spain
e-mail: clopez@ub.edu

E. Navarro-Jurado et al. (eds.), *Urban Dynamics in the Post-pandemic Period*,
The Urban Book Series,
https://doi.org/10.1007/978-3-031-36017-6_5

5.1 Tourism and the Pandemic

The global lockdown and the crisis attributable to the COVID-19 pandemic have had a serious impact on destinations where the tourism sector has established itself as one of the main drivers of the economy. In addition to these direct economic effects—the World Tourism Organization (UNWTO) estimates that the cancellations made by almost 1 billion tourists resulted in losses in global GDP that exceeded 2 trillion dollars putting 100–120 million jobs at risk (UNWTO 2020)—the enforced break in activity caused by the unprecedented health, economic and social crisis has provided a unique opportunity to reflect on the effects of tourism and, in particular, on the impact of urban tourism on consolidated city destinations. Within public institutions and at different levels of government, especially the local tier, a discourse has been generated regarding the need to rethink tourism in order that it might be made more sustainable, inclusive, resilient and fair (UNWTO 2020).

This twenty-first century is characterized by the express desire of many cities to become tourist destinations, highlighting the heritage of their historic centers and/or organizing cultural and/or sporting events (Crespi-Vallbona and Costa-Planells 2011). The destinations seek to stimulate their economic development (Miles and Paddison 2005; Smith 2007) to the point, that a city becomes a business (Cabrerizo et al. 2016), managed as a company to attract capital and tourists (Harvey 1989), prioritizing investment that adapts the city to tourism but which neglects the needs of its citizens (Hall 2009). In this way, city brand creation and positioning has become strategic (Eisenschitz 2016).

The concept of touristification was coined to describe the process by which the tourism industry transforms a city into a monopolized, hyperspecialized scenario (Lanfant 1994). This process is usually accompanied by the symbolic dispossession of the buildings and spaces of sociability of traditional residents (Benach Rovira 2016; Crespi-Vallbona and Mascarilla-Miró 2018; Martínez-Caldentey et al. 2020), which represents an obvious threat for a city's neighborhoods, as they are places of ties and attachment for the community where everyday lives are lived out and where the well-being of the neighbors is satisfied (Barañano 2021; López-Villanueva and Crespi-Vallbona 2022).

Since the first decade of the twenty-first century, the realities of mass, invasive and unwanted tourism (Pimentel de Oliveira 2020) have triggered a multitude of critical movements condemning the negative effects of tourism for cities and their neighborhoods. Such actions are embedded in the different social protest movements that develop strategies of criticism and rejection in the face of participatory deficits (Schneiberg and Lounsbury 2008) but, above all, those reclaiming their right to the city (Lefebvre 2017).

Even before the outbreak of the pandemic, but now once again with the upsurge in tourism, these movements, which have both local expression but also a global view (e.g. the Network of Southern European Cities Against Touristification or SetNet), denounce the following trends: (a) the reduction in and the precariousness of the right to housing caused by the rise in rental and purchase prices; (b) the transformation and increased cost of local commerce attributable to the specialization in tourism which makes supply incompatible with the demand of the population residing in the neighborhoods, resulting in the disappearance of the relationship of proximity, hitherto, the backbone of local commerce; (c) the overcrowding of public space that makes daily life difficult and which generates conflicts in the use of space by café terraces and rental vehicles (such as motorcycles, scooters or bicycles), hindering accessibility and mobility and generating problems of coexistence with neighbors, noise, dirt and uncivil behavior; (d) the saturation of public transport networks; (e) the precariousness of working conditions for those that work in the tourism sector; (f) the growing dependence on the tourism sector in the economic sphere and (g) the environmental problems associated with the intense traffic of airplanes and, especially, cruise ships, without mentioning the excessive consumption and waste of energy. In response, Fletcher et al. (2021) and Higgins-Desbiolles (2020), among others, call for limits on tourism aimed at changing the city and on the promotion of urban tourism as a form of merchandising (Andriotis 2018; Fletcher et al. 2020; Hall et al. 2020; Crespi-Vallbona and Domínguez-Pérez 2021).

The outbreak of the pandemic in March 2020, which confined the population to their homes and curtailed any type of mobility, left these social platforms with little activity. As we have seen, the crisis attributable to COVID-19 provided public institutions and local government with an exceptional opportunity to rethink tourism at both global and urban levels. In a similar vein, the UNWTO (2020) presented its recommendations for urban tourism structured around three main dimensions: (a) the integration of tourism into the urban development agenda—which, undoubtedly, appeals to the governance of tourism insofar as it facilitates the participation of government at different scales (national, regional and local) and gives the private sector and local communities a significant role (Suarez-Rosa and Fumero 2020; Crespi-Vallbona and Farran-Plana 2022); (b) sustainable policies and practices in urban tourism which take into account the four dimensions of sustainability—that is, environmental, sociocultural, economic and destination management (Olcina 2020; Crespi-Vallbona et al. 2019)—in keeping with the Sustainable Development Goals (SDGs) and (c) the construction of cities for all, that is, citizens and their visitors (Zapata 2020), ensuring that the benefits of tourism can contribute to wealth development and equal distribution, guaranteeing decent jobs and social inclusion. In short, governance, sustainability, and social inclusion are considered key factors for the development of urban tourism in the post-pandemic economy.

Despite the apparent commitment during confinement to the rethinking of tourism, prioritizing lines of sustainability, governance, ecological awareness and the participation of the agents involved, these initiatives have ended up as little more than good intentions. The rapid recovery of tourism to pre-pandemic levels has seen the reemergence of all the old problems and the re-voicing of longstanding citizen demands, reestablishing old resistances.

5.2 The City of Barcelona. Context of the Case Study

Barcelona is an excellent example of a city that has been exposed to these changes and dynamics because of its specialization in tourism. At the same time, it has a well consolidated trajectory in neighborhood movements since the 1979 transition to democracy in Spain, a country with a rich urban substratum (Jiménez 2011). In more recent decades, these neighborhoods of Barcelona have seen how the city has become one of the most valued destinations among international tourists, receiving more than nine million visitors according to the Global Destination Cities Index (2019) and ranking among the top 20 urban cities worldwide. Figure 5.1 shows this increasing demand for international tourism, which started with the Olympic Games in 1992 and which has been further promoted thanks to a municipal strategy of global urban renewal, both physical and social (Alabart Vila et al. 2015).

Here, we focus our study on one specific area of Barcelona—District III or Sants-Montjuïc—and analyze the effects of touristification and the residents' strategies of resistance to uphold their "right to the city". Sants-Montjuïc (Fig. 5.2) is Barcelona's largest district (with a total area of 2,090 hectares) and is made up of eight neighborhoods: el Poble-sec, la Zona Franca, la Marina del Port, la Font de la Guatlla, Hostafrancs, la Bordeta, Badal and Sants.

We have selected this district for our case study for a number of reasons: (a) because of its location, bordering the city's historic center of the city and the area that is home to Futbol Club Barcelona's stadium, the Camp Nou, two of the city's main tourist poles according to statistics published by Turisme de Barcelona; (b) because it houses some of the city's main facilities and infrastructure, most notably the largest railway terminal (including the high-speed train or AVE terminal) and the largest complex of trade fair buildings and conference facilities in the city (Fira

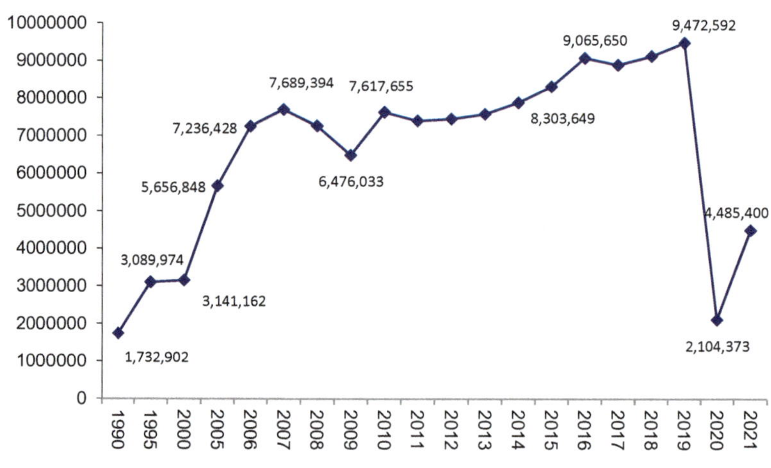

Fig. 5.1 International tourism arrivals in Barcelona, 1990–2021. *Source* Own elaboration based on data from Turisme de Barcelona

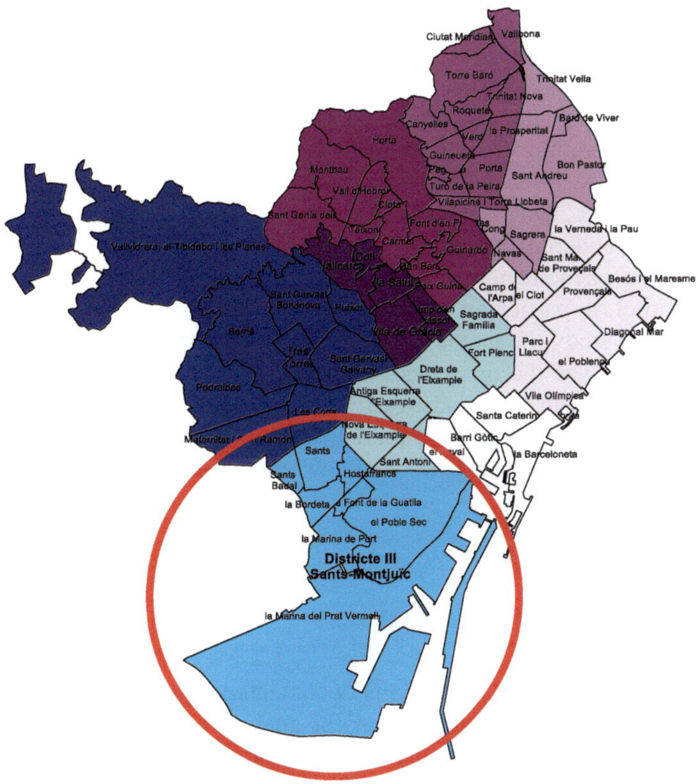

Fig. 5.2 Map showing location of District III (Sants-Montjuïc) in Barcelona. *Source* Own elaboration

de Barcelona and Palau de Congressos) as well as extensive leisure, culture and exhibition facilities; (c) because it has undergone a major urban transformation, which has served to embellish the space: the redevelopment of Montjuïc, the city's hill, to house the facilities for the 1992 Olympic Games (Estadi Olímpic and Palau Sant Jordi) and the reconversion of the old factory complexes (the district's industrial carries an enormous weight) into public facilities, including the burying of the train tracks and the landscaping of some areas of the district.

The district's significant industrial past is another fundamental element, underpinning its extensive and highly structured trajectory of trade union and cooperative movements (dating back to the nineteenth century). Indeed, workers organized themselves in cooperatives centered on socioeconomic relationships and based on cooperation and solidarity. But, as well as organizing consumption, the cooperatives contributed to create a cultural and neighborhood identity (Dalmau and Miró 2010).

Later, the decline in the city's industrial activity and the real estate boom fostered a period of speculation. Against this backdrop, major neighborhood protests were organized calling for the conversion of these spaces into facilities for the neighborhood.

Later, neighborhood associations took over the leadership of these protests (Alabart 2010). Neighborhood actions, which have taken on a variety of organizational and communicative forms, above and beyond those of the neighborhood associations (Fernández et al. 2019), have incorporated the negative effects of urban tourism into their protests. In sum, the district's cooperative and associative traditions were and continue to be the backbone of the strategy of citizen resistance (López-Villanueva and Crespi-Vallbona 2021).

The analysis undertaken here is structured into the following sections. First, we describe the phenomenon of tourism in the city, in general, and District III (Sants-Montjuïc), in particular, by analyzing the evolution in tourist accommodation. Second, we study the increase in housing for tourist uses vs. residential uses. And third, we analyze the resistance strategies adopted by organized society and local government in the case study area. This example serves to illustrate how the negative effects of the dynamics of touristification have reemerged after the interruption attributable to the pandemic and how citizen platforms of resistance have resumed their activity.

5.3 Sources and Methodology

The case study conducted here exploits documentary, qualitative and quantitative techniques of analysis. The information for analyzing tourist accommodation is drawn from the Cens d'establiments d'allotjament turístic (the Census of Tourist Accommodation) as maintained by Barcelona City Council and the Guild of Hotel Owners of Barcelona, while that concerning the residential use of the city's housing stock is taken from official Registres de la Propietat (Property Registers) as published by the Department of Territory and Sustainability of the Catalan Government, from the data maintained by the Institut Català del Sol (INCASOL) (Catalan Institute of Land Use), and from the Censo de Población y Viviendas (Census of Population and Housing) conducted by the Instituto Nacional de Estadística (INE) (National Institute of Statistics).

Our analysis of the resistance strategies employed is based on in-depth interviews with heads of local government and with leaders of neighborhood organizations in the Sants-Montjuïc District as well as in the rest of the city. Specifically, we interviewed the head of Economic Promotion and Tourism in the Sants-Montjuïc District (I1) and the councilor of Culture for the same District (I2); an activist from Lleialtat Santsenca (an association promoting neighborhood relations and cooperativism in Sants) (I3); a member of La Ciutat Invisible (a self-managed cooperative based in Sants) (I4); a member of Associació de Veïns d'Hostafrancs (a neighborhood association) (I5); the leader of the Platform Fem front al turisme (an anti-tourism pressure group) (I6); and the head of the Observatory of Economic, Social and Cultural Rights (ESCR) (I7).

5.4 Results

In what follows, we analyze the touristification processes in our case study area and identify the resistance actions implemented by its neighborhood/citizen organizations.

5.4.1 *Tourist Accommodation: Hotels and Apartments*

The number of hotel beds in the city of Barcelona shows an increasing trend, rising 21.5% from 63,952 in 2010 to 75,283 in 2021. However, in District III the trend is somewhat different, with the number being largely contained for much of this period until 2020 when a marked leap was detected. Thus, in this same period, the overall increase was 7% rising from 6,581 to 7,040 beds (Fig. 5.3).

This rise in the city's accommodation capacity is also apparent in tourist housing. The VUT (i.e. housing for tourist uses) licensing system was implemented in 2012 in response to the massive arrival of international tourism. The system introduced means that a licensed tourist home operates as a tourist company. In 2014, tourist homes accounted for 41,555 registered tourist beds, a figure that reached 57,442 in 2021, representing an average annual growth of 4.73%. This same increase, albeit with greater intensity, has also been observed in the District of Sants-Montjuïc (Fig. 5.4). Thus, while in 2021 the number of hotel beds in Sants-Montjuïc represented 9.35% of total accommodation in Barcelona, in this same year, accommodation in tourist homes in District III represented 10.16%.

The city's 2015 Pla Estratègic de Turisme (Strategic Tourism Plan) highlights government concerns for the management of tourist uses in Barcelona. Two years later, the Pla Especial Urbanístic d'Allotjaments Turístics or PEUAT (Special Urban Plan for Tourist Accommodation was introduced (gaining final approval in 2021) in an effort at regulating tourist accommodation, guaranteeing the fundamental rights of the city's local residents and establishing a system of accommodation zones that divided the city in a number of areas. Zone 1—comprising la Ciutat Vella, part of the l'Eixample, el Poblenou, la Vila Olímpica, el Poble-sec, Hostafrancs and Sant Antoni—was designated a negative-growth area forbidding the provision of

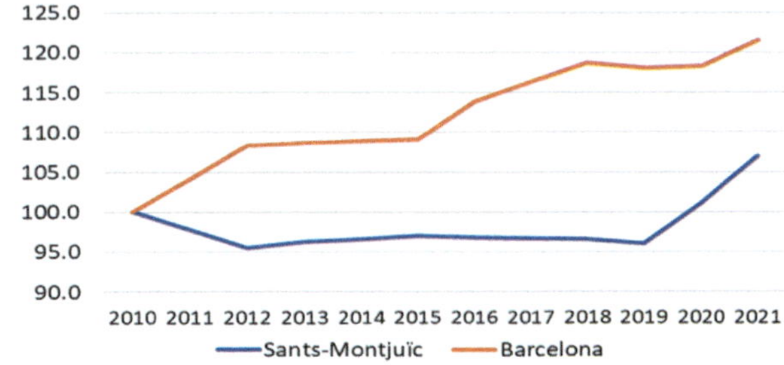

Fig. 5.3 Increase in number of hotel beds Barcelona, 2010–2021 (100 base index). *Source* Own elaboration based on data from Turisme de Barcelona and the Cens d'establiments d'allotjament turístic maintained by Barcelona City Council

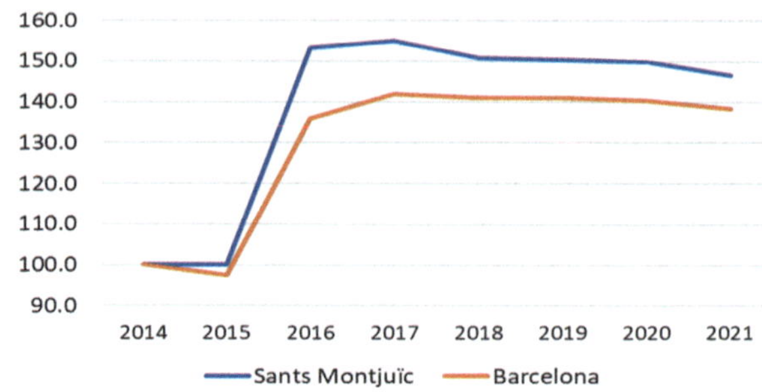

Fig. 5.4 Evolution undergone by tourist housing in Barcelona, 2014–2021 (100 base index). *Source* Own elaboration based on data from Turisme de Barcelona and the Cens d'establiments d'allotjament turístic maintained by Barcelona City Council

any new accommodation facilities and the issuing of any new licenses for tourist apartments. The PEUAT seeks to restrict the provision of hotel rooms and tourist apartments, thus easing the pressure of the temporary accommodation provision on plans to construct housing for purchase or rent by more permanent residents. The Plan also provides for the more rigorous inspection of illegal tourist homes on horizontal hosting platforms. However, restricting the issue of licenses for tourist housing and the closure of illegal tourist accommodation do not detract from the residents' real concern: the increase in the price of housing, both for purchase and for rent (Crespi-Vallbona and Mascarilla-Miró 2018).

One impact of the COVID-19 pandemic was that an indeterminate, but assuredly significant, number of tourist homes became residential apartments. However, with the lifting of the pandemic restrictions, "normality" in the Barcelona tourism sector has been quickly restored.

5.4.2 Evolution of Residential Housing vs Tourist Housing

The growth in tourism in Barcelona has sparked considerable real estate speculation and a concomitant increase in tourist apartments. This, not surprisingly, has generated marked discontent at the local level, with residents unable to afford the higher rental prices being forced to abandon these touristified neighborhoods (La Veu del Barri 2017). This silent but effective expulsion of residents is the counterpoint to the conversion of residential homes into tourist accommodation, which is accompanied by the disappearance of local businesses and the emergence of other types of establishment designed to cater to tourists and not to local residents (Frago 2021).

The average purchase price of housing (3,803.76 €/m2) and the average rental price (799.69 €) in District III are both below the average prices for the city of Barcelona as a whole (Table 5.1). However, the relative increase in purchase prices between 2014 and 2021 exceeds that of the city (90.80% vs 49.57%), pointing to a clear revaluation of the District. These property prices serve, moreover, to stress the diversity of neighborhoods that make up District III, above all the average price of 4,497.83 €/m2 recorded in el Poble-Sec, which exceeds the average for the city as a whole. Indeed, it is precisely in this neighborhood that the concentration of licensed tourist homes is highest, accounting for 42.33% of all such properties in the District.

Such housing costs have a direct impact on demand. Moreover, the expectation is that the higher the prices, the higher the annual disposable income of the resident population is likely to be. Yet, mean disposable incomes in District III are lower than those of the city as whole (18,949 € vs. 22,229 €) and, moreover, those of el Poble-sec and la Marina del Port are the lowest in the District (Table 5.2). This means that their populations are more vulnerable to the risks of "expulsion" caused by the rise in house prices.

The pressure of the VUT licensing system on the housing stock, especially in the case of the rental market, is a critical factor in upholding the right to housing. In this instance, its impact can be assessed by comparing the number of tourist apartments with the number of properties in the home rental market. In this case (Table 5.3), the housing given for free or at a low price has been added to the rental stock; thus, in the Sants-Montjuïc District for every 100 rental homes there are 4.8 tourist flats, a similar relationship to that found throughout Barcelona. In some of the District's neighborhoods, specifically, Hostafrancs and el Poble-Sec, this pressure rises to 7.9 and 7.7%, respectively. However, two limitations of these data should be mentioned. First, they are provided by Barcelona City Council's Cens d'establiments d'allotjament turístic and, as discussed, only licensed tourist apartments are listed. This means all unregulated apartments do not appear in these statistics. Second, the data source for the housing stock according to the tenure regime is the Censo de Población y Viviendas. However, the most recent edition corresponds to 2011, which means there is a substantial time gap between the two data sets; however, it does at least provide an approximation of the pressure to which the housing stock is under.

This tourist pressure is made especially evident if we observe the resident population in the District in relation to the total number of tourist beds available. In Sants-Montjuïc, for every 100 inhabitants there are 3.12 tourist beds registered in housing for tourist use (Table 5.4). In el Poble-sec and Hostafrancs, this proportion almost doubles that of District III and of the city of Barcelona as a whole. The intense tourist pressure in these neighborhoods is focused above all in the housing market, which is one of the main reasons accounting for the residents' discontent.

5.4.3 Resistance Strategies Employed by Organized Society

The dynamics of Barcelona's urban tourism has led to a rise in the protests and resistance platforms of its neighborhood residents. These platforms have materialized in different guises and are framed within broader social movements that seek to repel tourism's negative effects, especially its impact on house prices (both purchase and rental) (Gil and Sequera 2018); changes in residential dynamics, as well as in the commercial fabric and neighborhood relations (Huete and Mantecón 2018; Mansilla 2018; Milano et al. 2019) and the use and occupation of public space (Luque et al. 2019).

Table 5.1 Average prices and increases in housing for buying, selling and renting in the neighborhoods of the Sants-Montjuïc District, 2014–2020

	Buying and selling price			Relative Increase	Renting price			Relative Increase
	Average price € by m2		Increase		Euro by month		Increase	
	2014	2021	2014–2021	2014–2021	2014	2021	2014–2021	2014–2021
Poble Sec-Parc Montjuïc	1,607.2	4,497,8	2,890,7	179.9	582.6	780.1	197.5	33.9
Marina del Prat Vermell-Zona Franca	684.5	2,607.7	1,923.2	281.0	n/a	430.6	n/a	n/a
Marina del Port	1,876.8	3,448.0	1,571.2	83.7	527.3	743.1	215.9	40.9
Font de la Guatlla	2,188.7	3,754.9	1,566.2	71.6	644.6	826.7	182.1	28.2
Hostafrancs	2,174.7	3,922.2	1,747.5	80.3	593.9	789.8	195.9	33.0
La Bordeta	1,678.6	3,300.9	1,622.3	96.6	611.4	832.3	221.0	36.1
Sants-Badal	3,282.9	3,603.8	320.9	9.8	597.3	793.3	196.0	32.8
Sants	2,185.6	3,791.8	1,606,2	73.5	624.1	829.0	204.9	32.8
District Sants-Montjuïc	**1,993.6**	**3,803.8**	**1,810.1**	**90.8**	**598.4**	**799.7**	**201.2**	**33.6**
Barcelona	**2,754.4**	**4,119,8**	**1,365.4**	**49.6**	**688.2**	**918.8**	**230.6**	**33.5**

Source Own elaboration based on data from the College of Property Registrars (published by the Department of Territory and Sustainability of the Catalan Government) and INCASOL

Table 5.2 Household disposable income per capita (€/year), 2015–2019. Neighborhoods of the Sants-Montjuïc District

	2015	2016	2017	2018	2019
Poble Sec-Parc Montjuïc	14,664	14,939	15,082	16,091	14,664
Marina del Prat Vermell-Zona Franca	11,935	11,811	12,703	13,135	11,935
Marina del Port	16,119	16,379	16,244	16,997	16,119
Font de la Guatlla	18,850	19,034	19,147	20,011	18,850
Hostafrancs	18,397	18,533	18,736	19,413	18,397
La Bordeta	18,514	18,743	18,996	19,996	18,514
Sants-Badal	18,297	18,417	18,434	19,584	18,297
Sants	19,987	20,419	20,268	20,967	19,987
District Sants-Montjuïc	**17,528**	**17,793**	**17,827**	**18,710**	**18,949**
Barcelona	**20,055**	**20,733**	**20,956**	**21,484**	**22,229**

Source Own elaboration based on Barcelona Technical Programming Office. Barcelona City Council Statistical Institute

Table 5.3 Tourist apartments vs. Rental Homes in Sants-Montjuïc, 2021

	Rental housing + provided free or low price	Tourist use housing	Apartments & rental homes
Poble Sec-Parc Montjuïc	6,454	498	7.7
Marina del Prat Vermell-Zona Franca	–	3	–
Marina del Port	2,263	6	0.3
Font de la Guatlla	1,233	73	5.9
Hostafrancs	2,214	175	7.9
La Bordeta	1,665	39	2.3
Sants-Badal	3,165	103	3.3
Sants	6,884	244	3.5
District Sants-Montjuïc	**23,878**	**1,141**	**4.8**
Barcelona	**219,239**	**9,397**	**4.3**

Source Own elaboration based on the Cens d'establiments d'allotjament turístic (Barcelona City Council, 2021) and the 2011 Censo de Población y Viviendas

In Sants-Montjuïc there is both an extensive and dense associative network with a deep-rooted historical base that can be traced back to the workers' struggles of the last century and post-Francoism. This has contributed to forging the identity of the territory and its resistance to the capitalist appropriation of space (Dalmau and Miró 2010), which explains why several entities in the District have joined the struggle

Table 5.4 Tourist accommodation (tourist housing beds) vs. resident population in Sants-Montjuïc

	Tourist accommodation in tourist use housing	Population	Tourist accommodation beds in tourist use housing/population
Poble Sec-Parc Montjuïc	2,472	40,204	6,15
Marina del Prat Vermell-Zona Franca	10	1,356	0,74
Marina del Port	38	31,092	0,12
Font de la Guatlla	408	10,239	3,98
Hostafrancs	902	16,102	5,60
La Bordeta	210	19,451	1,08
Sants-Badal	490	24,903	1,97
Sants	1,310	43,679	3,00
District Sants-Montjuïc	**5,840**	**187,026**	**3,12**
Barcelona	**57,442**	**1,660,314**	**3,46**

Source Own elaboration based on the Cens d'establiments d'allotjament turístic (Barcelona City Council, 2021) and the 2021 Padró Municipal d'Habitants (Municipal Register of Inhabitants)

aimed at achieving the degrowth of urban tourism. The actions of this critical network converge on the Assemblea de Barri de Sants (Sants Neighborhood Assembly) as well as in the broader context provided by the city of Barcelona and even internationally. In 2015, the Assemblea de Barris per un Turisme Sostenible (ABTS) (Neighborhood Assembly for Sustainable Tourism) was founded in Barcelona as a popular counterpoint to combat the effects of tourism that threaten to undermine the character of the city's neighborhoods. This movement was set in motion by the Federació d'Associacions Veïnals de Barcelona (Federation of Neighborhood Associations of Barcelona) with the goal of coordinating the actions of those neighborhoods suffering most the problems of tourist overexploitation, actions that are further supported by SETNet.

These platforms seek, above all, to influence public opinion, give visibility to the problems of urban tourism and, generally, sound the warning to local residents. As such, they place an emphasis on high-impact collective actions that attract press interest, such as obstructing or disabling tourist buses and putting up posters identifying tourist apartments being operated without a license. Another strategic line adopted is more educational and sees the associations seeking to raise awareness by organizing workshops, debates and meetings. In addition, the platforms seek to establish dialogue with the government leaders with responsibility for addressing the problems that negatively impact local residents (Mansilla and Hughes 2021). Such actions acquire particular significance as they are based on two fundamental premises: the first, is their dependence on the participation of agents—public, private, social entities, etc.—that intervene in the territory, following a bottom-up approach and, the second, is their insistence on playing a key role in the management and decision-making of their territory. A good example of the latter is the participation of anti-tourism platforms in public policy design, including the aforementioned tourist accommodation plan, PEUAT.

In short, the residents of District III organize protest actions, interact with public institutions, empower their fellow citizens and cooperate with other neighborhood associations and movements across the whole city. Moreover, they act as intermediaries or spokespersons, whose role it is to communicate and provide information updates. However, local authorities are not always quick to respond to residents' demands nor are they always available when help is needed.

5.5 Conclusions

The chief problems resulting from the intensification of tourism in Barcelona's touristified neighborhoods can be summed up as follows: a reduction in housing stock, as residential uses are replaced by tourist uses; a rise in both the purchase and rental prices of these properties, leading to the "silent expulsion" of the most vulnerable neighbors; a shift in the nature of commercial establishments and a loss of local product stores as places of proximity and conflicts of social harmony and uncivil behavior given the new uses of the public space. The sources of contention of urban tourism are not unique to Barcelona, but rather are repeated in the world's other consolidated tourist cities—Berlin, Florencia, Venice, Lisbon, Prague, New Orleans, San Francisco and a long etcetera—where they have provoked various citizen movements whose protests are directed at policymakers, the tourism industry and the tourists themselves. Yet, at the same time, the growing resistance movements emerging in these tourist cities need to be understood as part of the wider struggles centered on urban restructuring, the transformation of urban governance and the "right to the city" (Novy and Colomb 2016).

What we are witnessing, therefore, are social movements that incorporate neighborhood protests against the effects of touristification and local residents voicing demands for their "right to the city". These platforms adopt various strategies of collective action to ensure that they, and the neighborhoods they represent, are included as key actors in the negotiation and design of public policies. Above all they seek to instigate singular actions with great media impact that can provide them with visibility and channels of dialogue with government bodies and other public institutions. The resistance movement targeting the touristification of Barcelona has focused its actions on reducing tourism on environmental grounds (by opposing the extension of the city's port and airport infrastructure and calling for limits to be imposed on cruise ship arrivals) and on demands for a more diversified economic model and the preservation of the right to housing and the city. In this respect, the initiatives taken are shared by a number of other cities in southern Europe (SETNet) which likewise call for the degrowth of tourism.

Following the pandemic and the rapid recovery of tourist activity, neighborhood and municipal collective action platforms have just as swiftly returned to their pre-pandemic protest actions. These social organizations claim that the pandemic has not led to any rethinking or diversification of economic models and condemn the fact that local government has dedicated public resources to meeting the needs of multinational companies (especially in the transport sector and accommodation market), whose large profit margins have contributed next to nothing to projects of social, economic and environmental improvement. Meanwhile, they argue that the growing encroachments of the restaurant sector in the public space (with the provision of bar and restaurant terraces in the open air), justified by the outbreak of COVID-19, have yet to be reversed even though the risks of infection have fallen. Indeed, the platforms condemn this commodification of urban space and its obvious impact in terms of both mobility and noise. Moreover, they denounce the resumption of cruise ship traffic and the re-launching of large infrastructures given their environmental impact in a context of climate emergency and energy crisis.

In short, all the former contradictions have reemerged in this post-pandemic era and, if anything, they are now vastly intensified, highlighting the need for new approaches to the management of urban tourism. As we have stressed in the course of this chapter, the following watch words need to be urgently adopted to ensure the good governance of tourist activity: sustainability; ecological awareness; participation and collaboration with all stakeholders; digitalization and a sharing economy; and a tourism of proximity and nature-based tourism.

Governance occupies a central place in tourism debates, where it is understood as the coordination of actors, social groups and institutions, whose objective is to obtain defined and deliberate results in a collective manner (Le Galès 2010). It is concerned with how to mobilize a set of critical actors—public and private (including interest groups, civic associations, NGOs, etc.)—to create new forms of collaboration out of existing situations of conflict, fragmentation and inconsistency (Iglesias Alonso 2021). In short, governance emphasizes such factors as network activity, citizen participation, public management and social capital (Conejero 2005; Aguilar 2007; Dredge and Pforr 2008; Trunfio and Della Lucia 2013, among others).

In this sense, it is important to take into account all potential actors, ranging from different levels of government, tourism businesses and private organizations and public transport to neighborhood associations, shopkeepers and residents. It is also critical that all the areas affected—economic, cultural, social and environmental, including aspects of mobility—are taken into account in open debates and participatory processes of future planning. One example that affects tourism is the need to reconsider the PEUAT and its exhaustive and real control of tourist licenses for housing, while as far as residents are concerned there is a need for fresh initiatives to regulate rental prices and the promotion of social housing.

In addition, the UNWTO and the World Travel and Tourism Council (WTTC) reiterate the need to factor in "sustainability" to the management of tourism policies, stressing just stress how important this is in the post-pandemic period. The post-COVID tourism paradigm highlights the pressing need to seek a balance between citizens and tourists in the world's urban spaces, as they inevitably compete for common uses.

References

Aguilar LF (2007) Gobernanza y Gestión Pública. Fondo de Cultura Económica, México

Alabart Vila A, Dominguez-Perez M, Crespi-Vallbona M (2015) Turismo y ciudad. In: Dominguez Perez M, Lopez-Villanueva C (eds) Madrid-Barcelona: procesos de transformación urbana y social. Similitudes y disimilaridades. Sintesis, Madrid, pp 247–274

Alabart Vila A (2010) Polítiques urbanístiques i moviment associatiu veïnal. Barcelona Societat (19):87–97

Andriotis K (2018) Degrowth in tourism. Conceptual, theoretical and philosophical issues. CABI, New York

Associació de Veïns (2017) Turismofòbia. Mite o realitat? La veu del barri d'Hostafrancs. nº 119

Barañano M (2021) Los hogares como espacios de arraigo y sus transformaciones en un contexto global. Encrucijadas. Revista Crítica de Ciencias Sociales 21(1)

Benach Rovira N (2016). ¿Ciudades en el mapa o en la guía turística? Venta de la ciudad y sentido del lugar/Cities on the map or in the tourist guide? Selling the city and sense of place. Revista CIDOB d'afers internacionals 113:89–105

Cabrerizo C, Sequera J, Bachiller PG (2016) Entre la turistificación y los espacios de resistencia en el centro de Madrid: Algunas claves para (re) pensar la ciudad turística. Ecología Política 52:78–82

Conejero E (2005) Globalización, gobernanza local y democracia participativa. Cuadernos Constitucionales de la Cátedra Fadrique Furió Cerriol 52–53:13–31

Crespi-Vallbona M, Domínguez Pérez M (2021) Las consecuencias de la turistificación en el centro de las grandes ciudades. El caso de Madrid y Barcelona. Ciudad y Territorio, Estudios Territoriales LIII:61–82

Crespi-Vallbona M, Farran-Plana M (2022) Fruiturisme: the boost of fruit tourism in Catalan rural areas. Tou Plann Dev. https://doi.org/10.1080/21568316.2022.2122072

Crespi-Vallbona M, Domínguez-Pérez M, Mascarilla Miró O (2019) Urban food markets and their sustainability: the compatibility of traditional and tourist uses. Curr Issue Tour 22(14):1723–1743

Crespi-Vallbona M, Mascarilla Miró O (2018) La transformación y gentrificación turística del espacio urbano. El caso de la Barceloneta (Barcelona). Eure, Revista Latinoamericana de Estudios Urbano Regionales 133(44):165–184

Crespi-Vallbona M, Planells Costa M (2011) Recursos Turísticos. Síntesis, Madrid

Dalmau M, Miró I (2010) Les cooperatives obreres de Sants. Traficantes de Sueños, Barcelona

Dredge D, Pforr C (2008) Policy networks and tourism governance. In: Cooper C, Scott N, Baggio R (Coord) Network analysis and tourism: from theory to practice. Channelview. https://doi.org/10.21832/9781845410896-008

Eisenschitz A (2016) Tourism, class and crisis. Hum Geogr 9(3):110–124

Fernández M, Hernández, M. H., & Barragan, V. (2019). Rompiendo el consenso: comunidades virtuales y discurso contra-hegemónico en el movimiento contra la turistificación en Sevilla. IC Revista Científica de Información y Comunicación, (16).

Fletcher R, Blanco-Romero A, Blázquez-Salom M, Cañada E, Murray I, Sekulova F (2021) Caminos hacia un turismo posa-capitalista. Serie Informes en Contraste, núm 18. Alba Sud

Fletcher R, Mas IM, Romero AB, Blázquez-Salom M (eds) (2020) Tourism and degrowth: towards a truly sustainable tourism. Routledge

Frago L (2021) Impact of COVID-19 pandemic on retail structure in Barcelona: From tourism-phobia to the desertification of city center. Sustainability 13(15):8215

Gil J, Sequera J (2018) Expansión de la ciudad turística y nuevas resistencias. El caso de Airbnb en Madrid. Empiria. Revista de Metodología de las Ciencias Sociales 41(3):15–32
Hall CM (2009) El turismo como ciencia social de la movilidad. Síntesis, Madrid
Hall CM, Scott D, Gössling S (2020) Pandemics, transformations and tourism: Be careful what you wish for. Tour Geogr 22(3):577–598
Harvey D (1989) From managerialism to entrepreneurialism: the transformation in urban governance in late capitalism. Geografiska Annaler Series B, Hum Geogr 71(1):3–17
Higgins-Desbiolles F (2020) Socialising tourism for social and ecological justice after COVID-19. Tour Geogr 22(3):610–623
Huete R, Mantecón A (2018) El auge de la turismofobia, ¿hipótesis de investigación o ruido ideológico? Pasos: Revista de Turismo y Patrimonio Cultural 16(1):9–19
Iglesias Alonso A (2021) La gobernanza local del turismo rural como respuesta a los efectos de la Covid-19. BARATARIA. Revista Castellano-Manchega de Ciencias Sociales 30:86–98
Jiménez IB (2011) "Las Asociaciones de Vecinos son la mejor escuela de formación de ciudadanos conscientes". El moviment veïnal durant el tardofranquisme i la transició. Plecs D'història Local 141:5–7
Lanfant M (1994) Identité, mémoire et la touristification de nos sociétés. Sociétés. Revue des Sciences Humaines et Sociales 46:433–439
Le Galès P (2010) Gouvernance. In: Boussaguet L et al. (eds) Dictionnaire des politiques publiques. Presses de Science Po, pp 299–308
Lefebvre H (2017) El derecho a la ciudad. Capitan Swing
López-Villanueva C, Crespi-Vallbona M (2021) Gentrificación y turistificación: dinámicas y estrategias en Barcelona. Encrucijadas, Revista Crítica de Ciencias Sociales 21(1):a2102
López-Villanueva C, Crespi-Vallbona M (2022) Cuidados y arreglos. La importancia del arraigo al barrio en un contexto de pandemia. El caso de la ciudad de Barcelona. Comunicación presentada al Congreso de la FES. 29 June–2 July. Murcia
Luque JB, Villa MÁ, Gámez AE (2019) El derecho a la ciudad y rescate del espacio público en zonas urbanas turistizadas. Una reflexión para Los Cabos, Baja California Sur (México). Aposta. Revista de Ciencias Sociales 80:109–128
Mansilla J (2018) Vecinos en peligro de extinción. Turismo urbano, movimientos sociales y exclusión socioespacial en Barcelona. Pasos 16 (2):279–286
Mansilla J, Hughes N (2021) "En dos años no nos vamos a acordar de la pandemia". Análisis del discurso sobre el decrecimiento turístico en Barcelona. Bataria 30(30–52). https://doi.org/10.20932/barataria.v0i30.623
Martínez-Caldentey MA, Blázquez-Salom M, Murray I (2020) Habitabilidad y airbnb: El alquiler de la vivienda en el distrito centro de Madrid. Cuadernos de Turismo 46:367–393
Milano C, Novelli M, Cheer JM (2019) Overtourism and tourismphobia: a journey through four decades of tourism development, planning and local concerns. Tour Plan Dev 16(4):353–357
Miles S, Paddison R (2005) Introduction: the rise and rise of culture-led urban regeneration. Urban Stud 42(5–6):833–839
Novy J, Colomb C (2016) Protest and resistance in the tourist city. Routledge, Abingdon
Olcina J (2020) Pandemia, cambio climático y turismo: acciones para lo inmediato y para lo próximo. In: Simancas M, Hernández R, Padrón N (eds) Turismo pos-covid-19. Reflexiones, retos y oportunidades. Cátedra de Turismo CajaCanarias-Ashotel de la Universidad de La Laguna, La Laguna, pp 31–42
Pimentel de Olivera D (2020) La sostenibilidad como apoyo en la desescalada de la COVID-19: una comparación al modelo de Doxey. In: Simancas M, Hernández R, Padrón N (eds) Turismo pos-covid-19. Reflexiones, retos y oportunidades. Cátedra de Turismo CajaCanarias-Ashotel de la Universidad de La Laguna, La Laguna, pp 271–283
Schneiberg M, Lounsbury M (2008) Social movements and institutional analysis. In: Greenwood R, Oliver C, Suddaby R, Sahlin-Andersson K (eds) The handbook of organizational institutionalism. Sage, London, pp 648–670
Smith MK (ed) (2007) Tourism, culture and regeneration. Cabi

Suarez-Rosa S, Fumero E (2020) La hora del turismo sostenible y de las personas. In: Simancas M, Hernández R, Padrón N (eds) Turismo pos-covid-19. Reflexiones, retos y oportunidades. Cátedra de Turismo CajaCanarias-Ashotel de la Universidad de La Laguna, La Laguna, pp 283–293

Trunfio M, Della Lucia M (2013) Social capital and governance for sustainable rural development. Stud Agric Econ 115(2):104–110. https://doi.org/10.7896/j.1220

UNWTO (2020) https://www.unwto.org/es/covid-19-y-sector-turistico-2020

Zapata VM (2020) De espectadora a protagonista: la sociedad en la reformulación del turismo tras el coronavirus. In Simancas M, Hernández R, Padrón N (eds) Turismo pos-covid-19. Reflexiones, retos y oportunidades. Cátedra de Turismo Caja Canarias-Ashotel de la Universidad de La Laguna, La Laguna, pp 293–304

Chapter 6
Neoliberalism, Collaborative Economy, and Short-Term Rentals Regulation in Andalusia, Spain

Ibán Díaz-Parra and María Barrero-Rescalvo

Abstract The emergence and rapid expansion of digital platforms has made the regulation of short-term or tourist rentals a key issue in current urban planning policy. As other fields, there are competing class interests over the regulation of this type of rental. This work seeks to examine the discourses of the different agents in the short-term market using the notion of ideology. It functions as an integrator of class practices and discourses that seek to legitimize the appropriation of rents generated by tourism revenues. The fieldwork used to support this idea consists of 30 interviews with short-term rentals landlords, managers, and companies in the main Andalusian tourist cities (Seville, Malaga, Granada, Cadiz, and Cordoba), carried out throughout 2020 and early 2021. Since then, a certain social alarm has been generated by the proliferation of this type of accommodation in their historic districts and forms of regulation have been proposed. In addition, we also analyze local and regional official documents, reports and urban and tourist plans. The chapter identifies the variety of agents, interests, and discourses around the issue and points out the complementarity between pro-market and pro-sharing economy discourses.

Keywords Short-Term Rentals · Ideology · Neoliberalism · Regulation · Sharing economy

I. Díaz-Parra · M. Barrero-Rescalvo (✉)
Department of Human Geography, University of Sevilla, Seville 41004, Spain
e-mail: mbrescalvo@us.es

I. Díaz-Parra
e-mail: ibandiaz@us.es

E. Navarro-Jurado et al. (eds.), *Urban Dynamics in the Post-pandemic Period*, The Urban Book Series,
https://doi.org/10.1007/978-3-031-36017-6_6

6.1 Introduction

Concern about the regulation of Short-Term Rental (STR) market has increased in recent years. This concern has been very evident in the case of Andalusia (southern Spain), a region with a growing specialization in tourism. The impact of p2p (peer-to-peer) platforms in making the accommodation market more flexible went hand by hand with the recovery of the tourist economy after the 2008 crisis. At that time, it was impossible to dimension the impact of these technical and management innovations on the real estate market. As a result, the Spanish Urban Rental Law (LAU) of 2013 excluded from its scope the regulation of STRs. Regional governments assumed this competence with exclusive jurisdiction on tourist policy in Spain, and in any case, the local governments in their land use planning. The 2016 Andalusian legislation on the so-called tourist rentals (officially, *Viviendas con Fines Turísticos* or VFT), was equally hasty, liberalizing, and full of contradictions. After a long debate and a tortuous procedure, the city council of Cadiz approved its own STR regulation at the end of 2021 as a punctual modification of the city's Master Plan. Seville's local government followed Cadiz with similar measure in 2022. A modification of the 2016 Andalusian decree is also on the horizon, in line with the regulations being established at the local level. STR regulation has raised a range of conflicts and ideological discourses, from the common free market speeches to the defense of STRs as a form of collaborative economy.

STR platforms such as Airbnb, Vrbo, Booking, and the like, have emerged largely under the auspices of the sharing or collaborative economy. It has been described as a type of economy that puts idle or underutilized goods into circulation, not guided by the desire for accumulation. Renting available rooms has been characterized as a complementary resource for households, around which altruistic motivations developed, such as the creation of social relationships between the temporary tenants and hosts.

Much of current critical scholarship on STRs has sought to deny the collaborative nature of this economy. Critical researchers have shown a trend toward concentration and monopolization of STR in both large cities with a strong presence of the sector (Gil and Sequera 2020), and cities with a lower rank (Katsinas 2021). Also, there is an ample evidence of a growing connection of STR platforms with large hotel chains and transnational financial investors. A limited number of professional hosts, including investors and property management companies, own or manage large property portfolios with multiple listings (Cocola-Gant et al. 2021; Arias Sans and Quaglieri-Domınguez 2016; Cocola-Gant and Gago 2021; Wachsmuth and Weisler 2018; Yrigoy 2019). Other scholars have identified a trend toward the creation of rentier coalitions that seek to politically influence regulatory decisions (Müller et al. 2021).

The opposite of professional and large agents would be ordinary or amateur guests who obtain an extra income from STRs (Dolnicar 2019; Dredge & Gyimóthy 2015; Guttentag 2015). These amateur agents still have a relevant presence, at least discursively, although their demographic weight has been greatly reduced. Semi and Tonetta

(2020), for instance, focus on marginal hosts, who engage in tourist rentals as a way of maintaining a certain middle-class standard of living by capturing these rents. In contexts of crisis, the extraction of rents from land property would become a recurrent resource, which is facilitated by the development of p2p platforms (see also Katsinas 2021; Yrigoy et al. 2022). According to Cocola-Gant et al. (2021), even in these cases, hosts could be mistakenly considered as outsiders to the dynamics of concentration in the sector, when in fact many of these properties end up being managed by intermediaries who optimize the income generated by the STRs. In any case, these small landlords seek economic benefit and do not tend to mobilize an idle resource, so it would be better to call them small rentiers rather than pioneers of the collaborative economy (Gil and Sequera 2020; Yrigoy et al. 2022).

Critical research on these topics coincides in presupposing a previous moment of a true sharing economy, which was subsequently displaced by the dynamics of professionalization and concentration. On the contrary, we propose to understand the STR sharing economy as an ideological device. From this, the following questions arise: What role does the discourse of the sharing economy currently play in the dynamics of professionalization and concentration? How are the contradictory discourses of the free market and the sharing economy articulated in the context of STR regulation? Do they represent different interests and strategies or are they complementary devices?

The following section develops a framework on urbanism and ideology, based on a critical reading of Henri Lefebvre and Slavoj Zizek, and focusing on the role of moralism in today's dominant ideology. Next, we explain the methodology used for this chapter, based on in-depth interviews and the analysis of public documents. The presentation of the results explains the analysis of the discourses and practices of the supply-side agents in relation to STR regulation. Finally, some elements of the theoretical framework are taken up and discussed in the light of the field work.

6.2 Ideology and Urbanism

Henri Lefebvre considers modern urbanism as an ideological device, part of a class strategy aimed at the reproduction of social relations of production (Lefebvre 1968). This implies that the practice and discourse of capitalist urbanization are oriented toward organizing the reproduction of society, naturalizing and legitimizing certain practices that are functional to the capitalist mode of production. The Marxist conception of ideology relates primarily to the legitimization of accumulation for the sake of accumulation and economic exploitation. Ideology refers to the way in which capitalist society represents its social conditions. For Lefebvre, relations between human beings are not self-evident; on the contrary, social practice produces mainly appearances, distorted representations of itself. In that sense, ideology is entirely determined by the practical life from which it emanates and helps to organize (Lefebvre 1966), so that we could find different and changing representations of well-established practices.

Lefebvre hardly speaks of neoliberalism in his writings on the city, and much of his critique is in fact devoted to the expose of centralized capitalist regimes that would rather coincide with the period dominated by Fordist-Keynesian economic policy. However, it will be a key element in the work of authors strongly influenced by the French philosopher, specially the geographer David Harvey (2001) since the 1990s, as well as the critical political geography of Neil Brenner (2004).

In contrast to the previous situation, neoliberalism is characterized by the attempt to make the market the main and almost the only social coordinator of land use and the production of the urban structure. This, in turn, implies a very defined role of the state as the promoter of the conditions for this market to function. It implies privatizations, but also the creation of conditions for private capitalist predominance and security (Jessop 2007, 2016). In general terms, the state must establish surveillance mechanisms, security conditions, and infrastructures that may require an even greater effort on the part of public agents than on the old interventionist policies (*Ibídem*). One of its characteristic elements has been the dismantling of productive development and territorial rebalancing policies of state economic planning, also the deregulation of land and housing markets and their progressive financialization through the indebtedness of households or their appropriation by investment funds. The great landmark of neoliberal deregulation in Spain was probably the Boyer Decree of 1985, which meant the elimination of protectionist policies in urban rents and the deregulation of the credit market, giving rise to the emergence of a version of capitalist development based on the real estate-financial sector. This has been interpreted as an exacerbation of the spatialization of the economy in the context of late capitalism (Lefebvre 2013; Harvey 2001). Urbanization and housing, practically turned into a financial asset, become a fundamental circuit for the circulation of capital, where huge speculative profits are generated, and which lead to increasingly violent periodic crises (1996, 2008, …).

Beyond the liberalizing and individualistic economic discourse, neoliberalism cannot be reduced to conventionally conservative positions. In fact, the opposite can often be the case. Radical discourses on architectural and urban theory from the 1970s have often been recycled to provide cover for neoliberal practices (Díaz-Parra 2015). The institutionalization of the environmentalist discourse is paradigmatic, with the idea of sustainability multiplying among documents and plans, in such a way that it ends up becoming mere noise, an empty word whose only function is to give moral support to urban development projects (Swyngedowu 2011). Certain types of progressive discourse and standard-bearers of a higher morality are the ideological complement par excellence of the most cynical neoliberal urbanism.

Ideology in its traditional sense is often understood as an inverted mirror of reality, intended to provide a moral representation of a fundamentally amoral (political-economic) practice. In Das Kapital, the notion of ideology is not named, subsumed in the idea of the commodity fetishism. Fetishization works by representing the market logics as its opposite, so that greed and usury become equality and freedom. However, the equality and freedom of economic agents conceals the subordination and coercion of wage labor (Marx 1990 [1867]). This idea is the basis of critical perspectives on capitalist society during the twentieth century. The critique of ideology allows us to go beyond the mere appearances of the market and observe the profound foundational injustice of capitalism, and to discover how the ideological affirmation of equality or freedom conceals its opposite.

Slavoj Zizek, following Sloterdijk, explains that ideology functions cynically in postmodern society: 'The cynical subject is aware of the distance between the ideological mask and social reality, but nevertheless insists on the mask' (1989, 56–57). In the face of this cynical reason, traditional critique of ideology would cease to function, entering (in Adorno's view) a post-ideological universe. Zizek denies this thesis. Ideology is not meant to be taken literally and continues to structure social reality in postmodern bias. There may be an intellectual awareness of the ideological character of liberal society representations, but the postmodern subject still prefers ideological and distorted representations. Ideology would not be so much a delusion as a fantasy (of liberty, equality, and fraternity) that structures social reality and enables its reproduction (*Ibídem*, 61).

Jodi Dean's (2009) central thesis, which she constructs from her reading of Zizek, is that the Western counterculture of the 1960s and 1970s has become the dominant culture. In that sense, rather than a cynical assumption of ideology, we might be looking at an ideology that has absorbed critical and cynical perspectives. The radical perspective of the 1970s returns in twenty-first century as a legitimizing discourse of liberal society, making capitalist and market contradictions invisible. Following Balibar, Zizek asks how Christianity became a dominant ideology. The answer is by incorporating fundamental motives and aspirations of the oppressed (Zizek 2000, 1994). For Zizek, the dominant ideology, in order to be operative, has to incorporate features in which the exploited majority can recognize their authentic yearnings (Zizek 2000; 201). Thus, George Soros or Bill Gates can rebuild as philanthropists what they destroy as financial speculators, the multinationals of the textile industry take cultural diversity as their banner in the same way that food franchises take the green discourse and fair trade as theirs. In a kind of 'symbolic overinvestment', the commodity itself embodies a compensation for the immoral character of its consumption (Zizek 2009). In this way, the traditional ideological critique can be integrated as a part of commodity fetishism, which allows this ideological fantasy to continue to function in today's informed and cynical society. According to this, STR platform's claim 'live like a local' is an ideological device that allows the STR commodity to embody a critique of the *mcdonalization* of culture that legitimizes the liberalization of the rental market.

6.3 Methodology

We focus on Andalusia, a traditionally farming region in which the service and construction sectors have become increasingly important. Although the tourism subsector has had an established presence since the mid-twentieth century, the explosion of cultural and urban tourism in recent decades up to the arrival of the COVID-19 pandemic has had a considerable impact on the current socio-spatial dynamics. For the study of this reality, the five large cities with the greatest weight of tourist rentals have been selected: Seville, Malaga, Cadiz, Granada, and Cordoba. They are the target of the research project funded by the Andalusian Plan for Research, Development and Innovation (PAIDI 2018, code P18-RT-2427).

Our qualitative methodology is based on 30 in-depth interviews with STR suppliers conducted between July 2020 and February 2021, who were contacted following the snowball technique. We selected informants with diverse profiles in the five cities studied: small rentiers, individual property managers, and professionals working in intermediary companies. Some of the informants also responded to questions as representatives of host associations (AVVA, at the Andalusian level; VEAN, from Seville; and CONVIVIR, from Cordoba). Overall, the interviews were divided into two parts. Firstly, we asked about the origin, management, and motivations in relation to ownership, and later, we focused on the platform, the technology, and the organization with an eye to the future. Most interviews lasted between 1 and 2 h and were semi-structured.

Interviewees are identified by a code that adds the following relevant information. The first two letters correspond to the city (SE, Seville, CA, Cadiz, MA, Malaga, CO, Cordoba, GR, Granada) followed by an identification number. Next digits indicate the number of dwellings they own or manage in each case (when it is a single room, it is indicated by 1/2). The last letters indicate the acronym of the association to which, if any, they belong.

Complementarily, the qualitative research was also based on the analysis of secondary sources such as official documents, reports, and local and regional urban and tourism plans. The elements of special interest were recorded in analysis sheets, according to the following categories: geographical references, agents involved, housing aspects, conflictive elements, and social effects.

6.4 Results

The Andalusian regulation on STR has been pointed out as a pro-market and *laissez-faire* regulation (Jover Báez and Berraquero Díaz 2020). The 2016 legislation faced the proliferation of STRs allowed by the development of p2p platforms. This problem was bequeathed by the exclusion of this type of uses from the LAU in its 2013 amendment, and virtually established an Andalusian Registry of STR (officially VFTs) that legalized at a stroke the installation of tourist uses over residential dwellings, knowing that this was in contradiction with local planning. The document is aligned with the neoliberal philosophy of minimal intervention of European legislation, specifically with regulations on collaborative economy (European Commission 2016), a soft law that seeks to regulate as little as possible and aims to influence lower-ranking regulations (Nieuwland and van Melik 2018). By not establishing a time limit, nor a limit on the maximum number of dwellings, nor any element that obliges landlords to use their properties for tourist purposes, the practical result of the regulation is that this type of rental evades the regulations for both conventional tourist accommodations and conventional urban rentals.

The idea that new technologies and online platforms generate conditions that cannot be regulated is echoed and reproduced by most interviewees. For example, one of them said, rejecting restrictive regulations, that 'they are putting up walls around Internet, which is like erecting walls in the countryside' (MA6_43_AVVA), which means that platform capitalism is *unregulatable*. A defense of the free market coincides with immediate interest of the interviewees.

Homeowners and STR professionals prefer tourist rentals to conventional rentals. The justification for this option is mainly of an economic nature, especially among specialized suppliers and management company professionals. However, as Yrigoy et al. (2022) also point out for the case of Palma, this reason is hardly mentioned in the first place, so they insist on the importance of other reasons. The first would be the difficult conditions supposedly established by the LAU, which 'does not exactly protect the owner' (SE3_260_AVVA) against non-payments and squatting: 'it gives you a bit of a break, especially now […] with the squatting issue […] and now rental agreements are automatically renewed for up to five years, and with [tourist rental] you don't have that problem' (GR1_30). Intermediaries emphasize the importance of the availability of the property for the landlord's use or the careful maintenance they enjoy in the STR. This is a way of asserting their specialized work and thus justifying their appropriation of part of the rent obtained. The only informant who clearly states the purely economic reason for his choice is tenant GR3_0, who chooses to sublet a room in his home: 'I did it to have some extra income, because of my situation, to earn some money that help me pay the rent'. Up to this point, agents' position is consistent with their interests as rentiers and specialized professionals, claiming free market. Supply side agents defend their right to rent maximization, as well as to legal security of ownership, arguing the pernicious nature of state regulations.

Another formula for legitimizing the STRs, without abandoning the free market discourse, is that of the more professionalized agents (especially intermediary companies) who represent themselves as a kind of decentralized hotel. This statement seeks to claim for the STR market the professionalism and prestige of the traditional accommodation sector, as a version adapted to the information society: 'we operate as a kind of hotel chain, that is, everything is very standardized, in that sense we are very similar to a chain such as a Meliá or Barceló, but in a small version' (SE3_260_AVVA).

With the issue of STR, collaborative economy has become a tourism policy concern. In this trend, Andalusian legislation, with exclusive competence on tourism sector, is particularly relevant. The central question here is the consideration of STRs as residential land use dedicated to tourist economy purposes, which implies two series of contradictions. First, the decree's choice of the term 'Viviendas para Fines Turísticos' ('dwelling for tourist purposes' in English) or VFT inherits a determining ideological component. It is the tourist use that makes it possible to extract these accommodations from the state legislation on urban land rental, where more restrictive regulations have recently been introduced (for example, compulsory extensions for conventional rental agreements). However, there is no obligation for STRs to be used exclusively for tourists. Secondly, tourist activity in these accommodations is considered an accessory activity, which justifies their tolerance on residential land uses. This latter argument is supported by the ideological discourse of the collaborative economy, mainly the idea that it is not a professional or full-time activity.

The collaborative economy discourse appears mainly among small landlords, but it is also used and serves the interests of large landlords and more professionalized agents. The discourse moves away from purely pecuniary interests to claim the opposite of what would be a neoliberal approach, i.e. the management of a resource (accommodation) in terms of its utility or use value and not exchange value. The use value of accommodation has a particular ideological component, as it refers to high cultural values, far from the culture of mass consumption, such as the way of life of Andalusian cities and the enjoyment of their 'idiosyncrasy', search for 'authentic' experiences, in a discourse of exclusivity that tries to move away from the idea of alienating mass tourism (for example, CO1_1/2_CONVIVIR). On the one hand, this is an ideological or a symbolic valorization of the product, which is based in moral issues. On the other hand, the discourse of the collaborative economy also offers resources for the positive representation of the owner's motivations: Entry into the STR market occurs in order to connect with other cultures, increase social contacts, etc.

Often, supply-side agents who defend neoliberal and free-market policies admit that discourses on the sharing economy in STRs do not match reality: 'My personal opinion is that this collaborative economy and "living like a local" is a myth, it is a product of Airbnb. But it is not true. I mean, it's no real, you [the tourist] live in a building, but you don't really interact with the locals' (SE5_23). However, other supply-side agents, without abandoning free market discourse, refer to the collaborative economy arguing against the regulation of the STR market. In their opinion, STRs allow a socialization of market benefits to agents who are not big capitalists, as a sort of spill-over effect (SE8_30_AVVA), or a non-abusive but rather symbiotic relationships between tourists, landlords, and intermediaries:

> Yes, this is collaborative economy, because I, without having 20 million euros, have been able to create a 200-bed business. Buying a hotel would have cost me a fortune. This is a collaborative economy in which the money stays where it is produced, the money does not go outside. For example, if you look to Malaga and certainly in Seville [...], there are not many Sevillians who own hotels. (MA6_43_AVVA)

Associations such as VEAN are dedicated solely to small landlords. They defend the collaborative economy discourse and are promoted and funded directly by Airbnb, according to the landlord and representative interviewed (SE11_1/2_VEAN). They elevate the collaborative economy discourse to institutions and public opinion, ultimately trying to influence regulation. These landlords accuse the platforms of not distinguishing between different hosts (i.e. between morally or socially motivated landlords and speculators who damage the image of the collaborative economy), while drawing axes of collaboration with the latter.

6.5 Discussion and Conclusions

Although neoliberal and collaborative economy discourses seem diametrically opposed, together they function as legitimizers of the same political-economic practices, these of the ultra-liberalization of the rental market, supported by development of new technologies and platform capitalism. Free market narratives reflect class interests of the capital invested in the sector; however, collaborative economy discourses offer an ideological complement that turns the pursuit of profit into a socially responsible business. Small homeowners sharing a room, economically marginal in current STR sector, play a key role in legitimizing the functioning of the whole market. Hence interest of the platforms in maintaining it and making it visible, as Yates (2021) points out and we have seen in our analysis. The humanistic practice of the small owner who sharing breakfast with a tourist couple legitimizes and sanctifies the profit of the investment fund that hoards housing and turns permanent accommodations into temporary rentals for visitors.

The ideological background of STRs has a mainly pragmatic and chrematistic core, which oscillates between the defense of its contribution to job creation (predominant in political parties with a government vocation) and the basically neoliberal discourse of free initiative. Here, following Zizek (2000), the collaborative economy functions as a moral complement which really makes the ideology of the free market work, absorbing society's legitimate pretensions of a fairer and more solidarity-based economy, which cares about small agents, about a human relationship between supplier and client, etc. In this sense, we could not say that the masses are cynical toward ideology; this moralistic component in fact shows the absence of a cynical attitude. On the contrary, it shows that the need of a distorted and moralistic representation of market relations persists in contemporary capitalist economy.

This ideological device does not only justify practices. It does not come a posteriori to justify an economic practice, but emerges as a constituent part of it, of the development of the sector, of the conflicts between agents and of those that operate around regulation. On the one hand, the moral complement of the collaborative economy is part of the valorization of the commodity. As Gil and Sequera (2020) point out, it represents an 'added value' to the model. These narratives give a symbolic plus to the accommodation activity, it is part of a symbolic valorization of the STRs that swell monopoly rents (Harvey, 2001) in a similar way to any marketing activity. On the other hand, the collaborative economy is a discourse and a practice that justifies and legitimizes the market interests in political struggle over regulatory policies, and determinates the greater or lesser appropriation of the global tourist rent by one or another group of agents.

Of course, as Gil and Sequera (2020) argue, STRs in p2p platforms do not fit into the theoretical scheme of the collaborative economy. The key is that the collaborative economy is neither an ideal that is sought to be realized nor something that has been perverted. The collaborative economy is an ideological device, part of the strategies of certain rentier classes, which makes a mystifying and moral representation of a series of economic activities that currently play a major role in capturing urban land rents.

Acknowledgements The fieldwork that supports this chapter was carried out in the framework of the projects 'Urban development and socio-spatial impacts of the tourism sector in large Andalusian cities' (PAIDI 2020, P18-RT-2427) and 'ITUCA. Impacts of Urban Cultural Tourism in Andalusia: Urban development and socio-spatial processes linked to the tourism sector in large Andalusian cities' (Centro de Estudios Andaluces, 3869/1047). Some of the interviews were also conducted by the researcher Julio José Parralejo.

References

Arias-Sans A, Quaglieri-Domínguez A (2016) Unravelling Airbnb: Urban perspectives from Barcelona. In: Russo P, Richards G (eds) Reinventing the local in tourism: travel communities and peer-produced place experiences. Channel View, London, pp 209–228

Brenner N (2004) New state spaces: urban governance and the rescaling of statehood. Oxford University Press, Oxford

Cocola-Gant A, Gago A (2021) Airbnb, buy-to-let investment, and tourism-driven dis-placement: A case study in Lisbon. Environ Plann A Econ Space 53(7):1671–1688. https://doi.org/10.1177/0308518X19869012

Cocola-Gant A, Jover J, Carvalho L, Chamusca P (2021) Corporate hosts: the rise of professional management in the short-term rental industry. Tour Manage Perspect 40.https://doi.org/10.1016/j.tmp.2021.100879

Comisión Europea (2017) Resolución del Parlamento Europeo, de 15 de junio de 2017, sobre una Agenda Europea para la economía colaborativa (2017/2003(INI)). Documento digital. Accesible en https://www.europarl.europa.eu/doceo/document/TA-8-2017-0271_ES.html. Consultado el 14 March 2022

Dean J (2009) Democracy and other neoliberal fantasies. Duke University Press, Durham

Díaz-Parra I (2014) El regreso a la ciudad consolidada. Ciudades. Análisis de coyuntura, teoría e historia urbana, 103.

Díaz-Parra I (2015) A back to the city movement by local government action: gentrification in Spain and Latin America. Int J Urban Sci 19(3):343–363

Dolnicar S (2019) A review of research into paid online peer-to-peer accommodation: launching the annals of tourism research curated collection on peer-to-peer accommodation. Ann Tour Res 75:248–264. https://doi.org/10.1016/j.annals.2019.02.003

Dredge D, Gyimóthy S (2015) The collaborative economy and tourism: critical perspectives, questionable claims and silenced voices. Tour Recreat Res 40(3):286–302. https://doi.org/10.1080/02508281.2015.1086076

European Commission (2016) Communication from the commission to the European parliament, the council, the European economic and social committee and the committee of the regions. A European agenda for the collaborative economy. Brussels, 2 June 2016. Available at: https://eur-lex.europa.eu/legal-content/ES/TXT/?uri=CELEX%3A52016DC0356

Gil J, Sequera J (2020) The professionalization of Airbnb in Madrid: far from a collaborative economy. Curr Issue Tour:1–20.https://doi.org/10.1080/13683500.2020.1757628

Guttentag D (2015) Airbnb: disruptive innovation and the rise of an informal tourism accommodation sector. Curr Issue Tour 12(18):1192–1217. https://doi.org/10.1080/13683500.2013.827159

Harvey D (2001) The art of rent: globalization and the commodification of culture. Spaces of capital. Routledge, London, pp 394–411

Harvey D (2014) Seventeen contradictions and the end of capitalism. Profile Books, London
Jessop B (2007) State power. Polity Press, Cambridge
Jessop B (2016) The state: past, present, future. Polity Press, Cambridge
Jover Báez J, Berraquero Díaz L (2020) ¿Habitantes o visitantes? El impacto del alquiler vacacional en el mercado de vivienda en Sevilla. Ciudad y Territorio. Estudios Territoriales 206(LII). https://doi.org/10.37230/CyTET.2020.206.07
Katsinas P (2021) Professionalisation of short-term rentals and emergent tourism gentrification in post-crisis Thessaloniki. Environ Plann A Econ Space 53(7):1652–1670. https://doi.org/10.1177/0308518X21988940
Lefebvre H (1966) Sociologies de Marx. Presses Universitaires de France, Paris
Lefebvre H (1968) Le droit à la ville. Anthropos, Paris
Lefebvre H (2013) La producción del espacio. Capitán Swing, Madrid
Marx K (1990 [1867]) Capital: A Critique of Political Economy Vol I. Penguin Classics, London
Müller N, Murray I, Blázquez-Salom M (2021) Short-term rentals and the rentier growth coalition in Pollença (Majorca). Econ Plann A Econ Space 53(7):1609–1629. https://doi.org/10.1177/0308518X211033307
Nieuwland S, van Melik R (2018) Regulating Airbnb: how cities deal with perceived negative externalities of short-term rentals. Curr Issues Tour. https://doi.org/10.1080/13683500.2018.1504899
Semi G, Tonetta M (2020) Marginal hosts: short-term rental suppliers in Turin, Italy. Environ Plann A Econ Space 53(7):1630–1651. https://doi.org/10.1177/0308518X20912435
Swyngedowu E (2011) ¡La naturaleza no existe! La sostenibilidad como síntoma de una planificación despolitizada. Urban 01:41–66
Wachsmuth D, Weisler A (2018) Airbnb and the rent gap: gentrification through the sharing economy. Environ Plann A Econ Space 50(6):1147–1170. https://doi.org/10.1177/0308518X18778038
Yates L (2021) The Airbnb 'movement' for regulation. How platform-sponsored grassroots lobbying is changing politics. The University of Manchester—Ethical Consumer, Manchester. Recovered from https://www.research.manchester.ac.uk/portal/files/192396608/Yates_2021_The_Airbnb_Movement_Corporate_Sponsored_Grassroots_Lobbying_in_the_Platform_Economy.pdf
Yrigoy I (2019) Rent gap reloaded: Airbnb and the shift from residential to touristic rental housing in the Palma Old Quarter in Mallorca, Spain. Urban Stud 56(13):2709–2726. https://doi.org/10.1177/0042098018803261
Yrigoy I, Morell M, Müller N (2022) Why do middle-class positions matter? The alignment of short-term rental suppliers to the interests of capital. Antipode. https://doi.org/10.1111/anti.12798
Zizek S (1989) El sublime objeto de la ideología. Siglo XXI, Madrid
Zizek S (1994) Mapping ideology. Verso, New York
Zizek S (2000) The ticklish subject: the absent centre of political ontology. Verso, New York, NY
Zizek S (2009) First as tragedy, then as farce. Verso, NY

Chapter 7
Changes in the Role of Heritage in Historic Centres: The Mutation of Historic Buildings into Tourist Accommodation

María Elvira Lezcano González and Margarita Novo Malvárez

Abstract Cultural attractions in the tourism sector are becoming increasingly important in many countries. In recent years, historic centres have undergone an intense process of "touristification", generating economic development and driving new urban dynamics that are rooted in the monumental wealth of cities. In this context, this chapter aims to determine the extent to which historic buildings are integrated into new urban processes and transformed into boutique hotels, in order to put the interest cultural tourists show in heritage to full benefit. To this end, we have analysed the case of the city of Palma (Mallorca-Spain), considering those characteristics common to this new type of accommodation, the protection of buildings, their location, degree of adaptation to their new function and the role that these establishments can play in the post-COVID future of tourism. Our findings highlight the possibility of bringing the buildings back into use and giving them a new life, although the expansion of boutique hotels influences destination dynamics, exerting high tourist pressure on the historic centre and causing conflicts with local residents.

Keywords Boutique hotels · Palma · Heritage dissemination · Historic buildings · Post-COVID tourism

M. E. Lezcano González (✉)
Departamento de Humanidades, Universidade da Coruña (UDC), A Coruña, Spain
e-mail: m.lezcano@udc.es

M. Novo Malvárez
Departamento de Ciencias Históricas y Teoría de las Artes, Universitat de les Illes Balears (UIB), Palma, Spain
e-mail: m.novo@uib.es

E. Navarro-Jurado et al. (eds.), *Urban Dynamics in the Post-pandemic Period*, The Urban Book Series,
https://doi.org/10.1007/978-3-031-36017-6_7

7.1 Introduction

The early twenty-first century has brought with it the revival of the historic centres of numerous cities, coinciding with the repurposing of cultural heritage. Generally speaking, they are urban areas with a high concentration of historical and cultural assets, many of which have been affected by a procession of functional upcycling (new domestic, institutional, cultural or accommodation uses) despite attempts to maintain those elements that are most representative of their origins and which form part of the collective memory. Francois Choay (2007), Pierre Nora (2009) and Mathieu Dormaels (2012) have considered this issue from the perspective of the connection between memory and heritage as complementary concepts, even though considerations regarding the extent and nature of this relationship have varied throughout the history of heritage studies (Zúñiga 2017, p. 189). Angeles Querol (2010) claims that memory, identity and territory are inextricably linked to cultural heritage, which in turn is directly associated with the identity and memory of a culture within the boundaries of a specific territory. In turn, Gilberto Giménez (2005) and Olga Lucía Molano (2007) believe that identity is expressed through cultural heritage and is linked to memory, whilst the hallmarks of culture are ingrained on the territory through the actions of the communities that live there. In Spain, Manuel de la Calle (2002, 2008) and Miguel Ángel Troitiño (1998, 2003) highlight the key role cultural assets may play in the territorial and strategic planning of cities, analysing tourist dynamics and their influence on the heritage of historic centres.

Taking these reflections as our starting point, we have observed how, since the turn of the twenty-first century, the city of Palma has undergone a reconfiguration of its space, attributable to the introduction of neoliberal policies that have paved the way for a new city whose rich heritage is not only considered a historical and artistic asset, but also an economic resource. Palma's walled historic quarter covers an area of just under 125 hectares and is home to 23,955 residents (2021), 21.03% of whom are of foreign citizenship. In turn, 34.86% of these are from EU-15 countries. Social change and the increasingly elite nature of foreign groups is a key indicator in explaining the changes to the historic heritage and the process of gentrification this area of the city is experiencing (González-Pérez 2020).

These new dynamics drive changes in the use of monuments, residents' profile and the shift from the traditional usage of land towards a spatial configuration based on private logic, which in turn lead to the banalisation and loss of identity of the affected districts that are swept up within a process of touristification. There is a wealth of scientific literature addressing the touristic gentrification of Palma (González-Pérez 2020; Franquesa 2010), in which authors attribute this development to the entry of foreign property investors from central and northern Europe (Novo 2019, p. 86). As in other locations, the touristification of Palma has meant the progressive encroachment of hotels and visitor amenities, sparking debate on the urgent need to reconsider a model that highlights the dilemma between economic growth and social sustainability (Vila 2016, p. 285). Fernando Carrión, who has studied these processes in the historic centres of Latin America, considers that one of the greatest dangers of applying these policies is the adoption of monumental measures that privilege physical heritage, yet which detract from its authenticity and therefore lead to the loss of its true essence (Carrión 2007, p. 11).

In Palma, one of the key indicators of the touristification of the historic centre is the proliferation of luxury hotels that are often housed in monumental buildings whose heritage recovery is seen exclusively from a commodified perspective: a business that caters to visitors' needs and their enjoyment. These hotels are generally referred to as "boutique hotels" and fall within an urban category of buildings with a heritage value that have been functionally upcycled. Common factors include their location in the most monumental districts, their wealth of built heritage and the luxury facilities they provide, whilst also representing a good example of reusage and sustainability, in contrast to new-build hotel developments. Eighty percent of all luxury hotels (four-star, four-star superior, five-star and five-star luxury) and all those classified as boutique hotels are located in the most emblematic districts (Fig. 7.1). Average occupancy stands at 70% and guests stay for an average of three days. Given their rapid spread, we can establish a connection between tourism-based gentrification and/or touristification and the architectural mutation of heritage. This association is no coincidence, as the recovery of the monumental heritage was a strategic goal of urban planning and policy aimed at revitalising tourism (Novo 2019, p. 86).

Despite the positive connotation implicit in the heritage recovery of the former fabrics that house these hotels, we question whether these rehabilitation actions

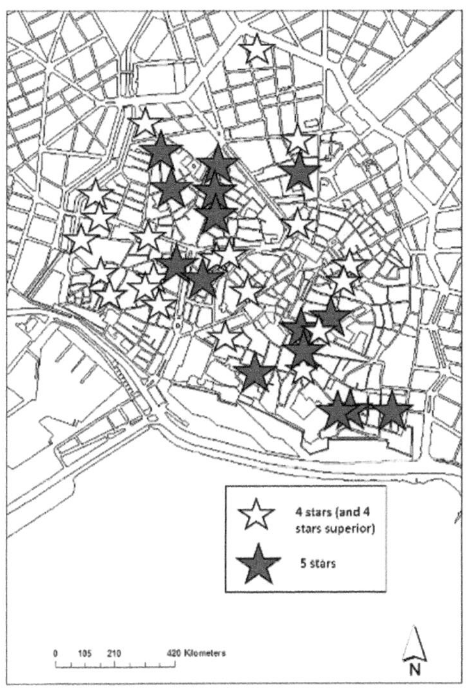

Fig. 7.1 Location of the 4- and 5-star hotels currently operating in Palma's historic quarter. *Source* Authors' own, drawn from the Catàleg d'establiments d'allotjament turístic de Mallorca (Vicepresidència i Conselleria d'Innovació, Recerca i Turisme, 2018)

have destroyed the original essence of these buildings. A further concern is whether they have undergone all phases of the heritage management process: research, cataloguing, protection/conservation and finally dissemination. The common characteristics these buildings share also point to a third issue for consideration, namely the difficulties implicit in their lack of definition as a homogenous group, as Balearic Island legislation does not contemplate a specific category for "boutique hotels".

In the light of the questions we have posed, the aim of this communication is to determine the extent to which historic buildings are integrated into new urban processes and transformed into boutique hotels. Further objectives include identifying their characteristic elements, shared values and whether their developers encourage their cultural dissemination. We end by considering the role these properties may play in the post-COVID future of tourism, which would appear to prioritise more intimate settings that shun massification.

Our study of these issues allows us to conclude firstly that the repurposing of these buildings is essential for their continuity, even though this may imply the loss of their traditional functions. A further conclusion is the benefit of repurposing for their external dissemination, which to date has focused exclusively on their promotion as an accommodation option.

7.2 Methodology

For the purpose of our research, we conducted an initial qualitative study in order to define the characteristics of properties of this type, followed by a quantitative study based on official data.

The scope of the study comprises Palma's intramural historic quarter, as this is where practically all the properties of this type are located. The timeline extends from 2002, when the first of these hotels appeared in the city, until the present day and a scenario in which the debate surrounding the challenges of expansion has moved beyond academia into social and political spheres.

The initial intention was to locate all boutique hotels in order to determine the characteristics of this type of properties and their action frameworks. The next stage consisted of compiling data relating to the projects, their architects and the media coverage given to their rapid expansion. The analysis of the project reports revealed that the architects sought advice from trans-disciplinary sources, as well as the progress and evolution of heritage conservation knowledge and awareness, resulting in interventions that demonstrate the necessary criteria and responsibility required for intervention on historic buildings.

The corresponding fieldwork was also carried out, based on direct contact through a series of semi-structured interviews with managers and staff at these properties, as well as the President of the Palma's Hoteliers' Association and architects involved in the restoration and refurbishment projects, in order to discover the nature of the conservation work in keeping with the principles of authenticity and identity, as well as to determine whether the construction programmes were conditioned principally by investors' profitability goals or the real needs of the buildings. The aim of the interviews was to obtain information related to the adaptation of the buildings, management of the properties and the dissemination of their historical and artistic values.

7.3 Results and Discussion

7.3.1 The Precedent of the Paradores Nacionales de Turismo

Palma's boutique hotels are the true reflection of the ties that bind heritage and tourism, as they are housed in buildings of historical and artistic value. From 2010 onwards, their number has risen considerably, due to the rapid growth of urban tourism.

In Spain, the first references to the convergence of both sectors date back as far as 1911 and the creation of *Comisaría Regia de Turismo y Cultura Artística y Popular*. One of the greatest achievements of this Royal Commission for Tourism and Artistic and Popular Culture was the creation of a "type of hotel for travellers tailored in accordance with the regional styles and needs of each location, known as a *parador*" (Lorenzo 2017, p. 674). Since 1928, the *paradores nacionales* have formed a growing network of hotels housed in historic buildings, preferably in old quarters (Eslava and Ontañón 1997, p. 12). This pioneering initiative, which successfully combined hotel and heritage management (Cupeiro 2019, p. 118), attempted to converge the protection and conservation of monumental heritage sites and the promotion of the tourism industry (Lorenzo 2017, p. 674), forging ties between both spheres which had previously shared very little common ground (Cupeiro 2019, p. 127).

The *parador* model can be considered the forerunner of boutique hotels in the sense that they both commonly reuse historic structures and that the revenue generated is the result of the commodification of heritage. However, there are also major differences between these two business models: the former are a state-owned hotel chain (Paradores de Turismo de España S.A.) whilst the latter are private enterprise.[1] Another fundamental difference lies in the manner in which their heritage is promoted and disseminated: the *paradores* also tend to include collections of movable property (Lezcano 2019, p. 218). In recent years, the hotel industry has also witnessed the emergence of other regional private initiatives for the reconversion of numerous types of historic buildings (convents, manor houses, lighthouses, country estates, factories, etc.) into various types of accommodation which are generally promoted online under the umbrella term of "singular properties".

[1] Paradores de Turismo (2022) is a Spanish state-owned hotel chain that operates just under one hundred hotels located throughout Spain and, since October 2015, franchises in Portugal. https://paradores.es/.

The evolution of both options has benefitted from favourable international tourist regulations, with an increasing focus on the potentially beneficial impact of tourism on heritage conservation. Examples in this sense include the International *Cultural Tourism Charter* (1999), the *Global Code of Ethics for Tourism* (2001) or the *World Charter for Sustainable Tourism +20* (2015) (Adams 2021, p. 3).

7.3.2 Boutique Hotels: Concept, Values and Characteristics

The first boutique hotels appeared in England in the 1980s (Agget 2007) as a reaction against the standardisation of accommodation (Elliott and Chan 2012) together with the demand for differentiated services. They were marketed as unique properties housed in singular buildings, which Jones et al. (2013) define as small hotels offering first-class services and authentic cultural and historical experiences. They are characterised by their prime locations, outstanding architecture, welcoming settings and a high standard of service. In addition, the architecture and brand focus create a unique identity for each property, setting it apart from the rest (de Oliveira 2018). Following Anhar (2001), architecture and design are key elements within the boutique hotel concept, whilst Mcintosh and Siggs (2005), as well as Khosravi et al. (2012), highlight singularity as the main differentiating feature. In turn, Wheeler (2006) and Rogerson (2010) argue that in addition to the core concepts of architecture and elegance, a prime urban location is also a crucial factor, together with the reduced number of rooms, which allow the hotels to deliver personalised customer service and unique experiences.

In short, the review of literature in this area enables us to conclude that the key characteristics that set boutique hotels apart are their size, design, location and personalised service. Architecture is another principal differentiation factors.

7.3.3 The Problems Surrounding the Expansion of Boutique Hotels in Palma

Boutique hotels are an increasingly essential part of Palma's image as a tourist destination. Their common characteristics and design criteria shed light on how the urban tourist territory is constructed and the architectural conditions, which form a clearly differentiating element for tourism products. Renovation and repurposing actions are the response to a strategy driven by the private sector, which is increasingly providing a dynamic model for urban and heritage renovation. Albeit with varying degrees of success, these hotels have contributed to the renovation of the city based on the use of its built heritage.

The rise in the number of these hotels located in Palma's historic centre dates back to 2010, following the introduction of new urban policies designed to promote the city as a tourist destination (Melo et al. 2021, p. 98) that triggered complex processes of touristification and gentrification (Novo 2019, p. 92). However, just a few years later and due to the dramatic rise in applications for opening licenses, the public administrations that had indirectly encouraged their appearance decided to limit future growth by imposing a moratorium on the urban plan between 2017 and 2019, based on a joint agreement with Palma's Hoteliers' Association. Today, licences are subject to a series of conditions: the hotels must be located in listed buildings that have been declared Assets of Cultural Interest (BIC in the Spanish initials), hold a five-star category and have no more than twenty rooms. Furthermore, they must comply with the terms and conditions stipulated in the modified General Urban Ordinance Plan (PGOU in its Spanish initials).

A further complication is their category, which is not contemplated in the legislation currently in force. In other words, in accordance with the current regulations (Decree Law 3/2022 of 11th Febrero on urgent measures for the sustainability and circularity of tourism in the Balearic Islands) instead of being listed as Boutique Hotels, the majority are classified as City Hotels (H) or Inland Tourism Properties (TI) or, albeit to a lesser extent, under small-sized categories with a high degree of product differentiation (Casañas and Blázquez-Salom 2021, p. 101). City Hotels provide better quality services and amenities, and generally occupy listed buildings in the historic quarter. The General Urban Ordinance Plan (PGOU in its Spanish initials) determines where these hotels may be located (Casañas and Blázquez-Salom 2021, p. 105). These classifications are an indicator of how tourist legislation has sought the reactivation of the historic quarter through the commodification of its heritage.

In 2022, 40 City Hotels and 2 Inland Tourism hotels are operating in the historic quarter, although just 26 of these can be considered truly singular and comply with all the characteristics that define boutique hotels. They include five-star, five-star luxury, four-star luxury and four-star properties, offering a total of 1,320 beds. More than half of these are included in the upper-upscale category: 16 have five stars and two are classified as five-star luxury, accounting for a total of 816 beds. All the properties are listed and a number have also been declared Assets of Cultural Interest (BIC in the Spanish initials). In chronological terms, this phenomenon is extremely recent, as the majority of these hotels have opened in the last five years. Nine opened following the passing of the moratorium in 2017 (Table 7.1) and a further 8 are scheduled to open between 2022 and 2023. These correspond to licences granted previously and will add a further 313 rooms.

These hotels are relatively small and have an average of 25 rooms and 50 beds per property. Of the five-star hotels, the only one that exceeds the average is the Es Princep Hotel, which has 68 rooms and 163 beds, yet which is included in this classification as it complies with all the conditions that define boutique hotels (Fig. 7.2).

The guest profile for these hotels is an international tourist aged between 45–54 and 55–64, with a high purchasing power and awareness of cultural issues, combined with the desire to experience the city. They stay mainly from Thursday to Sunday (city breaks); from Sunday to Thursday there are growing numbers of cruise tourists who opt to spend one or two nights in the city before embarking or after disembarking in Palma's port. As for nationalities, Germans comprise the largest group, followed

Table 7.1 Five-star and five-star luxury boutique hotels by year of opening (Palma)

City hotel	Category	Number of rooms	Number of beds	Year of opening
Hotel Ca Sa Galesa	5-star luxury	12	24	1995
Can Bordoy Grand House and Garden	5-star luxury	24	48	2018
Es convent de la Missió	5 stars	27	54	2003
Can Cera	5 stars	30	14	2011
Calatrava Mediterranean Sea House	5 stars	32	16	2017
Can Alomar Urban Luxury Retreat	5 stars	32	16	2014
Sant Francesc Hotel Singular	5 stars	42	88	2016
Boutique Hotel Sant Jaume	5 stars	36	72	2017
Summum Prime Boutique Hotel	5 stars	18	36	2017
Hotel Gloria de Sant Jaume	5 stars	14	28	2018
Es Princep Hotel	5 stars	68	163	2018
Hotel Mamá	5 stars	32	64	2018
Antigua Palma	5 stars	27	54	2019
El Llorenç Parc de la Mar	5 stars	33	66	2019
Concepció by Nobis	5 stars	31	62	2021
Nivia Born Boutique Hotel	5 stars	24	48	2021
Palma Riad	5 stars	11	22	2021
Inland tourism				
Palacio Can Marqués (T	5 stars	13	32	2018

Source Authors' own, based on data from the General Directorate of Tourism. Open Data Team. Government of the Balearic Islands. https://catalegdades.caib.cat/Turisme/Mapa-Allotjaments-turistics-Mallorca/4wg5-per5

by Britons and Scandinavians, although over the last year there has been a sharp rise in the number of French visitors. In 2022, bookings by US visitors are expected to rise, following the start of three weekly flights between New York and Palma. Staff numbers in these hotels are high, with an average ratio of 1.5 per room.

Fig. 7.2 Es Princep Hotel. *Source* Authors' own

7.3.4 The Architectural Singularity of Palma's Boutique Hotels

Palma's boutique hotels are housed in emblematic and singular historic buildings (former stately homes, palaces, convents or other types of fine residential buildings), in which contemporary interior design blends with an intimate and historic setting, producing a sense of contrast that is further enhanced by the inclusion of avantgarde furniture, occasionally designed by the architects themselves or specialised interior designers (Fig. 7.3). Their historic nature provides added value and is part of the experience guests seek.

The developers' objective was to ensure a harmonious balance between the existing architecture and the definition of the new use through the inclusion of the cutting-edge techniques and technologies demanded by modern-day society, whilst at the same time minimising their impact on the buildings' identity (Gutiérrez 2013, p. 722). In this sense, the greatest difficulty lay in harmonising the traditional fabrics, one of the greatest assets of this new business model, with the new needs, whilst also taking into account the singular features of the urban setting.

Our analysis of the works' studies and reports revealed that the architectural restoration projects have striven to conserve as far as possible the buildings' historical memory and their authentic significance, on the understanding that the priority was to combine this objective with the demands of their repurposing as hotels. An example of these efforts is the corridor leading to the rooms in the new wing of the Hotel Es Convent de la Missió. Renovated in 2019, the current setting is fairly similar to that of the original. The same is true of the current state of the Casa Misión de San

Fig. 7.3 Hotel Icon Rosetó. *Source* Authors' own

Vicente de Paúl, which is adjacent to the hotel, and specifically the corridor where the monks' cells were located (Fig. 7.4).

All the properties in the city that have been converted into boutique hotels have set up complex integral restoration programmes that include the use of contemporary materials and recent building innovations. Most of the alteration work has been carried out inside the buildings, in contrast to the less invasive architectural solutions applied to the façades and exteriors that have mostly conserved their original appearance.

The interventions have turned the architecture into one of the tangible attributes guests value most highly. The intangible characteristics include the amenities,

Fig. 7.4 Casa Misión de San Vicente de Paul and Hotel Es Convent de la Missió. *Source* Authors' own

customer service and character, atmosphere or charm, which in the majority of cases is also associated with the architecture and its personality (Fuentes et al. 2014, p. 635). After cataloguing and analysing the properties in Palma, we have identified a number of concurring elements (Table 7.2) that position them as locations of major value, attributable to a considerable extent to the actual buildings. The interviews with the hotels' managers and staff provided us with detailed insight of the terms and circumstances surrounding their cultural dissemination.

The greatest alterations to the original building fabrics were the structural changes necessary for the luxury facilities the categories of the new establishments require, such as lifts, swimming pools and sun terraces, function rooms and the design of large guest rooms (Fig. 7.5).

Actions to disseminate the built heritage target the hotels' clientele, stemming mainly from guests' requests for information. Actions to disseminate the cultural values vary in intensity, although some properties are now attempting to connect with the general public. An example of this is the Summum Prime Boutique Hotel, whose employees provide information on request to the public as a complement to the information included on Perspex panels and the photographs that can be seen at various points of interest around the building. This hotel has also created an

Table 7.2 Elements common to Palma's boutique hotels and their dissemination characteristics

A quality architectural legacy is a key element for these hotels; it also offers guests the opportunity to experience emotions related to the buildings' idiosyncrasies
They are all located in the historic centre
Each hotel has architectural features that define and differentiate it, although the contemporary design of the furniture and fittings and the communal facilities (sun terraces, swimming pools and meeting rooms, etc.) contribute a globalised character to this type of property. Despite the differentiating features, the end product is standardised
The décor and furnishings contrast with the architecture and tend to be exclusive to each particular hotel. Efforts are also made to acquire contemporary works of art and the majority take part in the *Nit del Art* event by organising exhibitions
All the buildings are listed and some have even been declared Assets of Cultural Interest (BIC in the Spanish initials), highlighting the fact that they are properties with character and added value
The hotels blend into the quarters and the urban landscape. There is no volumetric or formal-aesthetic impact, at least in terms of the exterior façades. The swimming pools, sun terraces, patios and other new elements cannot be seen from the exterior
Sustainability is sought by optimising resources and energy self-sufficiency. This has required the inclusion of geothermal resources and the use of renewable energies, recyclable and prefabricated locally sourced materials, suitable plants, thermal insulation, natural light and LED lighting, etc.
Most of the alterations to the original structures were designed to be reversible. This is a requirement for listed buildings, and in many cases required a reduction in the number of rooms originally planned
Dissemination of this architecture and its associated heritage is targeted exclusively at customers and there are practically no external communication actions in this sense

Source Authors' own

Fig. 7.5 Solarium Hotel Cort. *Source* Authors' own

Experiences Department which, despite focusing on sales and marketing actions, has proposed the publication of a monographic study on the former palace which houses this hotel. Nevertheless, in spite of these advances, it can generally be claimed that there is a need for new formulas capable of combining the hotels' private nature with the external dissemination of their heritage.

7.4 Conclusions

The construction of boutique hotels in historic buildings in the city of Palma is the direct consequence of the introduction of public municipal policies for the development of cultural and urban tourism, which in turn have driven the recovery of the local heritage. In this case, tourism as a contemporary activity has contributed to the continuity and improvement of a considerable number of buildings, some of which had fallen into disuse and oblivion, through a process of repurposing and a new form of inhabitation. In all likelihood, the most pressing topic of urban debate refers to the new uses of this heritage, and therefore the impact of touristification on the city, displacing other issues such as the environment (1980s) and territory (1990s and 2000s) which were predominant in previous decades.

Actions programmed in certain buildings in Palma's historic centre to convert them into boutique hotels are in line with the conservationist philosophy of the *paradores nacionales*, a model that drove monumental intervention in order to repurpose them as hotel accommodation. As with the *paradores*, the intention was to create an image of exclusivity based on the singular nature of the architecture, which in turn has become a source of revenue. In 2021, the results exceeded pre-pandemic performance levels, indicating a favourable post-COVID tourist scenario that prioritises more intimate and less crowded settings.

From a historical and artistic perspective, we consider that the restoration actions have had a positive impact, revitalising historic building fabrics, which in some cases were in a severe state of deterioration. However, we are aware that the new uses also imply a loss of significance and the commodification of heritage. This is the price that must inevitably be paid if we consider that the architects' aim was not to restore a convent, factory or former palace, but rather to construct a hotel by taking advantage of the historic structures and/or recovering existing ones as far as possible, considering that the directives are not determined by the actual buildings but rather by the requirements of a hotel.

Nevertheless, and despite the losses and disappearances, this heritage retains its documentary nature, from which we can glean the messages transmitted by the original architecture and indirectly grasp its intangible messages, albeit to varying degrees of intensity. Ideally, these messages, which can be perceived by the guests, should be shared with the local community through the introduction of mechanisms that would not interfere with the logical privacy demanded by guests staying at this type of accommodation. This would not only enhance the documentary value of the monument in question, but would also complete the management process that should form the part of the trajectory of all cultural assets.

We also consider that this dissemination could forge links between the hotel, tourists and the local community. This spirit of cooperation would not only prove beneficial for the heritage assets, which would become more familiar and more appreciated by citizens, but it would also strengthen the tourist image of the brand in particular and the city in general. The dissemination strategy deployed by *paradores* could provide a model for boutique hotels, as their commitment in this sense has allowed for the promotion and later appreciation of the inherited architecture.

References

Adams C (2021) Arquitecturas reinventadas. Hoteles sostenibles en construcciones industriales. Cuadernos de Turismo 48:351–368. https://doi.org/10.6018/turismo.492991

Aggett M (2007) What has influenced growth in the UKs boutique hotel sector? Int J Contemp Hosp Manag 19:169–177

Anhar L (2001) The definition of boutique hotels. Retrieved from http://www.hospitalitynet.org/news/4010409.html

Carrión F (2007) El financiamiento de la centralidad urbana: el inicio de un debate necesario. In: Carrión F (ed) Financiamientos de los centros históricos de América Latina y el Caribe. FLASCO/Lincoln Institute of Land Policy, Quito/Cambridge, MA, pp 9–24

Casañas F, Blázquez-Salom M (2021) De casal a hotel boutique al Centre Històric de Palma (2012–2020). Kult-ur 8(15):99–122. https://doi.org/10.6035/Kult-ur.2021.8.15.3

Choay F (2007) Alegoría del Patrimonio. Editorial Gustavo Gili, Barcelona

Cupeiro P (2019) Viajar a través del tiempo: el reto cultural de Paradores. Estudios Turísticos 217–218:117–132

De la Calle M (2002) La ciudad histórica como destino turístico. Ariel, Barcelona

De la Calle M (2008) El turismo en las políticas urbanas: Aproximación a la situación de las ciudades españolas. In: Ivards JA, Vera JF (eds) Espacios Turísticos. Mercantilización, paisaje e identidad. Instituto Universitario de Investigaciones Turísticas, Universidad de Alicante, Alicante, pp 507–529

de Oliveira FA (2018) Hostels e hotéis boutique. Análise de fatores de seleção na cidade de Oporto. Tesis Doctoral, Universidade de Vigo, Vigo

Dormaels M (2012) Identidad, comunidades y patrimonio local: una nueva legitimidad social. Alteridades 22(43):9–19

Elliott S, Chan M (2012) Lodging subsector report: boutique hotels. International Tourism Marketing and Development. Recuperado de: https://dspace.lib.uoguelph.ca/xmlui/bitstream/handle/10214/3587/Lodging_subsector_report_boutique_hotels.pdf?sequence=7

Eslava J, Ontañón F (1997) Paradores históricos. Lungwerg Editores, Barcelona

Franquesa J (2010) Sa Calatrava mon amour. Etnografia dún barri atrapat en la geografía del capital. Edicions Documenta Balear, Palma

Fuentes L, Villacé T, Muñoz A (2014) Estudio de los atributos distintivos de los alojamientos *boutique* y su gestión desde la perspectiva del género del emprendedor: aplicación a los alojamientos *boutique* en España. ESIC Mark Econ Bus J 45(3):633–662

Giménez G (2005) La concepción simbólica de la cultura. In: Giménez G (ed) Teorías y análisis de la cultura. Conaculta, Mexico, pp 67–68

González-Pérez JM (2020) The dispute over tourist cities. Tourism gentrification in the historic Centre of Palma (Majorca, Spain). Tour Geogr 22(1):171–191. https://doi.org/10.1080/14616688.2019.1586986

Gutiérrez ML (2013) La revitalización de la ciudad histórica a través de la rehabilitación patrimonial. Arte y Ciudad - Revista de Investigación 3(I):719–732 (Extraordinario)

Jones DL, Day J, Quadri-Felitti D (2013) Emerging definitions of boutique and lifestyle hotels: a Delphi study. J Travel Tour Mark 30(7):715–731

Khosravi S, Yussof I, Bahauddin A, Mohamed B (2012) Potentials of boutique hotel: The case of Penang Island, Malaysia. OIDA Int J Sustain Dev 4(8):71–78

Lezcano ME (2019) Paradores y comunicación turística del patrimonio. Estudios Turísticos 217:217–226

Lorenz A (2017) Monumentos convertidos en hoteles: el sacrificio de la memoria arquitectónica. El caso de Santo Estevo de Ribas de Sil. PASOS. Revista de Turismo y Patrimonio Cultural 15(3):673–685. https://doi.org/10.25145/j.pasos.2017.15.045

Mcintosh AJ, Siggs A (2005) An exploration of the experiential nature of boutique accommodation. J Travel Res 44(1):74–81

Melo L, Barrera M, Franconetti J (2021) Boutique hotels as a form of active recovery of the historical-cultural heritage in Palma de Mallorca, Spain. J Tour Herit Res 4(4):87–102

Molano O (2007) Identidad cultural, un concepto que evoluciona. Opera 7:69–84

Nora P (2009) Les lieux de mémoire. LOM, Santiago

Novo M (2019) Los nuevos usos del patrimonio: la expansión de hoteles boutique en Palma (Mallorca). Estoa. Revista de la Facultad de Arquitectura y Urbanismo de la Universidad de Cuenca (Ecuador) 8(16):83–95. https://doi.org/10.18537/est.v008.n016.a07

Paradores de Turismo (2022) Misión, visión y valores. Web oficial de Paradores, Madrid, España. Recuperado de https://empresa.parador.es/quienes-somos/bienvenida-pedro-saura/

Querol MA (2010) Manual de Gestión del Patrimonio Cultural. Akal, Madrid

Rogerson JM (2010) The boutique hotel industry in South Africa: definition, scope, and organization. Urban Forum 21(4):425–439

Troitiño MA (1998) Turismo y desarrollo sostenible en ciudades históricas. ERIA 47:221–228

Troitiño MA (2003) La protección, recuperación y revitalización funcional de los centros históricos. Colección del Mediterráneo Económico 3:131–160

Vila F (2016) La (in)esperada gentrificación cultural. El caso de Barcelona. Periférica Internacional. Revista Para El Análisis De La Cultura Y El Territorio 17:285–297. Recuperado de https://revistas.uca.es/index.php/periferica/article/view/3283.

Wheeler DF (2006) Understanding the value of boutique hotels. Doctoral dissertation, Massachusetts Institute of Technology

Zúñiga P (2017) Patrimonio y Memoria: una relación en el tiempo. Tribuna 36:189–194

Part II
Urban-Tourist Spaces on the Coast: New Urbanised Spaces and Resistance

Chapter 8
Urban-Tourism Spaces on the Coast: Transformations, Challenges and Globalisation in Uncertain Scenarios

Rosario Navalón-García

Abstract The variety of activities related to tourism generates differentiated transformation processes in cities, in the rural world and in coastal spaces. Of all the territorial mutations generated by tourism, those occurring in coastal areas are particularly interesting as these are fragile and complex environments which have suffered the greatest territorial transformations. The importance and significance of the changes taking place in them are directly related to the effects of the urbanisation and occupation of the land, due to tourism real estate development. As a result, the coastal areas are often scenarios of landscape degradation and pollution, social segregation, saturation or infrastructure deficit, among other effects.

Based on the uniqueness of the urban-tourist spaces of the coast, this article will offer a description of the most representative transformation processes in the tourist areas of the Spanish Mediterranean, as an example that can be extrapolated to other spaces. The mechanisms that have made it possible and which have given rise to the different models and implementation sequences will be analysed. Then, the effects related to the urbanisation process will be addressed in order to identify common development patterns, enabling a global diagnosis to be made. Finally, some of the future trends and challenges that will be faced by the coastal urbanised spaces will be examined.

Keywords Real estate development · Coastal tourist area · Mass tourism · New lifestyles · Globalisation

R. Navalón-García (✉)
University of Alicante, Alicante, Spain
e-mail: r.navalon@ua.es

E. Navarro-Jurado et al. (eds.), *Urban Dynamics in the Post-pandemic Period*, The Urban Book Series,
https://doi.org/10.1007/978-3-031-36017-6_8

8.1 Introduction

Since the universalisation of mass tourism in the mid-twentieth century, the spaces affected by this activity have been transformed drastically from both a socioeconomic and environmental and territorial perspective. Historically, the coast has been subjected to strong demographic pressure, leading to more than half of the world's population living on a narrow strip of land (less than 60 km from the sea), due to urbanisation, industrialisation and infrastructure concentration processes. In addition, due to its environmental features, the coast is the principal receiving space and the one that generates the most tourist displacements on a global level of both international tourists and the internal flows of each nation. The coast is the principal tourist space in many countries. It constitutes the leading economic activity of many coastal populations, island spaces, such as the Balearic Islands and even countries, such as Cyprus or Malta, for example, in the Mediterranean; but also in Florida, in the Caribbean or Southeast Asia, among many others.

Therefore, analysing the urban-tourism transformations that take place in these spaces is of enormous interest and, due to their breadth and relevance, this is a challenging task.

The tension arising from the concentration of activities in the small coastal space means that this space is undergoing constant change. This reflects, like no other space, the global trends of the moment, crisis situations or the uncertainty regarding global challenges, such as climate change or sustainability.

The different interests, often opposing, that come into play in the different process of the activities in this narrow strip of land, enable these areas to be characterised as spaces of potential conflict. This is aggravated considerably when the pressure of tourism real estate development is incorporated into the territory.

8.2 Characterisation of the Urban-Tourism Space of the Spanish Mediterranean

The dynamism of the tourism sector and the positive and negative impacts that it generates make this the principal transforming agent of the territorial reality of the places in which it is implanted (Vera et al. 2011). The Mediterranean is the world's leading tourism region, accounting for a fifth of international tourism (WTO 2020), to which a large volume of internal tourism should also be added. Coastal tourism is the most common and differential form of tourism, despite the considerable increase in urban and interior tourism in recent years. Specifically, the sun and beach destinations of Spain, from Girona to Huelva, including the archipelagos, represent 65% of the total supply and tourism demand of the second most visited country in the world (Exceltur, 2019) (UWTO, 2019).

The relevance of this activity is beyond all doubt, as it represents 70.5% of Spain's total tourism GDP and more than six out of every ten jobs in the sector

(Exceltur 2021). According to the tourist expenditure data of Egatur (INE) and International Tourism Highlights (UNWTO 2019), some of the Spanish destinations on the Mediterranean coast and the islands generate an income from international tourism higher than other large competing foreign sun and beach destinations. For example, Catalonia earns the same amount as Egypt and Cyprus together or Andalusia earns the same as the Puerto Rico, Cuba and Costa Rica together (Exceltur 2021). The size of the coastal tourism activity is also visible in the fact that of the 20 urban and holiday destinations with the most number of hotel beds in Spain, 15 correspond to coastal municipalities, above some of the most important Spanish cities. This does not contemplate the data for the extra-hotel supply, which are not registered on a destination scale by the INE (2019). Only Madrid, Barcelona, Seville, Valencia and Granada are in this ranking.

From the 1960s, there has been a high consumption of land in coastal municipalities basically related to the development of the tourism supply. This activity undoubtedly constitutes one of the principal driving factors of the coastalisation process of the country, which, between 1987 and 2011, has led to the artificialisation of 47% more land in these areas (Carvalho Cantergiani and Jiménez Herrero 2007). This has also generated an increase in the construction, infrastructure and demographic pressure in coastal municipalities. Although these municipalities represent 7% of Spain's surface area, 44% of the country's population live in them. To this, the significant volume of seasonal population related to tourism should also be added.

The tourism boom in the mid-twentieth century was due to a series of comparative advantages, such as the existence of attraction resources, including the climate or the beaches, the proximity to tourist issuing countries from central and northern Europe and their favourable socioeconomic position with respect to Spain. Furthermore, the partial opening up of the country from the 1950s and the existence of an uncritical environment with the possible impact that could be derived from the development of the necessary accommodation supply, among other factors, fostered a favourable context for the take-off of the tourism activity. This laid the foundations of a fundamentally receiving country and the tourism and currencies that it attracted enabled this activity to constitute a lever of development and a vehicle for the tertiarisation and urbanisation of the coastal spaces that attracted the demand.

The functional specialisation as a mass tourist destination, which began six decades ago and which was based on the climate factor and the good connections with issuing countries, prevails today, with a clear concentration in the sun and beach product, which accounts for 75% of total foreign demand visiting Spain (according to the hotel accommodation data of the INE) and a clear spatial polarisation in the Mediterranean and archipelagos. The relevance of internal tourism in the shaping of the Spanish tourism territory should be taken into account, which has tended to be given a secondary role due to its late incorporation. As indicated by Gutiérrez et al. (2005), internal demand played a relevant role in the development of the second home supply, with the precedent of returning visitors in the places of origin and the perception of acquiring these homes for vacation use as an investment for families and the less wealthy classes.

This aspect is undoubtedly relevant in the characterisation of the coastal tourist territory, if we take into account that in the 733 municipalities of the Mediterranean coast, together with the Balearic and Canary Islands, 89.7% of the total accommodation capacity is made up of homes for residents and second residences for tourists. Particularly, striking is the fact that the principal destinations of Spanish tourism demand coincide with those where there is a predominance of second residence accommodation supply (Exceltur 2021). Furthermore, in these municipalities, only 10.3% are considered as regulated accommodation (in the form of hotels, registered apartments, campsites or rural houses). All of this helps us to understand the features that define the Spanish coastal tourism supply, as shown in Table 8.1.

As already mentioned, Spanish tourism is characterised by an undifferentiated supply, mainly coastal, based on its proximity to the sea and the good climate. It is polarised on a narrow strip of land and with a time concentration in the summer

Table 8.1 Sun and beach destinations with the greatest volume of tourism accommodation beds. Proportion between regulated and residential supply (not regulated). Data of 2019

Municipality	Regulated accommodation beds + residential supply	% of regulated accommodation	% of residential beds
Torrevieja	249,080	2.7	97.3
Orihuela	111,523	4.2	95.8
Santa Pola	98,773	5.4	94.6
San Javier	95,643	6.1	93.9
Cullera	69,039	6.5	93.5
Cartagena	126,699	14.0	86.0
Estepona	83,930	15.7	84.3
Gandía	91,712	16.6	83.4
Mijas	67,431	18.0	82.0
Roses	74,235	19.8	80.2
Marbella	130,225	22.0	78.0
Fuengirola	67,447	23.0	77.0
Cambrils	75,895	34.0	66.0
Roquetas de Mar	68,937	34.4	65.6
Benidorm	176,504	44.8	55.2
Salou	127,664	47.5	52.5
Calvià	115,388	56.5	43.5
Arona	72,158	57.9	42.1
Adeje	65,974	79.0	21.0
San Bartolomé de Tirajana	95,473	80.2	19.8

Source Exceltur, based on the Occupation Survey (INE) and data referring to second residences in the Housing Census of the INE and Ministry of Development

months, when 58% of sun and beach displacements are made. In the coastal destinations, in the summer the population density is similar to that of medium-sized cities, with the resulting problems of saturation and congestion, waste generation and water and electricity consumption needs and the costs of cleaning, health, etc. All of these factors must be addressed with the budgets assigned in accordance with the resident population, which obviously does not correspond to the needs of these destinations for maintaining public spaces and infrastructures, for basic service endowment and to finance the investments required to satisfy the local and tourist population. The question is that the income that the municipalities receive is not aligned with the constructed housing stock or with its tourism supply. In fact, the average income per person is 38.7% less in tourist municipalities than in other municipalities with a similar but permanent population. For example, the summer population of Marbella is similar to that of A Coruña and that of Torrevieja is similar to that of Valladolid (Exceltur 2021). This situation of underfunding affects the satisfaction of the tourist, but also the quality of life of the residents and the feasibility of maintaining and creating adequate and sufficient amenities. In this way, the municipalities find a way to compensate this insufficient budget based on extraordinary income derived from the new construction of housing in the form of taxes and rates. All of this favours the development of new residential tourism complexes which, based on the argument of creating new supply for tourism, mask and justify real estate-tourism development operations more focused on urbanisation than on the creation of an attractive supply for the tourist destination.

This situation is the result of an accelerated growth, initially driven by the pressure of European demand and the mediation of international tour operators and also interior tourism and the national initiative. In many cases, tourism was implanted in the territory according to similar and standardised patterns, based on an undifferentiated supply focused on a climate and coastal product and, therefore, highly seasonal and with a strong concentration and saturation in the coastal space. The result has been the development of new urban-tourism spaces with a relevant component of accommodation supply, both regulated and second residences, which has enabled the creation of benchmark destinations, such as Benidorm, Torremolinos or Palma de Mallorca, among many others. However, also over the last 60 years, there has been an extensive and intensive occupation of the territory, which has ignored urban planning guidelines but followed those marked by the real estate promoters and investors. The result is the generation of strong territorial impacts and imbalances which the destinations must address with reorientation and restructuring formulas. This would help slow down the decline of these areas and adapt to a changing and more exacting demand. A competitive paradigm must be adopted more focused on quality, sustainability and which responds to the challenges of the planet, which the Covid-19 pandemic has only aggravated.

8.3 Transformation Processes in the Coastal Areas

In parallel with the take-off of democratic tourism, in the 1960s certain models began to take shape and the dynamics of development and the evolution of the tourism activity took place through different approaches and in different areas (Europe, North America, Australasia or Japan) (Pearce 1993). Of these, the most interesting focused on evolutionary aspects that formed part of the tourist areas, given that they enable us to understand similar inertias and dynamics, which can help to identify explanatory factors of the transformations that have taken place over time and also those that may occur in the future, according to the patterns observed in other environments. Therefore, learning about the models not only helps us to understand how development occurs, but also helps us to predict the future.

Based on the idea of concentric belts described by Miossec (cited by Pearce 1993), in turn, based on the idea of pleasure peripheries (Turner and Ash 1991), several models emerged which can be principally applied to the shores of the Mediterranean for the issuing spaces of north-western Europe and the Caribbean for North America. According to these space–time evolutionary models, from the mid-twentieth century, tourism developed and expanded in the coastal spaces, globalising the activity in successive tourist peripheries, as coined by Gormsen (1981, 1997), as a formula for reproducing the capitalist system (Navarro-Jurado et al. 2015). Taking this model as a reference, the sequence of the emergence and development of new tourist spaces not only takes into account the territorial and time evolution of the coastal destinations but also incorporates a diachronic transformation in the interior of each of these peripheries. This can be identified in successive implementations over time in other places, based on the evolution of three elements: the amount and variety of the available accommodation services, the local and exogenous participation in the development processes and the social structure of the demand in each case (Fig. 8.1).

The Gormsen model was defined in 1981 and had a strong Eurocentrist perspective. It distinguished proximity peripheries (first and second), which were developed by local agents and with diverse types of supply to respond to a varied demand. They were initially concentrated in cold beach areas and, subsequently on the southern European coasts; with comparable peripheries for the United States. These were followed, in time, by the development of intermediate and distant peripheries, characterised by a predominantly foreign investment and a minority high-income demand, located on the coasts of North Africa (third) and the Caribbean, the Pacific Ocean, the Indian Ocean, Southeast Asia and South America, for the fourth and successive peripheries.

In order to better understand the transformation processes of the coastal destinations, it may also be useful to review some of the premises of Butler's life-cycle model of tourist destinations (1980). It is with good reason that this model is based on the fact that destinations are dynamic and the analysis of their development, as indicated in the sub-title of this theory, is useful for understanding the implications that this has for the management of the resources of an area. According to Butler (2012), the shaping of the destinations is the result of the tension between inertia, as

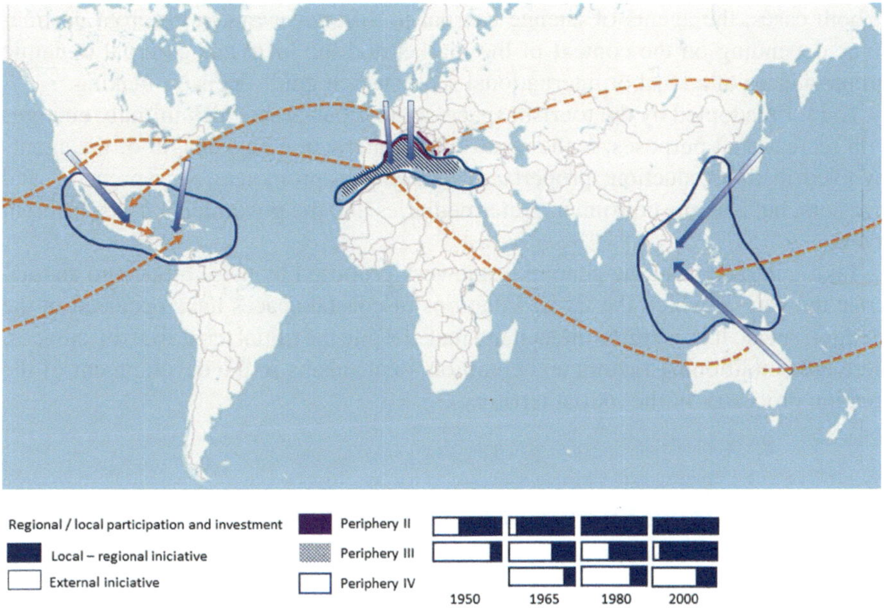

Fig. 8.1 Tourist peripheries coastline in the twentieth century, based on space–time evolution model of international coastal tourism. *Source* Own elaboration based on Gormsen (1981). The spatio-temporal development of international tourism; and Gormsen (1997). The spatio-temporal development of seaside tourism

a tendency to remain and grow in the same way over time and dynamism, as a need to seek new ways of development and orientation in order to increase the attractiveness of the destinations. In both cases, the tourist areas are affected by endogenous and exogenous forces that promote permanence or transformation in the tourism sector. Furthermore, as the destinations grow, they are subject to changes as a result of internal and external pressures that condition the different paces of transformation and the associated consequences. The usual impression is that growth is necessary and irreversible and that if it is initially successful, it should be replicated as a process of repetition and accumulation. In this way, new units are added to those that already exist, with results that are often unplanned and unexpected and generate impacts which have to be subsequently addressed when it is difficult to counteract them.

Change is constant in tourism and, in itself, it is not good or bad. Therefore, it is relevant to focus on the pace and type of transformations, which can be as important as development itself. According to Butler (2012), the change occurring in destinations can be gradual, based on pre-existing structures and driven by endogenous forces, maintaining the same form and functions. Alternatively, the change can be revolutionary, sudden and variable, usually caused by exogenous forces as a result of external investment, with unpredictable effects due to its nature and scope.

In both cases, the agents of change can act in different ways in different destinations, depending on the context of the market and the local and external dynamic circumstances of a local or international scale, which guide decision-making.

The form adopted by the tourism space, as an urban area which initially emerges with recreational purposes, is the result of the interaction between all of the agents involved in its production: property owners, urban promoters, entrepreneurs and investors, but also the consumers, intermediaries and the government at its different levels.

Taking into account the general framework proposed by these models to characterise the way in which the transformations of coastal spaces have occurred in the Mediterranean, it is useful to understand that, despite the initial territorial differences, there are conditioning factors and common mechanisms in the development of the creation processes of the tourist territory.

8.4 Conditioning Factors of the Transformations of the Tourist Coast in Spain

In this section, some of the explanatory factors are described. These can be used as a base to learn about the starting factors and the mechanisms that have intervened in the development and profound transformation of such an emblematic coastal tourist destination as Spain throughout the twentieth century and beginning of the twenty-first century (Fernández-Tabales and Navarro-Jurado 2022). Understanding the processes of these areas enables us to determine the origin of some of the more relevant problems and also acknowledge the challenges arising from the current internationalisation processes of tourism. Despite the territorial differences in the different regions, a series of conditioning factors and common mechanisms can be identified that have enabled the generation of the transformations resulting in the development of a tourism landscape marked by a tourism real estate supply that is polarised in terms of territory and time and specialised in the sun and beach product.

Evidently, the starting point for the process is the presence of the comparative advantages of the physical, ecological, climate and coastal factors, which are the principal attraction elements of the demand, together with proximity and connectivity with the issuing spaces of northern and central Europe, which explain the initial success of these areas. These factors must be accompanied by other aspects that link the creation of competitive advantages, such as improvements in the physical accessibility of the attraction spaces, with the construction of road and airport infrastructures, a key aspect for explaining the location of the tourism strategies in certain spaces on the coast. However, in addition, and above all, we should consider the relevance of other financial and legal aspects that largely explain the pace, dimension and way of implantation in the territory (Navalón-García 1995; Antón Clavé and González Reverte 2008; Vera et al. 2011; Rullán 2011; Fernández Tabales and

Cruz Mazo 2011; Navarro-Jurado et al. 2015; Ivars-Baidal and Vera Rebollo 2019; among many others).

In relation to these last factors, for the Spanish case, the typification of the key aspects carried out by Rullán (2011) and Fernández Tabales and Cruz Mazo (2011) is highly useful. First, these authors consider that a context favouring the availability of capital and finance for acquiring property and real estate investment forms the basis of the enormous transformations that have taken place in recent decades. The good conditions for accessing these loans which were granted at very low rates of interest and an increase in the repayment terms were not only enjoyed by individuals, but also enjoyed by promoters and construction companies, which could more easily bear the financial costs of the real estate operations.

The increase in the tourist or speculative demand explains the increase in the price of housing and the increase in the profits for promoters who supported the successive construction waves. Furthermore, the arrival of the euro in 2001 favoured the entrance of black money in money laundering processes, through real estate sales transactions. Within this context, the attraction of capital and the specialisation in the real estate financial dynamics of the coastal spaces constituted a trigger for the transformations in the territory. The business does not reside in the value of the land and the resulting constructed space but in the expectations of the rapid increase in value of the investment. In this way, with the excuse of promoting the development of tourism and the creation of employment and wealth for the whole of the host society, the real estate business operates for the benefit of a very few.

Closely related to this conditioning factor is the traditional Spanish fiscal policy which favours the acquisition of properties and the investment in real estate. To this we should also add the scarce promotion of the rental market and the afore-mentioned attractive conditions to obtain loans for purchasing second homes, not only for their use as holiday homes, but also with which to earn a profit through their subsequent sale. We should also remember that the Spanish coasts have been a space of interest for the purchase of property by foreign citizens. They are tourists who come to stay (Huete 2009) and who, as well as investing their capital, find monetary, health and commercial security in the country. We should also add that it is particularly attractive due to the price differential and quality of life in Spain. This foreign demand, who can be considered as climate immigrants, together with the arrival of international capital, can be considered as some of the principal invigorating factors of the Spanish real estate sector, but they are not the only ones.

Another key aspect that explains the scale of the transformations of the coast is the regulatory framework referring to land legislation, given that the simple classification of land as urban or developable land implies the revaluation of the land and the source of tension for its execution. While the authorities responsible for this issue are the Regional Governments in Spain, when the territorial reality and scope of the urbanised and constructed spaces are analysed, no major differences are found between territories, despite the fact that the chronology and the inspiration of the norms that mark the patterns of territorial organisation are based on different sensitivities. The recent land laws have introduced regulatory mechanisms more focused on sustainability and more precise in terms of the processes and effects. However, in general terms, from the mid-twentieth century, these documents have been based on the premise that the increase in the available land for building lowers the prices and, therefore, benefits the population. Thus, these regulations have not slowed down the transformation of large areas of land. Rather, the contrary has occurred. They have favoured the liberalisation of the land market, as they conceive as developable all the land that is not specifically classified as protected.

The public authorities have been conditioned by the macroeconomic context that promotes the construction and sale of property. At the same time, they must also provide the necessary amenities and facilities for these new urban areas, within a context (already mentioned) of budgetary underfunding in these municipalities with a large seasonal population. In this way, a solution to the problems has been found in the instrumentation of the tourist real estate development in the form of local taxes. It is important to emphasise the explanatory relevance of the structural insufficiency of the local financing system in the mature coastal tourist destinations. It should be remembered that the income of these municipalities is derived from the participation of the State in the local budgets, depending on the censored population, which, are clearly insufficient. These are complemented by the rates derived from the constructed housing stock (in accordance with the IBI [Property Ownership Tax]) and the execution of urban developments and the different rates and revenue from licences related to construction. Within this context, it is logical to understand that, in order to satisfy the service needs of a floating tourist population that is much higher than the censored population, the income connected to the construction in the municipal territory must be maximised.

In this way, the urban planning subordinates the objective of the appropriate organisation and planning of the territory to the need to use urban development as a tool to compensate for budgetary weaknesses. Therefore, the reclassification of all types of land is promoted, frequently under the argument that this enables the development of initiatives that fulfil the "function of social interest" (similarly to the expansionary land law of 1956). However, in many cases, the proposals refer to urban-tourism development projects in the form of large residential complexes, only sometimes accompanied with sports facilities, such as golf courses or marinas.

During the first decades of the development of coastal tourism, the encouraging context of economic development at any cost led to much more building construction than urbanisation. This generated many problems for the local governments, which had to equip a posteriori these areas with a sufficient percentage of completed buildings so they may be classified as urban spaces. Therefore, the need to guarantee, prior to building, the endowment of infrastructures and basic amenities for the new urbanised areas was the initial argument used in the LRAU (land law) of the Region of Valencia (1994) and subsequently in successive land laws in Spain (1998, 2008, 2015), in order to promote the development agents. These were the guarantors of the preparation and urbanisation of the territory, prior to the building of the properties. In this way, the local revenue authorities would not have to assume the cost of the urbanisation of already constructed and sold spaces, as was the case after the first and second construction waves on the coast.

However, the innovative system of the developer has not always fulfilled the initial objective of avoiding individual speculation and the control of the urbanisation processes of the territory. Often, the collusion between the municipal authorities and the developers has left the owners of the land and pre-existing buildings defenceless, leading to a high number of lawsuits in different national and European courts, as they lost the right to decide over their properties. According to this mechanism, the appropriation of the capital gains derived from the transformation process was transferred from the owners of the land to the owners of the companies accredited to urbanise it, who were those who obtained the majority of the profit. In this way, far from promoting sustainable and fair developments, the developers have promoted the transformation of thousands of hectares and the construction of large urbanisations based on initiatives only possible for agents with access to large sums of capital or promoters connected to international financial flows. This has given rise to the involvement in these change processes of large national or international promoting companies which will take the initiative from now.

The planning goes against the plan, which should order the uses and define the social function and public interest, together with the private interests of the development initiative. Therefore, the development of certain macro-projects often responds more to speculative property purchase actions than to developments for tourism use and a medium to long-term commitment to the territory in which they are located. Given that the focus is the real estate business, the relationship with the space used is temporary, hence its execution (in some cases) has not been exempt from controversy, due to the questionable quality of the construction, the scarcity of public amenities or the lack of respect for the environment.

Finally, land ownership affects both the urbanisation and real estate promotion processes, as an accelerator or restraint to the change of use or as a conditioning factor in the way the new tourist urban space is adopted. The pre-tourism structure (the size of the properties, the type of ownership, use and profitability of the land) conditions the form and dimension of the resulting actions and the pace of the transformation. In the early years of the tourism real estate boom, in irrigated areas with a small cadastral framework, under the land law of 1956, when the size of the property was less than five hectares, it was possible to develop small urbanisations without prior planning. Due to the juxtaposition of small actions, the result of this has been the generation of a continuous stretch of residential territory, made up of small, disperse and poorly connected pieces. This has led to the unregulated occupation of large areas that are barely urbanised and poorly endowed in the plains opposite the coast or its catchment area. This process has also occurred in areas with a predominance of small dry land properties on terraced hillsides overlooking the Mediterranean Sea. Meanwhile, the large rain-fed estates, managed as latifundios in the southern regions of Spain or the areas for forestry use, due to their large size, have been subject to the elaboration and approval of plans and end up being transformed in a planned way. In these cases, the change of use from agricultural or mountain use to residential tourism use is based on large-scale partial plans, sometimes thousands of hectares, which can arise from the grouping together of several properties or their segregation. In any event, the transformation of these plots requires access to large amounts of capital, the approval of plans and the preparation and urbanisation of the space for their execution over time in different sectors (Navalón-García 1995, 1999, 2001), which, in some cases, are still being executed today.

8.5 The Effects of the Real Estate Instrumentation of the Urban-Tourism Developments

The combination of all of these factors has conditioned the development model of coastal tourism, in the majority of cases leading to a mono-specialisation of the sun and beach product, which renders it seasonal and territorially polarised. Although there are compact destinations, interwoven into the pre-existing population nuclei, with a significant hotel supply and synergies that generate positive dynamics, the supply focused on residential urbanisation and real estate promotion is predominant. The complexes are often barely planned from a global planning perspective as they respond to independent juxtaposed projects in the territory. Furthermore, it is not surprising that there are some cases of corruption in some way related to the figure of the developer (Fig. 8.2), which has grown at the same pace as the tourism sector itself.

Given that the initial objective is not specifically the tourism business, but rather the real estate business, the chosen model of land occupation is usually extensive, not concentrated or dynamic, and ignores criteria of efficiency in water and energy consumption or territorial occupation. Considerable environmental impacts are generated, due to the invasion of sensitive spaces, deforestation or the aggravation of the erosion caused by construction on hillsides with steep slopes. To this, we should also add the lack of development of complementary activities, able to increase the attractiveness of these areas and favour deseasonalisation and an increase in expenditure. All of this leads to a lower level of tourist expenditure and a lower profitability of investments and efficiency from an economic and social point of view (Baños Castiñeira 1999).

This has occurred in very different areas in the coastal regions of Spain with similar transformation processes and results but also in areas beyond the Spanish borders. Specifically, Spanish legal and fiscal issues have conditioned the speed and size of the changes of the autonomous regions of the Mediterranean and islands, but it is possible to describe similar effects in other environments corresponding to the peripheries of pleasure, such as those indicated in the afore-mentioned evolutionary models of Gormsen or Butler, among others.

Taking into account what has been described for the traditional sun and beach tourist spaces which have reached a stage of maturity or decline, according to Butler, there is a need to reorientate and reconvert these destinations. This is particularly the case if we bear in mind the coordinates of the international market and the emergence of new competitors on the shores of the Mediterranean and more distant peripheries, such as the Caribbean.

Fig. 8.2 Map of corruption in Spain by municipality. *Source* Corruption map by municipality and political party. Developed by Corruptódromo http://wiki.nolesvotes.org/wiki/Corrupt%C3%B3dromo by @PixSpain, created with Google my maps. Creative commons Attribution-ShareAlike 3.0 Unported (CC BY-SA 3.0)

In recent years, the shrinking of profit margins in coastal destinations with a hotel supply, together with the low revenues obtained from residential use, has reduced the possibility of reinvesting and reorienting the destinations. Therefore, it is more difficult to resolve the problems of urban saturation, the obsolescence of tourist facilities and amenities or the deterioration of public spaces and attraction resources, which would be ideal so as to retain a demand who travels a lot and more cheaply to anywhere in the world.

An attempt to compensate this loss of liquidity to resolve the shortcomings of the destination resides in the revenue derived from construction, but this leaves very little margin for the incorporation of renovation processes and public investments which would address the stagnation or decline of the traditional destinations. This deepens the vicious circle (Perelli 2012) which seeks for the solution to finance the correction of the problems of saturation in the income derived from construction, while creating new urbanised spaces difficult to maintain with the local budgets; therefore, the urbanisation and construction cycle must continue, creating vast urban spaces and undifferentiated destinations.

This is a phenomenon with a global scope backed by the dominant financial policies which have given rise to a Spanish specialisation in the revaluation of urban assets, that is, construction as an investment for speculation. The result is the irreversible occupation and building up of the territory, sometimes with unfinished buildings, which generates enormous costs of unproductivity and inefficiency when the buildings are empty not to mention the enormous landscape and environmental impact.

8.5.1 The Globalisation of the Instrumentalisation of Tourism

As contemplated by Gormsen, the tourism activity advances by colonising new spaces for tourism in successive "peripheries" which are evolving rapidly in their life cycles in other regions of the world. Following the capitalist logic, the promoters operate in other environments geographically apt for developing tourism in which the income differential enables the optimisation of investments. Authors such as Rullán (2008) or Navarro-Jurado et al. (2015), among others, have compared different environments following the patterns of the afore-mentioned model in identifying the mechanisms and effects of urban-tourism urbanisation in the so-called fourth periphery, located on the shores of the Southern Mediterranean (Chaime 2016) and in areas of Central America and north-east Brazil (Buades 2006) or Southeast Asia.

As previously described, in these cases, the scope of the transformations depends on the pre-existing conditions; these can be environmental due to the existence of favourable climate features or tourist attraction resources; economic, such as the availability and low price of land or the differential value of the currency; or social, with a population (and leaders) prone to changing the uses of the land and activity out of pure need. However, the pace, size and form are fundamentally related to the objectives of the capital (Buades 2006; Antón Clavé and González Reverte 2008) and the integration or otherwise of the tourism development strategy in the global planning of the territory. In this way, as described by Blázquez et al. (2011), in these cases, it is the transnational capital that chooses the development model, in the form of integrated tourist enclaves, which are offered as "all inclusive" in hotels and closed urbanisations, with the backing (and even subsidies) from the governments of these countries which seek in tourism a way towards development and financing their needs.

Although it could be understood as an opportunity for economic growth, it also represents a potential aggravation of the socio-environmental impacts, particularly if the initiatives are developed in fragile areas, from an environmental and social point of view. The problem is greater if the capture of these foreign investments implies the diversion of public funds to favour tourism real estate growth policies, in detriment to policies to improve the living conditions of the population (Navarro-Jurado et al. 2015). The costs of creating infrastructures and basic amenities are externalised and assumed by the local governments who wish to attract the investment and the profits which barely reach the residents are privatised.

8.6 Challenges of Globalisation

Tourism acts as a principal agent for transforming the coastal territory with similar processes and sequences in very different places. In the more recently developed peripheries, the model repeats the patterns of traditional tourist destinations, although in a more accelerated and polarised way. Although the mistakes made in these spaces should be taken into account, similarly to their predecessors, the territory is sold, its natural resources are altered and its local population is unable to enjoy it and, in these cases, barely receives a return on the investment.

There is a homogenisation of the proposals in terms of the approach taken by the projects, which clone the urban initiatives and proposals of other regions, although initially the theoretical intention is based on differentiation criteria. The prevalence of criteria related more to the real estate business than to tourism and the presence of the same companies that provide the capital (Blázquez et al. 2011) explains the similarity of the urban complexes. The residential and hotel complexes are repeated with similar sports facilities (basically golf and water sports), areas of well-being and health, gardens and bathing areas. Standardised and banal urban environments are generated, in which the resources of the place itself are relegated to a secondary level, or simply provide an exotic scenario. Given that the initial approach is not rooted in the overall planning, the urbanisations are isolated from the traditional urban nuclei and are poorly connected, which aggravates the fragmentation of the territory.

When the approach to the new developments is embedded and closed, with specific leisure functions and own endowments, different spatial segregation processes are aggravated which end up privatising the space which becomes reserved for tourism use and its guests and prevents society from enjoying its resources. A cultural homogenisation process by country may even occur in the form of closed complexes of tourist ghettos which have very little contact with the context.

Elitisation or gentrification processes also occur in the coastal areas of the space whereby certain social groups are replaced with others with a higher purchasing power. The residents are deprived of the use and enjoyment of places due to the resulting increase in prices and traditional retail establishments or local customs are replaced with new uses and global brands.

8.7 Future Challenges for Coastal Destinations and Planning Proposals

In recent years, the urban spaces emerging from tourist real estate development have continued to adapt to the social changes and to the environment in which they are located. After a long period of incessant growth which led to the saturation of the receiving spaces, only altered by occasional crises, the needs of those who use them have changed and the COVID-19 pandemic and the post-pandemic period have opened a new phase of great interest. This situation has raised a series of latent questions which have now become highly relevant. Issues such as mobility and connectivity, the widespread use of technologies, the reinforced sensitivity to health safety, the renewed appreciation of natural and unpolluted spaces or the way in which movements are understood (López Palomeque et al. 2022) have become key aspects in the choice of destinations and living spaces.

The tendency for tourists to become citizens and construct new places of coexistence in regions far from their country of origin is not new (Huete and Mantecón 2011) but, after the pandemic, the appreciation of and demand for spaces with a better quality of life have risen. The new residents, attempting to redirect their lives in places distant from the productive function, either temporarily or permanently, appropriate these places and call for new endowments and amenities which were initially not contemplated or, at least, not on the scale that the new users require. We are referring to health services, retail areas and even education centres, typical of residential areas that are interwoven with the residential tourism supply. In relation to this and with the new lifestyles (remote working and digital nomads), the destinations should adapt to the changes in mobility, both residential and leisure, and adapt to the new needs, incorporating innovation into the transport networks and systems of these areas, with sustainable mobility proposals that incorporate technology to improve efficiency and which also cater to all people irrespective of their capacities.

Other challenges are those associated with the adjustment of the destinations to demographic changes, such as the ageing of the population, which requires specific facilities; also the new healthier leisure and lifestyle patterns or the more demanding users related to their greater experience in tourism consumption. The incorporation of the platform economy in the relationship between supply and demand and the mutations in the forms of communication and marketing are aspects that can be considered as being disruptive. This is also the case of the penetration of technology in all phases of the tourism experience and in practically all areas of its development, which poses the challenge of the digital transformation of companies and destinations. Finally, although there are many other aspects that are not covered here, we should also refer to the relevance of energy measures and prevention measures related to climate change and its impacts. Also important is the sustainable planning and management of the urban areas and spaces that should offer quality of life and address the environmental impacts of urbanisation and saturation.

Bibliography

Antón Clavé S (1998) La urbanización turística. De la conquista del viaje a la reestructuración de la ciudad turística. Documents d'Anàlisi Geogràfica 32:17–43

Antón Clavé S, González Reverte F (coord) (2008) A propósito del turismo. La construcción social del espacio. UOC, Barcelona

Baños Castiñeira CJ (1999) Modelos turísticos locales. Análisis comparado de dos destinos de la Costa Blanca. Investigaciones geográficas 21:35–57

Carvalho Cantergiani C, Jiménez Herrero LM (2007) Artificialización del suelo y sostenibilidad en las costas. Indicadores del observatorio de sostenibilidad de España. Ambienta, November: 66–68

Blázquez M, Cañada E, Murray I (2011) Búnker playa-sol. Conflictos Búnker playa-solBúnker playa-solderivados de la construcción de enclaves de capital transnacional turístico español en el Caribe y Centroamérica. Scripta Nova XV:368, July. Universidad de Barcelona. Available at http://www.ub.es/geocrit/sn/sn-368.htm

BOE (2015). Royal Legislative Decree 7/2015 of 30 October, approving the modified text of the Land and Urban Rehabilitation Law number 261. https://www.boe.es/buscar/pdf/2015/BOE-A-2015-11723-consolidado.pdf

Buades J (2006) Exportando paraísos: La colonización turística del planeta. La Lucerna, Palma de Mallorca

Butler R (1980) The concept of tourism area cycle of evolution: implications for the management of resources. Can Geogr 24:5–12

Butler R (2012) Matura tourist destination: can we recapture and retain the magic? In: Vera JF, Rodríguez I (eds) Renovación y reestructuración de destinos turísticos en áreas costeras. PUV—Universitat de València, Valencia, pp 19–36

Chaime S (2016) Turismo globalizado en Marruecos. El caso de Saidía. Doctoral thesis, University of Malaga.

Exceltur (2021) Solytur 2019. Monitor de competitividad turística de los destinos de sol y playa español. https://www.exceltur.org/solytur/monitor-de-competitividad-turistica-de-los-destinos-de-sol-y-playa-espanol/

Fernández Durán R (2006) El tsunami urbanizador español y mundial. https://www.ecologistasenaccion.org/4824/el-tsunami-urbanizador-espanol-y/

Fernández Tabales A, Cruz Mazo E (2011) Territorio y actividad constructora: del «tsunami» a la crisis. Factores explicativos y propuesta de indicadores a escala municipal en Andalucía. Boletín de la Asociación de Geógrafos Españoles 56:79–110

Fernández-Tabales A, Navarro-Jurado E (2022). Nuevos enfoques en la geografía del turismo: gobernanza e inteligencia territorial como últimas fases en la evolución de la gestión turística del litoral. Documents d'Anàlisi Geogràfica 68(3). https://doi.org/10.5565/rev/dag.747

Gormsen E (1981) The spatio-temporal development of international tourism: attempt at a centro-periphery model. In: La Consomation d'Espace par le Tourisme et sa Préservation. Centre des Hautes Études Touristiques, Aix en-Provence, pp 150–170

Gormsen E (1997) The impact of tourism on coastal areas. GeoJournal 42:39–54. https://doi.org/10.1023/A:1006840622450

Gutiérrez J, Viedma A, Callejo J (2005) El proceso de constitución de España en una sociedad turística. Política y Sociedad 42(1):151–168. https://revistas.ucm.es/index.php/POSO/article/view/POSO0505130151A

Huete R (2009) Turistas que llegan para quedarse: Una explicación sociológica sobre la movilidad residencial. University of Alicante, Alicante

Huete R, Mantecón A (2011) Introducción: sobre la construcción social de los lugares. In: Mazón T, Huete R, Mantecón A (eds) Construir una nueva vida. Los espacios del turismo y la migración residencial. Milrazones, Santander, pp 11–19

INE. Instituto Nacional de Estadística (2019). Movimientos turísticos en frontera. Número de turistas según comunidad autónoma de destino principal. Madrid, INE. https://www.ine.es/jaxiT3/Datos.htm?t=23988#!tabs-tabla

Ivars-Baidal JA, Vera Rebollo JF (2019) Planificación turística en España. De los paradigmas tradicionales a los nuevos enfoques: planificación turística inteligente. Boletín de la Asociación de Geógrafos Españoles 82(2765):1–31. https://doi.org/10.21138/bage.2765

López Palomeque F, Vera Rebollo JF, Torres Delgado A, Ivars Baidal JA (2022) El turismo ¿fin de época? Desafíos de España como destino turístico en un nuevo scenario. Publicacions de la Universitat de València, Valencia

Navalón-García R (1995) Planeamiento urbano y turismo residencial en los municipios litorales de Alicante. Instituto de cultura Juan Gil-Albert, Alicante

Navalón-García R (1999) Caracterización del espacio turístico residencial del litoral valenciano. Boletín de la Asociación de Geógrafos Españoles 28:161–178

Navalón-García R (2001) Agricultura y turismo en la franja costera de la Comunidad Valenciana. Fundació Bancaixa - Fundación Bancaja, Valencia

Navarro-Jurado E, Thiel-Ellul D, Romero-Padilla YR (2015) Periferias del placer: cuando el turismo se convierte en desarrollismo inmobiliario-turístico. Boletín de la Asociación de Geógrafos Españoles 67:275–306

Pearce D (1988) Desarrollo turístico. Su planificación y ubicación geográfica. Trillas, Mexico, 168 pp

Pearce D (1993) Geographie du tourisme. Nayhan Université, Paris

Perelli O (2012) La reconversion de los destinos maduros del litoral en el actual contexto de cambio global: una reflexión sobre los nuevos instrumentos de impulso. In: Vera JF, Rodríguez I (eds) Renovación y reestructuración de destinos turísticos en áreas costeras. PUV—Universitat de València, Valencia, pp 37–54

Rullán O (2008) Reconversión y crecimiento de las zonas turísticas. Del fordismo al post-fordismo. In: Vinuesa T, Marchante G, Hernández G (coord) Destinos turísticos: viejos problemas, ¿nuevas soluciones? X Coloquio de Geografía del Turismo, Ocio y Recreación (A.G.E.). Universidad de Castilla-La Mancha, Cuenca, pp 587–626

Rullán O (2011) Urbanismo expansivo en el Estado Español: de la utopía a la realidad. Geografía retos ambientales y territoriales XXII:165–209. Congreso de Geógrafos Españoles. Universidad de Alicante

Thiel-Ellul DF, Navarro-Jurado E (2018) Medición y análisis de la sostenibilidad. Indicadores sintéticos a través de métodos multicriterio y su relación con el turismo en el litoral de Andalucía. Síntesis, Madrid

Turner L, Ash J (1991) La horda dorada: turismo internacional y la periferia del placer. Endymion. Turismo y Sociedad, Madrid

UNWTO (2019) Panorama del turismo internacional, edición 2019. https://www.e-unwto.org/doi/book/10.18111/9789284421237

UNWTO (2020) Barómetro del turismo mundial. Enero 2020. Año estadístico. https://catedratim.files.wordpress.com/2020/02/unwto_barom20_01_january_statistical_annex_sp.pdf

Vera JF, López F, Marchena M, Antón S (2011) Análisis territorial del turismo. Tirant lo Blanch, Valencia

Vera Rebollo JF, Baños Castiñeira CJ (2010) Renovación y reestructuración de los destinos turísticos consolidados del litoral: las prácticas recreativas en la evolución del espacio turístico. Boletín de la Asociación de Geógrafos Españoles 53:329–353

Yunis E (2001) Sustainable development and management of tourism in coastal areas. World Tourism Organization, Madrid, October. http://www.invemar.org.co/redcostera1/invemar/docs/518Management%20of%20Tourism.pdf

Chapter 9
Landscape Preservation on the Mediterranean Coast; Do Social Movements Play a Decisive Role? The Case of the Costa Brava (Spain)

Joan Vicente Rufí, Sergi Nuss-Girona, Margarida Castañer Vivas, Jaume Feliu Torrent, Javier Martín-Uceda, and Oriol Porcel Montané

Abstract Social movements are at the origin of environmental protection of the coastline in tourist destinations. Consequently, it is crucial that this role will continue in the future, if urbanisation pressures persist and in a context of energy transition. The various reasons for this starting hypothesis are historical, sociological, political, economic and cultural, so that the geographical variable is relevant, as it is related to the specific conditions of the territory.

The analysis of these social resistances must be complemented with the more environmentalist evolution of transversal policies that affect the tourist coastline, such as urban planning strategies or laws for the protection and management of the coast.

This chapter analyses the social mobilisations for territorial protection of the Costa Brava coastline over the last fifty years, mapping more than 230 mobilisations. The results are relevant maps regarding the actors and their motivations and, of particular interest, the negative impacts.

J. V. Rufí (✉) · S. Nuss-Girona · M. C. Vivas · J. F. Torrent · J. Martín-Uceda · O. P. Montané
Department of Geography, University of Girona, Girona, Spain
e-mail: joan.vicente@udg.edu

S. Nuss-Girona
e-mail: sergi.nuss@udg.edu

M. C. Vivas
e-mail: mita.castaner@udg.edu

J. F. Torrent
e-mail: jaume.feliu@udg.edu

J. Martín-Uceda
e-mail: javier.martin@udg.edu

O. P. Montané
e-mail: oriol.porcel@udg.edu

E. Navarro-Jurado et al. (eds.), *Urban Dynamics in the Post-pandemic Period*, The Urban Book Series,
https://doi.org/10.1007/978-3-031-36017-6_9

Keywords Urbanisation pressures · Social movements · Landscape · Transversal policies

9.1 Introduction

What role do mobilisations and social movements play in protecting landscapes in highly anthropised coastal areas? To what degree do they achieve positive results in response to their demands? These are the fundamental questions this paper asks and tries to answer. The starting point is a hypothesis inspired by Gramsci's "optimism of the will": these results are quantitatively and qualitatively more positive than they might seem. With these premises, this paper explains the objectives, theoretical foundations and some first results for the province of Girona (Catalonia-Spain) of a research project started in 2018 by the University of Girona research group APTA (Territorial and Environmental Analysis and Planning). The significance of this work stems from its generation of a very extensive database about movements in the last fifty years and from the verification of the hypothesis that their consequences, as previously mentioned, are of much greater significance than one might think. In other words, without these actors the environmental situation of the landscapes in this territory would be much worse than it might already be (Greenpeace 2019). Demonstrating this usefulness is a basic way of giving value to this social heritage, indispensable for the present and for a future that will presumably be equally or even more complex.

The work is organised into three parts. The first is a summary of the theoretical basis of research and methodology. Notable among them are the reflections on social movements, specifically in defence of the territory, and especially on the fact that we focus on their impacts (or results) more than on their motives, characteristics, arguments and strategies. A focus on the outcomes of the movements allows us to reflect on another typical and key aspect of studies on social movements, the "not in my backyard" (NIMBY) concept. Here, too, the hypothesis should be recalled because it is assumed that while the initial drive and development of the mobilisations very often responds to this particular interest, their outcomes can frequently surpass it (even without intending to), giving rise to results that go beyond the usual "Not here!"

The second part is the case study and contains the results obtained for the coastal province of Girona, and more specifically the three coastal regions that make up the so-called Costa Brava.[1] The Costa Brava is a well-known, first-class international tourist destination that has been subject to the pressures this creates for almost a

[1] Conflicts that have taken place in the three regions of the province of Girona—Alt Empordà, Baix Empordà and La Selva—that make up the coastal strip of the Costa Brava are included. The immediate Gironès hinterland has also been included because this is where the large infrastructures that serve the coastal area pass through or are located, including the airport, motorways, railways and electricity lines.

century. It has also been the scene of persistent social mobilisations demanding protection of the landscape, thus making it an especially interesting case.

Finally, the third part contains the conclusions, where we link the results of the case analysed with the general objectives and hypothesis of the research, while acknowledging that the Costa Brava may have certain characteristics that make it the exception rather than the rule in the set of Spanish coastlines with which the wider project is concerned.[2]

9.2 Social Movements in Defence of the Territory and Their Impacts: A Brief Theoretical Framework

A theoretical framework focused on the fundamental concepts of "social movement", "territory" and "impacts/outcomes" underpin the research in this paper. In terms of primary sources, theories around the so-called new social movements (NSM), since the last quarter of the last century, have attempted to explain the changes in ways of identification and social organisation in post-industrial societies characterised by globalisation. Fundamental for this line of research are authors like Touraine (1981), Tilly (2004) and Castells (1997). The latter two are of particular interest in that, when characterising NSM, they propose that space (or territory, landscape, place, etc., all concepts that suggest spatiality) has a relevant role. These authors are, of course, not the only ones concerned with this topic. Since the 1980s, others have used the term "spatial turn" or various forms of "topophilia" (Tuan 1974; Nogué 1991, 2015), almost as a paradoxical consequence of globalisation, which in its earliest and most primitive theorisations (Amin 2002) seemed to "annihilate"—in the words of David Harvey (1989)—diversity and geographical distances. In short, global–local dialectical tensions obliterate previously hegemonic theoretical frameworks and reignite the debate on the significance of space in all its imaginable scales, from the body to the planet.

To put it geographically, it is precisely among these tensions that borders, where the reality of conflict appears, are destroyed and rebuilt (Brenner 2004). In some way, inherent to "new" social movements are "new" social conflicts (although not all authors would agree; for Harvey, the central conflict continues to be capital/labour). These new conflicts usually have a territorial motive, scenario, instrument or consequence. Countless authors have studied and proposed interpretations of these territorial conflicts from very different standpoints, often opposed in one way or another. Since the last two decades of the twentieth century, in geography as well as in political science, history and economics, different theories have been formulated in which space has a central role (the aforementioned spatial turn)

[2] This study began on the Girona coast and has now been extended to the entire Spanish Mediterranean coastline and involves experts from the Autònoma de Barcelona, Rovira i Virgili, Valencia, Alicante, Murcia, Granada and Malaga universities. The project, with the acronym MOVXPAIMED, is funded by the programme I+D+i PID 2020-116850RB-I00.

(Warf and Arias 2009). The "rebirth" of sub-state nationalisms, models of local development and the new regional geography clearly clash with globalist models and more structuralist views, such as those of political economy and political ecology (some authors of reference are Massey 1991; Paasi 2002; Dematteis 1985; Nogué, 2015; Keating 2007; Robbins 2004).

In turn, this meant, and has meant, a growing interest in territorial conflicts, or environmental conflicts, although these have a genealogy of their own in part due to the emergence of environmental problems, demands and movements from the 1970s onwards (Melucci 1994; Brulle and Rootes 2015; Giugni and Grasso 1995).[3] It is an interest focused on different areas of study that could be divided into three main lines: the reasons for the movements (contexts, problems they generate, etc.); the ways of organisation, action and communication; and the analysis of their impact (outcomes) and political ramifications. From the area of geography at least, the major focus has been the first, seeking both the "structural" and "specific" causes of the conflicts. They include the impacts and transformations stemming from the alienation and subjugation of the territory (the "places" and their populations); the superstructures, flows and demands of globalisation and its decision-making networks (Nel·lo 2003); and NIMBY rationales that have dominated many of the most mainstream explanations, including selfish opposition to preserve specific and privileged interests (Driscoll 2013; Dear and Taylor 1982; Dear 1992; Nel·lo 2003; Bobbio and Zeppetella 1999).

The most extensive research currently underway on these movements is conducted by the Universitat Autònoma de Barcelona's Institute of Environmental Science and Technology. It is the Environmental Justice Atlas (EJATLAS) collaborative project,[4] which consists of compiling and mapping environmental-based conflicts. According to recent data, they amounted to 3,700 cases by the beginning of July 2022. One conclusion of this is that the global increase in environmental activism is highly correlated with the acceleration of social change experienced since the middle of the twentieth century (Scheidel et al. 2018; Temper et al. 2018).

Even within the analysis of the motivations and characteristics of the movements, another relevant perspective is the one that focuses on the impacts of tourism and the reactions that they generate. It is a field of study and reflection that also has a long history in our country with absolutely pioneering works such as *The Human Landscape of the Costa Brava* by the French geographer Yvette Barbaza (1966), also focused on the area of this article or, much more recent, that of Maria Kousis (2000) analysing various Mediterranean cases and much closer to the approach of this text. But it is in the last decade and a half that newer and essential approaches have been generated for contemporary real estate/tourism dynamics and, even, with an innovative vocabulary—turismophobia, touristification, overtourism (Colomb and Novy 2016, 2021; Milano et al. 2019; Navarro-Jurado et al. 2019).

---

[3] In the text, the terms "territorial" and "environmental" have been used indistinctly to refer to the conflicts. Without doubt, this decision is debatable in theoretical terms, or at least warrants serious reflection (Walter 2009). Nonetheless, the research is not based on concepts and models but quite the opposite, on the cases analysed and on this reality that "environmental" and "territorial" are largely interchangeable, including for the protagonists themselves.

[4] See Temper et al. (2018).

As mentioned in the introduction, the interest of the research is in the third area of focus, in territorial movements and conflicts and in their outcomes and impacts, a perspective undoubtedly much less developed than the others (Ibarra et al. 2021). As a result of this choice, which is justified by the hypothesis that the Costa Brava is a unique case in terms of the positive impacts of these movements, this research not only delves into the outcomes achieved ("saved" spaces, etc.) but also attempts to find explanations for the "why" of these impacts. To this effect, political opportunity theory is undoubtedly useful in interpreting when movements take place and the results they obtain (Tarrow 1998; Scheidel et al. 2018). On the one hand, it looks at the weaknesses of governance in each context to understand the movements' windows of opportunity (Meyer and Minkoff 2004; Alfama et al. 2007). On the other, it looks at the resources (moral, cultural, human, material) that each movement has and exploits in its action (Sribman 2021). As Aguilar (2021) highlights, the study of social movements needs to adopt a relational approach (with the media, institutions, etc.). Dynamic relational processes give rise to paradoxical situations. It is often when governments take a more protective stance that the movements acquire greater force (Alfama et al. 2007).

However, for this theoretical framework, we must go back to the NIMBY label previously mentioned. One of the research hypotheses involves putting the importance of this concept into perspective when explaining the movements and, above all, their impacts, at least in the area under study here. While it often acts almost like an automatically applied adjective to virtually any conflict that comes about, we believe that the reality is more complex. Even without denying that this type of reaction can be at the origin of many of the conflicts that feature in our study, in line with Eranti (2017), we discuss labelling territorial defence movements as NIMBY, with private interests as their only and main driver. As mentioned previously, we believe that place attachment—i.e., strong emotional ties to places and objects—and, even more so, general environmental justice and sustainability advocacy demands (the common good), also lie behind these social movements, with societal gains as the return. In fact, from the analysis of the movements on the Costa Brava over the last fifty years and the "events" arising from their mobilisations (as will be seen later), some outcomes can clearly be described as "structural", either in the form of very stable organisations or as rules and procedures that go far beyond the specific cases of conflict. In other words, the movements do not have to remain in the phases (or models) of resistance (Castells 1997), but can move on to "project" or "propositional" phases (Castells 1997; Nel·lo 2003). At that point, they generate more consolidated forms of territorial governance, either in an institutional and explicit way (as all types of official declarations rhetorically affirm[5]) or in a hybrid and implicit way.

[5] An example of this is the 2019 United Nations Human Rights Council's unanimous recognition of the indispensable role of social movements for environmental sustainability and protection (UN 2019). Another example is that since 2015, the Sustainable Development Goals (SDG) promoted by the United Nations repeatedly call on civil society and its entities as key elements to achieve their aims. Similarly, article 45 of the Spanish Constitution includes the obligation to preserve the environment on the part of the public authorities, but also and explicitly on the part of the citizens. More specific territory and landscape governance or planning instruments "necessarily"

It is a formula in which movements constitute a force for dialogue—underpinned by important resources of social legitimacy, technical capacities and political pressure—that makes them key and unavoidable actors for the protection/management/transformation of the territory.

9.3 Methodology

This research compiles and analyses social movements opposed to projects that activists considered harmful to the natural or cultural heritage in the regions of Girona since the 1970s. The information (about their identification and analysis) contained in this paper is quantitatively and qualitatively processed for the primary purpose of constructing a broad database and a portfolio of cases of environmental conflict and ecological movements. The information comes from collaborative work with environmentalist advocacy groups and documentary research[6] (literature, organisations' archives, online searches).

To analyse and understand the case of the Costa Brava, a cartographic-type database was generated to compile the different social movements in defence of a place, a thoroughfare, a natural space or a particular landscape against specific projects, as well as any opposition to land-use planning instruments or regulations in a wide territorial area. Starting from a pre-existing, provincial-level database that covers the years 1970–2020 (Nuss-Girona et al. 2020); the data were then refined for the Costa Brava study area. The provincial database began in 2018 with 60 cases and was progressively expanded and improved thanks to documentary research and an ongoing process of communication with activists (meetings, emails, telephone consultations and interviews). The database currently has more than 250 cases at the provincial level, of which almost 200 make up the case study limited to the coastal regions. The aim of this process was to generate a representative sample of diversity, continuity and geographical dispersion of movements throughout the study period, not a comprehensive one.

The movements, which synthesise a process that is often complex in terms of space and time, were given geographical coordinates. In other words, they were mapped to the exact place or to a barycentric point of the disputed space. The entry

---

include appeals to this central role of the movements. For example, there is the European Landscape Convention, which was a major boost to participation in the definition of landscape policies, with the Landscape Approach (Arts et al. 2017) representing the most recognised theoretical and practical trend in application. The European Green Deal currently emphasises that "the transition towards sustainability must be fair and inclusive, putting people first and bringing together citizens in all their diversity" (Figueiredo Nascimento et al. 2016).

[6] Main sources of information: (a) Document, Archives and Publications Management Service of the City of Girona (https://www.girona.cat/sgdap/cat/premsa.php); (b) Territorial Yearbook of Catalonia, published between 2003 and 2015 by the Catalan Society of Land Management (http://territori.scot.cat/cat/anuari.php); (c) Urban Map of Catalonia of the Ministry of Territory and Sustainability of Catalonia (http://dtes.gencat.cat/muc-visor/AppJava/home.do); and (d) Ministry of Justice of Catalonia, which has a search engine for legally registered civil society organisations.

Table 9.1 Status of the environmental conflicts in the Girona regions. Description

Saved	The plan or project triggering the conflict was cancelled by either an administrative decision or court ruling. Eventual changes in the policies or legislation that no longer permit certain plans and projects (e.g. fracking) are sometimes included. Cases can also be rejected before the administrative procedures begin, because knowledge of their possible implementation brings about pressure and reports from the movements, leading to their public rejection by the authorities
Provisionally/ partially saved	Provisionally saved cases are those stopped due to social pressure (the slowing down of administrative procedures, interim injunctions, etc.) and often their particular combination of external factors (elections, economic crises, developer company bankruptcy, etc.), leaving the projects on indefinite "stand-by". However, economic and political interests are so strong that the projects reappear some years later Partially saved are cases with relevant environmental impact corrective measures, applied thanks to social pressure
Open conflict	Cases that have not completed the administrative and/or court procedures and where there is a social movement opposed to them involving public actions, rallies, complaints, fund-raising, contracting legal services, etc.
Defeat	Cases that, despite social opposition, have been approved and developed

Source Own compilation

generated for each conflict had the following basic descriptive information: name of the conflict; reason for the conflict; name of the group or association that led the mobilisation; and status of the conflict.

Lastly, and as has been mentioned from the beginning, for this paper "Status of the conflict" was the main factor of analysis to cross with the rest of the parameters studied. Consequently, an important decision was the definition of the respective categories. Starting from the two results at either end of the spectrum—defeat or success, from the movement's point of view—three more categories were established. They respond to common situations in this type of conflict and in these territories, situations that correspond to long, intermittent processes that frequently have outcomes that, rather than being black or white, fall into a grey area. Table 9.1 describes the "Status" categories established.

9.4 The Emergence, Persistence and Socialisation of the Valuation of the Costa Brava Landscape

The Costa Brava is the brand by which the coastline of the province of Girona, in northeast Catalonia and bordering France, is known. At around 200 kms in length, it is characterised by rocky landscapes and pine groves that plunge straight into the sea, small coves and two large bays with wide beaches. It has been inhabited for millennia—it is where the Greeks and Romans arrived on the Iberian Peninsula—and is currently made up of towns with long histories that now specialise in tourism and other small fishing villages that have also become tourist spots. There are 22 municipalities with a population of 244,255 (2021), a hotel capacity of almost 80,000 beds and campsites with an additional 114,000 places (2021). In 2019, the last year of a certain normality in terms of tourism, the number of visitors was almost 5 million. Furthermore, the proximity to the metropolitan area of Barcelona and the South of France explains the very high proportion of second homes, which can exceed 50% of the housing stock in some municipalities, according to the last official census of 2011.

These municipalities are so integrated into three regions—Alt Empordà, Baix Empordà and La Selva, totalling 2,954 km^2—that an interior region, originally agrarian and wooded, and that also has small mediaeval villages, must be added to the coastline. They are part of the landscape imaginary and are populated with national and international tourists, which is an important point in terms of the research presented in Map 9.1.

The Costa Brava is the oldest tourism brand in Spain, dating back to 1908. That year, the coastline of the province of Girona received this name (various sections of the coast had other names). It was a sign that the, by then, regular presence of visitors deserved a "commercial" treatment similar to that of other locations of the Mediterranean such as the Cote d'Azur in France and the Italian Riviera. The name also revealed the relevance of landscape—not just the simple fact of the colour of the sea and the sky or the obvious seaside location—as a defining element of this territory, which was of an obviously Mediterranean character as imagined from the Grand Tour, both from a natural and a cultural perspective.

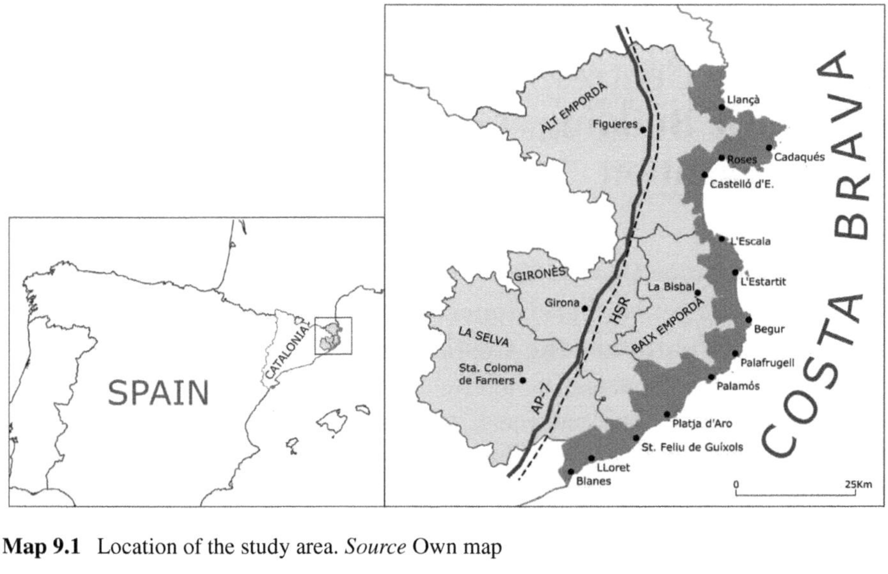

Map 9.1 Location of the study area. *Source* Own map

At practically the same time that the Costa Brava emerged as a tourist brand, it also became the object of another (and apparently) quite different view and reading. In the cultural and political movement known as *Noucentisme*, the landscapes of this coastline and the immediate interior rural space, especially the places linked to the Greek, Roman and mediaeval settlements of the regions of Alt and Baix Empordà, became "models" for the country. Catalonia was fully immersed in the process of constructing its modern national identity, which for *Noucentisme* incorporated an idealisation of the Mediterranean precisely as a paradigm of classical civilisation and order (Nogué and Vicente 2004). This qualitative leap also brought about a certain recognition of these landscape values (being, e.g. one of Salvador Dalí's iconic landscapes), both by the population itself and by groups outside the territory, and, above all, by economic and cultural elites from Barcelona and progressive international ones (Yanes 2020).

What makes this case interesting is that this heavy load of real and symbolic values very quickly generated some unique developments that made it stand out from other coastlines in the processes of territorial transformation/preservation.[7] In fact, a first consequence of this appreciation for the landscape was, from the 1920s onwards, voices warning of the risk of degradation of the natural and urban spaces (and their ways of life) due to the growing presence of visitors and ad hoc residential areas (Yanes 2020; Camps 2018). The first residential areas with a clear approach to landscape came into being, such as S'Agaró and avant-garde urban developments (Camps 2018). Much more explicitly, in 1935 the Catalan government convened the First Costa Brava Conference, started drafting a regional planning proposal and created the Costa Brava Tourism Board, which did not renounce tourism development despite its resounding protectionist rhetoric (Yanes 2020; Musquera 2015) (Figs. 9.1 and 9.2).

This brief account of the development of tourism along the Girona coastline would be difficult to understand without considering the impact of the Spanish Civil War (1936–1939). For tourism on the Girona coastline, the victory of the Franco regime and its almost 40-year dictatorship[8] meant a kind of "schizophrenia" that was, in many ways, lethal for the territory. This long period, from the mid-1950s onwards, saw the leap from elite tourism to mass tourism, framed by the effects of the welfare state on the European middle and working classes (Donaire 1997; Pack 2013). Given this context, which overwhelmed the Spanish reality, the government produced urban planning legislation comparable to international models, while also allowing and facilitating massive, speculative property developments that pushed urban planning to its very limits. Consequently, from the second half of the 1950s to the end of the 1970s, the destruction of natural and cultural heritage due to urban growth, especially caused by tourism, proved to be disastrous (Nuell 2001; Valdunciel 2011). The legacy of this period is still tangible in various ways: in spaces yet to be developed but with "acquired rights" by owners; in territories with a very precarious quality of development and construction; and, without doubt, in a deeply rooted

[7] It can be compared, for example, to the territories of the south of Catalonia (Saladié 2016).

[8] The first democratic elections since 1936 took place in June 1977.

Fig. 9.1 The beginnings: promotion of s'Agaró (1934). *Source* Arxiu Històric Municipal de Sant Feliu de Guíxols. Ibàñez-Espuña collection. Unknown author

Fig. 9.2 The beginnings: beach of Sant Pol at 1920. *Source* Municipal Historical Archive of Sant Feliu de Guíxols. Newspaper library

logic of speculative behaviour and a strong dependence on the construction sector (Valdunciel 2011; Vicente 2018). In all these characteristics of development, the Girona coastline fits in with what happened in much of the Spanish coastline and the logic of the "urban growth machine" (Blàzquez 2014; Romero-Martínez et al. 2015;

Navarro-Jurado et al. 2019),… although with some peculiarities that stem from the experience just reported in the first decades of the twentieth century.

Although the initiatives of the 1920s and 1930s were frustrated or isolated phenomena, awareness of the territorial quality remained partly implicit and partly explicit. Proof of this is that even in the 1940s and early 1950s; the Franco regime promoted debates and instruments that were unthinkable for other coastlines. In 1949, the *General Regulations for Constructions and Urban Planning in the Province of Girona* were approved pending a general management plan, as were the *General Urban and Landscape Protection Regulations for the Area of the Province of Girona, called "Costa Brava"* in 1954. All this was nonetheless useless vis-à-vis the power of economic interests and development needs that surpassed, distorted or ignored plans of any kind, and which, furthermore, had the approval of the administration itself.

It would not be until the mid-1970s when, after almost two "glorious" decades for tourism and two disastrous ones for landscape, and the demise of the dictatorship and the economic crisis, that the first actions came about: not only denunciations but also mobilisations (illegal, given that it was still not a democracy) against specific projects. The most notable among them were the protests to stop the development of the remaining coastal marshes of Alt Empordà, and the Debat Costa Brava, both in 1976 (Fig. 9.3). These two cases were milestones of much broader activism. The first gave rise to the first natural park since the return to democracy in 1985, and the second was an event of such a civic, technical and political nature (due to the opposition to Franco) that it has coloured all subsequent reflection on the Costa Brava since (Cals 1982; Sànchez et al. 2018). They are also the starting point for the research on social and landscape protection movements on this coastline.

9.5 Movements, Conflicts and Their Outcomes: Mobilising Is Useful

The number of conflicts analysed was 198, with an apparently balanced distribution throughout the study area, although highly concentrated in the coastal strip (see Map 9.2), especially in the municipalities with higher population densities and tourist activities. In comparison with the rest of the 134 inland municipalities, the narrow strip of 22 coastal municipalities had almost half (42%) of the coastal conflicts. Also notable is the fact that the vast majority of open conflicts today are concentrated precisely in these coastal municipalities, which shows the current pressure of urbanisation on some territorial resources already stretched to their limit.

From south to north, the area comprising Blanes and Lloret de Mar, together with Tossa de Mar, stands out for the conservation of coastal areas of high natural and landscape value. It has the highest population concentration and number of hotel beds on the Costa Brava and a particular prevalence of conflicts relating to the construction of road infrastructures (extending the C-32 and new routes of the N-2).

Fig. 9.3 Empuriabrava Marina. *Source* Arxiu Territori

It is important to highlight that, because of its proximity, this area is undergoing socio-territorial processes linked to the progressive integration into the metropolitan area of Barcelona, which accounts for some of these conflicts. The municipalities of Vall d'Aro (Santa Cristina d'Aro, Sant Feliu de Guíxols and Castell-Platja d'Aro) immediately stand out, with conflicts linked to tourist facilities on the coast and other economic activities in the interior. This area is where the most important active conflict on the entire Girona coast can be found: the second phase (the first is from the end of the 1970s) of Port d'Aro's navigable (and residential) marina.

The next hot spot consists of a group of municipalities with many largely active conflicts relating to the construction of second homes in coastal areas with very vulnerable landscapes. They are the municipalities on the central Costa Brava, including Palamós, Palafrugell, Begur and Pals (Fig. 9.4). Together they account for more than 25 conflicts. There is a twofold geographic and social explanation for this higher pressure on the central Costa Brava. First, the southern Costa Brava has a very steep orography, except at the southern end where the largest tourist resort in Girona has already been built (the Blanes-Lloret complex), so the capacity for expansion is to some extent limited to specific areas and very specific projects. Furthermore, from Pals northwards, there are three maritime-terrestrial natural parks of the Costa Brava, which "in their own way" have acted as retaining walls against construction (with the involuntary support of military facilities from different eras). Second, the coastal strip of the Baix Empordà region, which encompasses the entire central Costa Brava, is the cradle of many very local environmental organisations (up to 14 in 10 municipalities), while in the Alt Empordà and La Selva regions, there

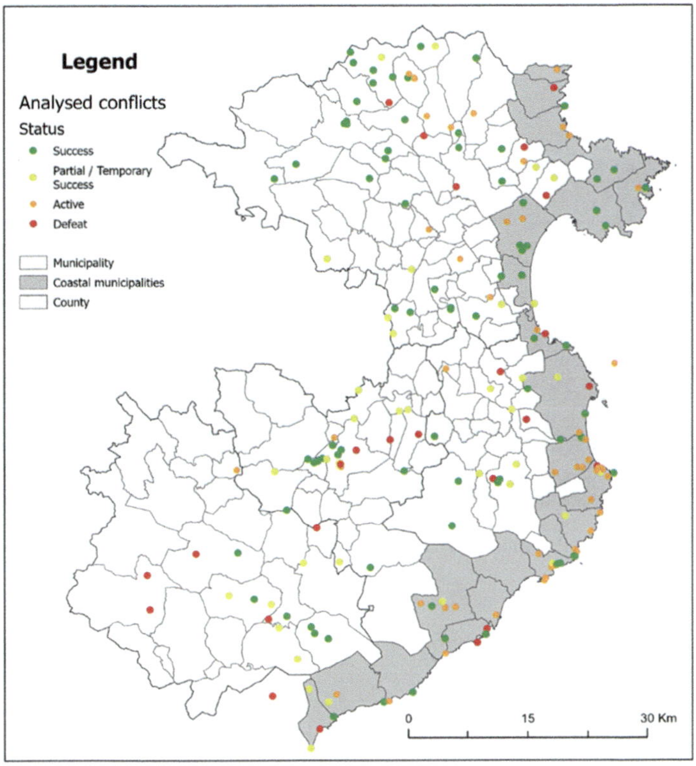

Map 9.2 Location and classification of the conflicts analysed. *Source* Own map

are five established movements that cover the remaining 12 coastal municipalities. The main advantage of this pattern is that the landscape is defended "inch by inch", but it also has a disadvantage. The movements are much more reactive and subject to the dynamics of the expansion and contraction of the economy and not so much to an actor with strategic vision and action. However, this weakness underwent a significant change of course in 2018, as will be explained later.

Continuing towards the border with France, in municipalities like Torroella de Montgrí, l'Escala, Castelló d'Empúries and Roses, there is a certain diversity of conflicts characterised by mobilisations against housing developments and tourist facilities, and especially in favour of the creation or protection of protected natural spaces. Among the many successful fights in defence of natural spaces that have taken place on the Costa Brava, the one that stands out the most is the defence of Aiguamolls de l'Empordà (1976) (Fig. 9.5).

Aside from the coastal spaces, some interior areas also stand out, including the capital of the province, Girona, because it is the location of controversial infrastructure that affects its entire territory (high-speed rail, AP-7 and A-2 motorways, the MidCat gas pipeline, the cross-border Very High Voltage line, etc.). There is

Fig. 9.4 Castell beach (Palamós), saved of urbanisation by social movements. *Source* Arxiu Territori

Fig. 9.5 Mobilisation for the defence of the Aiguamolls, blocking trucks. 1978 *Source* Aiguamolls de l'Empordà Defence Group

also a cluster of conflicts in the border section around La Jonquera, with demonstrations against the installation of various wind farms where wind and natural landscape values converge in an area of the Pyrenees that is also crossed by in strategic international infrastructures.

In line with the main research hypothesis, the results on the status of the conflicts are very important, particularly the high percentage of successful and partially or temporarily successful cases (61%) and their relatively equal distribution throughout

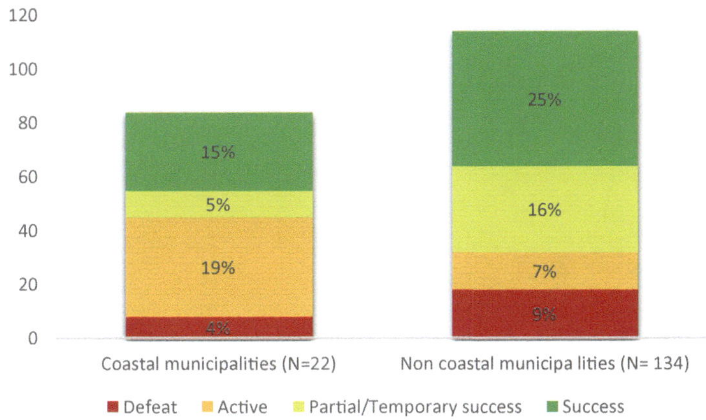

Fig. 9.6 Conflicts according to their location and status (% of the total number of conflicts). *Source* Own compilation

the area in both inland and coastal municipalities. Furthermore, this result also indicates that the mobilisations with satisfactory outcomes (for the movements) are particularly concentrated in the municipalities of Alt Empordà. This fact is probably due to the strong, spatially-intensive and long-standing establishment of certain organisations such as IAEDEN-Salvem l'Empordà (founded in 1980). Some organisations, including IAEDEN-Salvem l'Empordà and the Associació de Naturalistes de Girona (ANG), have a noteworthy organisational capacity (Nuss-Girona et al. 2020). Both entities have more than 40 years of history, including all the social, economic and political changes during the return to democracy after the Franco regime (Fig. 9.3). These groups have a solid social base, a very carefully prepared discourse, and a capacity for social, media and political influence that has become an asset of governance in this territory. Moreover, they are like nodes in a network of movements, making any cause with their "stamp" automatically more high profile and influential (Figs. 9.6 and 9.7).

As can be seen in the timeline (Fig. 9.8), in the last 50 years, there has been a kind of dialogue between the most iconic movements and the political and legal developments that have affected the Costa Brava. From a process of reclamation manifesting itself in different conflicts and campaigns, a set of political and legal decisions have changed the previous territorial paradigm. New elements, such as protected natural spaces and planning instruments with a "conservationist" purpose or designed, have rationalised town planning now based on a more complex vision of the territory in socio-ecological terms. For their part, the cycles of economic expansion and contraction act as fuel for and a "natural" brake on conflicts. Each recession leads to a period of fewer mobilisations due to lower public and private investment in infrastructure projects, property complexes and tourist facilities.

A good example of this is the reactivation of the construction sector after 2018 (Fig. 9.10) after the financial crisis of 2008 and a period of inactivity. Various groups

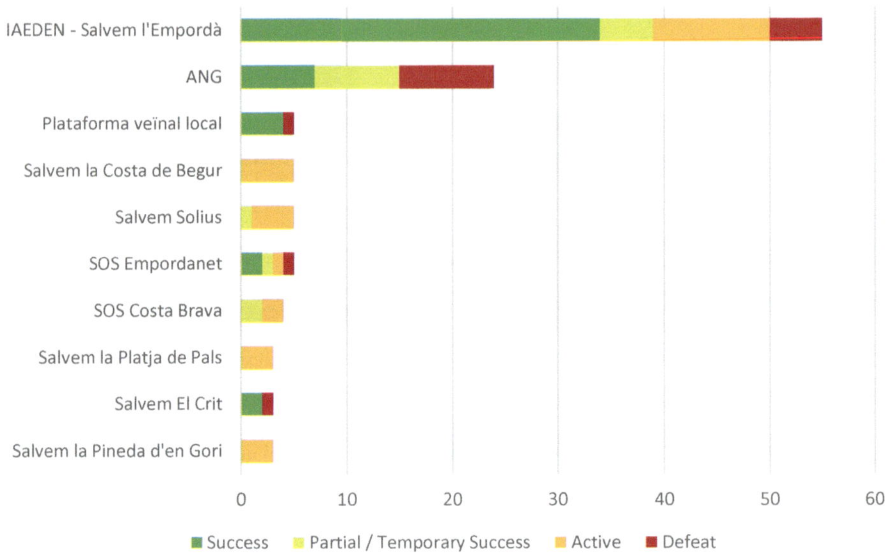

Fig. 9.7 Main leadership movements. *Source* Own compilation

responded with a number of currently active mobilisations (Figs. 9.7 and 9.9) and this new wave of projects sparked the formation of various groups (Salvem Begur, Salvem Solius, SOS Empordanet, Salvem la Pineda d'en Gori, Salvem la Platja de Pals, among others). They became part of the umbrella organisation SOS Costa Brava (first a platform, and since 2021 a federation), which is now the "Coordinator" of 25 entities, all on the Costa Brava. In only four years, this federation's political advocacy has become a reference point in Catalonia. One example is the campaign to create the Catalonia Coastal Conservancy, based on the French *Conservatoire du littoral*[9] and currently in a stage of institutional, jurisdictional, financial and legal development. Other examples include cases of high landscape impact and media coverage (promoted by its affiliated groups), such as the Sa Riera Living development in Begur.

Finally, an amalgamation of conflicts around urban development, recreational and tourist infrastructures, basic resources (water, energy and waste) and mobility (Fig. 9.4) leaves no doubt that the social metabolism of a consolidated tourist destination like the Costa Brava has far-reaching territorial and environmental implications, which is evident in the number and type of conflicts identified.

Figure 9.10 shows that the effectiveness of the movements (success or partial/temporary success) is about 50% or even more in almost all types of conflicts. Urban development without doubt stands out as the main threat to territorial sustainability, due to the Costa Brava's strong tourism sector and the response capacity of the groups. The movements also focus on impacts in the primary sector (water pollution

[9] https://www.conservatoire-du-littoral.fr (founded in 1975).

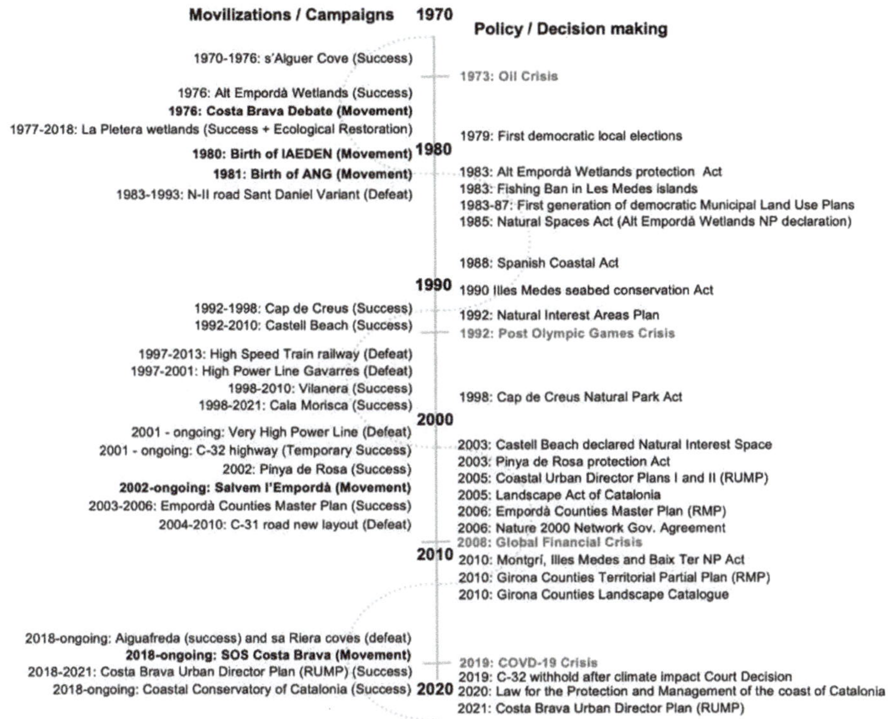

Fig. 9.8 Timeline of iconic conflicts/movements and political and legal decisions (Costa Brava, 1970–2021). *Source* Own compilation. NP: Natural Park; RMP: Regional Master Plan; RUMP: Regional Urban Master Plan

Fig. 9.9 Salvem l'Empordà mobilisation, 2004. *Source* AIEDEN

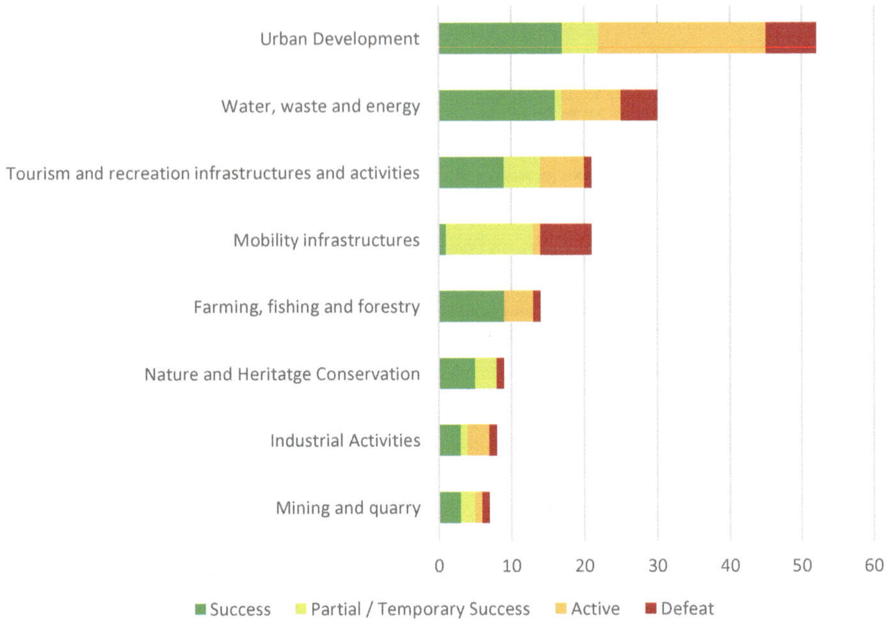

Fig. 9.10 Conflicts according to the trigger. *Source* Own compilation

due to slurry dumping, overfishing, logging in riparian or protected forests, etc.) and natural and cultural heritage protection (Fig. 9.11).

On the other hand, large international infrastructures that pass through the territory has led to a cluster of facilities such as industrial and logistics parks, quarries and mines that are the subject of permanent debate. The percentage of "success" in

Fig. 9.11 Urbanisation in Begur 2021. *Source* SOS Costa Brava

movements opposing them is lower due to rationales, interests and transportation routes that go beyond territorial limits. The maximum outcome of these movements is often an "improvement" of the planned project and hardly ever its cancellation.

9.6 A Socio-Spatial Trialectic?

The results and conclusions of this paper focus on the aspects that are of more general interest. One undoubtedly relevant result is that 61% of the cases result in total, temporary or partial success with respect to the aims of the organisations and individuals involved. This is a high enough percentage to challenge widely held views that nothing can be done to combat speculation and capital interests. The success stories, each with their own nuances, include not only isolated cases without any repercussion, but others with sufficient qualitative and/or structural significance to influence decisions elsewhere at other times, whether on the Costa Brava itself or in Catalan territory more generally (Fig. 9.12).

And these success stories have been significant. The first natural park created by the autonomous government was on the Costa Brava. The first Catalan law for popular consultations was based on a conflict in Baix Empordà. The first Territorial Master Plan was also in Empordà. The first urban master plan for the review of "non-sustainable land" was for this same coastline. In other words, this section of the coast has had many pioneering landscape protection and land management decisions since the 1980s, as it did avant-garde positions in the 1920s/1930s and even in the two decades under Franco. The relationship between different eras also seems to be relevant. Only in the post-transition democratic period have mobilisations had an undeniably positive impact. Not only have the movements had an indisputably positive effect on democracy; without the precedents of the first half of the twentieth

Fig. 9.12 SOS Costa Brava. *Source* SOS Costa Brava

century, it is most likely that the public would have recognised and appreciated the valuation of the coastal landscape considerably less. It remains to be seen to what extent the existence of an explicit, consolidated and, we insist, shared landscape "narrative" is a determining condition for the presence and, above all, the strength of protectionist movements in the other coastal areas in which we are currently working.

Another noteworthy and relevant aspect of the case analysed and the lessons learned from it is the persistence of the mobilisations—albeit intermittently, depending on the (re)appearance of cases—and of some of the players (AIEDEN, Naturalistes de Girona, etc., and even individual actors who set themselves up as collective referents). This persistence is a decisive factor in conflicts that often go on for years and appear and disappear depending on economic crises, political and regulatory changes and so on. In this way, mobilisation becomes social capital, despite periods of inactivity.

A possible model to explain the dynamics of transformation/protection of the landscape and the territory on the Costa Brava is that of a "trialectic" relationship of three poles (Fig. 9.13). One revolves around property/developers interests (which would encompass from the capital to the labour that drives this vast sector); another around institutional technical/political actors; and the third, and this would be the innovation, revolves around protectionist movements. A dynamic of tension, in which the strength of this third pole, although unstable, is relevant and characterised largely by the level of pressure it exerts on the real estate sector (which is also unstable by nature) and its capacity to influence public institutions.

This brings to the last section of these brief conclusions, which refers to NIMBYs. The basic analysis underlying this label is neither useful nor sufficient for the case studied. The results of the mobilisations show that the percentage of total or partial "success" is very relevant (this may only be true of NIMBY situations). Also, however, some of the most important results mean new forms of protection on a regional scale or, and even more evident, regulatory changes in the field of land-use

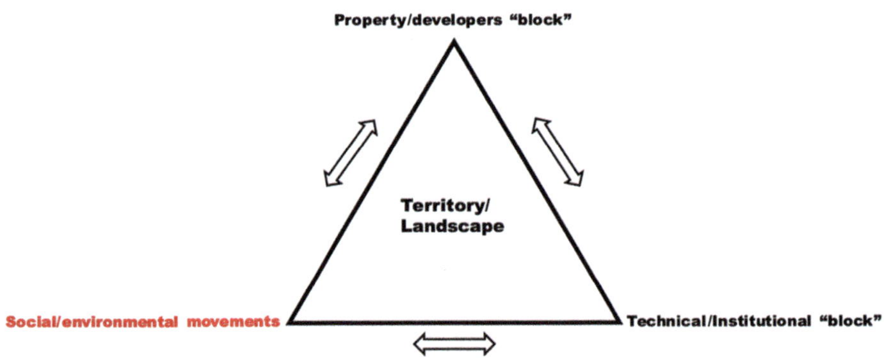

Fig. 9.13 Trialectic model of the territorial/landscape dynamics of the Girona coastline. *Source* Own compilation

planning that end up being applicable to the whole country. Many outcomes transcend municipal and private interests and logic: the first natural park decreed by the autonomous government of Catalonia; the Coastal Protection Master Plans and the declassification of land eligible for development and the reduction of land zoned for building; coastal laws that stem from the mobilisations on the Costa Brava; and the first local referendum in Catalonia.

Likewise, the existence and persistence of entities with forty years of history and with a structured, complex discourse on general and local environmental and territorial models are both an indication and a condition to drag unquestionable NIMBY postures towards more relevant positions of questioning and making proposals of broader ambition and impact. This balance does not exclude NIMBY attitudes from maintaining an important and even growing presence, and the current debates on projects related to the energy transition may require a re-evaluation of the hypothesis. These are emerging problems with a huge impact on societies where it is much easier to argue for the defence of the landscape than for its transformation in the name of a greater cause (from the environmental point of view), but which are not exempt from speculative logic, alternative models and so on.

From the beginning of this paper, we have insisted that this case may be unique among other coastlines under study by the group. The new challenge is to understand the reasons behind the similarities/differences and possible transfers of models and methods between territories that share problems but not conflict/resolution strategies.

Bibliography

Aguilar S (2021) Movimientos sociales y relevancia política. In: Ibarra P et al (eds) Impactos. ¿Qué consiguen los movimientos sociales? Bellaterra edicions, Manresa, pp 49–56

Alfama E et al (2007) Per una nova cultura del territori? Mobilitzacions i conflictes territorials. Icària, Barcelona

Amin A (2002) Spatialities of globalisation. Environ Plan A: Econ Space 34(3):385–399. https://doi.org/10.1068/a3439

Arts B, Buizer M, Horlings L, Ingram V, van Oosten C, Opdam P (2017) Landscape approaches: a state-of-the-art review. Annu Rev Environ Resour 42:439–463

Barbaza Y (1966) Le paysage humain de la Costa Brava. A. Colin, Paris

Blázquez M (2014) Un torcebraços pel decreixement urbà-turístic. Via [on-line], 6. http://journals.openedition.org/viatourism/780; https://doi.org/10.4000/viatourism.780

Bobbio L, Zeppetella A (1999) Perché proprio qui?: grandi opere e opposizioni local. Franco Angeli, Milano

Brenner N (2004) New state spaces: urban governance and the rescaling of statehood. Oxford University Press, Oxford

Brulle RJ, Rootes C (2015) Environmental movements. In: Wright JD (ed) International encyclopedia of the social & behavioral sciences. Elsevier, Amsterdam, pp 763–768. https://doi.org/10.1016/B978-0-08-097086-8.91016-X

Cals J (1982) La Costa Brava i el turisme. Estudis sobre la política turística, el territori i l'hoteleria. Kapel, Barcelona

Camps A (2018) Evolució del model turístic a Catalunya (1920–1930) Els exemples de s'Agaró i el Poblat d'Estiueig al municipi de Castell d'Aro. Final Degree Project, Department of Geography, University of Girona. Unpublished
Castells M (1997) The power of identity, the information age: economy, society and culture, vol II. Blackwell, Oxford
Colomb C, Novy J (eds) (2016) Protest and resistance in the tourist city. Routledge, London
Colomb C, Novy, J (2021) Making sense of (new) social mobilisations, conflicts and contention in the tourist city: a typology. In Fregolent L, Nel·lo O (eds) Social movements and public policies in southern European cities. Urban and landscape perspectives, vol 21. Springer, Cham. https://doi.org/10.1007/978-3-030-52754-9_4
Dear M (1992) Understanding and overcoming the NIMBY syndrome. J Am Plann Assoc 58(3):288–300
Dear MJ, Taylor SM (1982) Not on our street. Pion Ltd., London
Dematteis G (1985) Le metafore della Terra. La umana tra mito e scienza. Feltrinelli, Milano
Donaire JA (1997) El turismo a los ojos del postmodernismo una lectura desde la dialéctica socioespacial: la Costa Brava, Tunicia y los malls. Doctoral thesis, Universitat Autònoma de Barcelona
Driscoll A (2013) NIMBY movements. In: The Wiley-Blackwell encyclopedia of social and political movements. Wiley, Hoboken, NJ
Eranti V (2017) Re-visiting' NIMBY: from conflicting interests to conflicting valuations. Sociol Rev 65(2):285–301
Figueiredo Nascimento S et al (2016) Citizen engagement in science and policy-making. EUR 28328 EN. https://doi.org/10.2788/40563
Giugni M, Grasso MT (1995) Environmental movements: heterogeneity, transformation, and institutionalization. Annu Rev Environ Resour 40:337–361
Greenpeace (2019) A toda costa. Análisis de los ecosistemas naturales costeros vulnerables a la urbanización masiva del litoral. https://es.greenpeace.org/es/wp-content/uploads/sites/3/2019/06/Costas-2019-LR.pdf
Harvey D (1989) The condition of postmodernity an enquiry into the origins of cultural change. Wiley and Blackwell, New York
Ibarra P et al (2021) Impactos. ¿Qué consiguen los movimientos sociales? Bellaterra edicions, Manresa
Keating M (2007) New regionalism in western Europe: territorial restructuring and political change. Edward Elgar, Cheltenham
Kousis M (2000) Tourism and the environment: a social movements perspective. Ann Tour Res 27(2):468–489
Massey D (1991) A global sense of place. Marx Today 38:24–29
Melucci A (1994) ¿Qué hay de nuevo en los "nuevos movimientos ambientales"? In: Laraña E, Gusfield J (eds) Los nuevos movimientos sociales. De la ideología a la identidad. Sociological Research Centres, Madrid
Meyer DS, Minkoff C (2004) Conceptualizing political opportunity. Soc Forces 82(4):1457–1492. https://doi.org/10.1353/sof.2004.0082
Milano C, Novelli M, Cheer JM (2019) Overtourism and degrowth: a social movements perspective. J Sustain Tour 27(12):1857–1875
Mora P (2012) El moviment ecologista a Catalunya: el seu origen, evolución i inserció a la societat catalana. Doctoral thesis, Universitat Autònoma de Barcelona
Musquera S (2015) Les imatges d'un nou paisatge creat per al turisme. La fotografia com a testimoni del paper de l'arquitectura en la transformació del paisatge de la Costa Brava als anys 60. Doctoral thesis, Universitat Politècnica de Catalunya. http://hdl.handle.net/10803/326747
Navarro-Jurado E, Romero-Padilla Y, Romero-Martínez JM, Serrano-Muñoz E, Habegger S, Mora-Esteban R (2019) Growth machines and social movements in mature tourist destinations Costa del Sol-Málaga. J Sustain Tour 27(12):1786–1803
Nel·lo O (ed) (2003) Aquí, no! Els conflictes territorials a Catalunya. Empúries, Barcelona

Nogué J (1991) Els nacionalismes i el territori. El Llamp, Barcelona

Nogué J (2015) Sentido del lugar, paisaje y conflicto. Geopolítica(s). Revista de estudios sobre espacio y poder 5(2):155–163

Nogué J, Vicente J (2004) Landscape and national identity in Catalonia. Polit Geogr 23(2):113–132, February. https://doi.org/10.1016/j.polgeo.2003.09.005

Nuell H (2001) Evolució dels usos del sòl a la Costa Brava (1957–1993). Doctoral thesis

Nuss-Girona S et al (2020) 50 years of environmental activism in girona, catalonia: from case advocacy to regional planning. Land 9:172. https://doi.org/10.3390/land9060172

Paasi A (2002) Place and region: regional worlds and words. Prog Hum Geogr 26(6):802–811

Pack S (2013) Turismo en la Europa de la postguerra: de la diplomacia esterliniana al consumismo de masa. TST (24):138–166

Robbins P (2004) Political ecology: a critical introduction. Wiley, Hoboken, NJ

Romero-Martínez JM, Romero Padilla Y, Navarro-Jurado E (2015) Atributos urbanos contemporáneos del litoral mediterráneo en la crisis global: caso de la Zona Metropolitana de la Costa del Sol. Scripta Nova. Revista Electrónica de Geografía y Ciencias Sociales 19(515):1–10. Universidad de Barcelona. https://doi.org/10.1344/sn2015.19.15119

Saladié S (2016) Paisatge i conflictes territorials a les comarques meridionals de Catalunya. Doctoral thesis, Universitat Autònoma de Barcelona

Sànchez M et al (2018) Escola de radicals: quaranta anys defensant el territori de l'Empordà. Brau Edicions, Figueres

Scheidel A, Temper L, Demaria F et al (2018) Ecological distribution conflicts as forces for sustainability: an overview and conceptual framework. Sustain Sci 13:585–598. https://doi.org/10.1007/s11625-017-0519-0

Scheidel A et al (2020) Environmental conflicts and defenders. Glob Environ Chang 63:1–15. https://doi.org/10.1016/j.gloenvcha.2020.102104

Soja E (1989) Postmodern geographies: the reassertion of space in critical social theory. Verso, London

Sribman A (2021) Conceptualización de los impactos. Aspectos teóricos. In: Ibarra P et al (eds) Impactos. ¿Qué consiguen los movimientos sociales? Bellaterra edicions, Manresa, pp 59–69

Tarrow S (1998) Social movements and contentious politics. Cambridge University Press, Cambridge

Temper L, Demaria F, Scheidel A, Del Bene D, Martinez-Alier J (2018) The Global Environmental Justice Atlas (EJAtlas): ecological distribution conflicts as forces for sustainability. Sustain Sci 13:573–584. https://doi.org/10.1007/s11625-018-0563-4

Tilly C (2004) Social movements 1768–2004. Routledge, New York

Touraine A (1981) The voice and the eye. An analysis of social movements. Cambridge University Press, Cambridge

Tuan Y-F (1974) Topophilia. Prentice-Hall, Englewood Cliffs

UN (2019) Human Rights Council resolution 40/L.22. Recognizing the contribution of environmental human rights defenders to the enjoyment of human rights, environmental protection and sustainable development. https://documents-dds-ny.un.org/doc/UNDOC/GEN/G19/088/48/PDF/G1908848.pdf?OpenElement

Valdunciel J (2011) Paisatge i models urbans contemporanis. Les comarques gironines (1979–2006): del "desarrollismo" a la globalització. Doctoral thesis

Vicente J (2018) La destrucció creativa i l'urbanisme. In: Sala i Martí P, Puigbert L, Bretcha G (eds) (Des)fer paisatges. Observatori del Paisatge de Catalunya-Aigües Ter Llobregat, Olot

Walter M (2009) Conflictos ambientales, socioambientales, ecológico distributivos, de contenido ambiental... Reflexionando sobre enfoques y definiciones. Boletín Ecos (6):2–9, febrero–abril

Warf B, Arias S (2009) The spatial turn. Interdisciplinary views. Routledge, London

Yanes S (2020) La construcción social de la Costa Brava. Una aproximación bibliográfica. Pasos. Revista de turismo y patrimonio cultural 18(5). https://riull.ull.es/xmlui/handle/915/22011

Chapter 10
The Construction of Sustainable Territorial Models in the Aftermath of the Covid-19 Pandemic: Lessons from Medium-Sized Mediterranean Cities

N. Ruiz Moya, L. M. Sánchez Escolano, and Álvaro Navas González

Abstract In recent decades, medium-sized cities have become the setting for new, complex urban development processes, which result from new forms of land use produced by improvements in technology, infrastructure and mobility. These processes have given rise to tensions and conflicts between competing land uses which are vying for control of the limited available resources. In many cases, this can deepen pre-existing spatial imbalances or give rise to new ones. Planning, in its different forms, has yet to provide an effective framework of reference in response to these new scenarios that can transcend the purely technical aspects of urban planning to provide a truly integrated form of territorial management.

These processes also apply to medium-sized cities in the Mediterranean region, and to coastal cities in particular. Many of these cities enjoy great demographic, economic and social dynamism, but they are also the scene of major territorial imbalances which are reflected in the exorbitant growth proposed in current urban development plans. In this research, we study three representative examples of such cities—Vélez-Málaga, Motril and Roquetas de Mar (Andalusia-Spain)—analysing and describing the complex processes currently underway. We also propose new planning strategies and obtain a number of important lessons for the sustainable management of these areas.

N. Ruiz Moya (✉) · L. M. Sánchez Escolano
Department of Human Geography, University of Granada, Campus Universitario de Cartuja, 18071 Granada, Spain
e-mail: noeliaruiz@ugr.es

L. M. Sánchez Escolano
e-mail: lmsescolano@ugr.es

Á. Navas González
University of Granada, Campus Universitario de Cartuja, 18071 Granada, Spain
e-mail: alvarong@correo.ugr.es

E. Navarro-Jurado et al. (eds.), *Urban Dynamics in the Post-pandemic Period*, The Urban Book Series,
https://doi.org/10.1007/978-3-031-36017-6_10

Keywords Coastal tourist area · Urban planning · Medium-sized cities · Sustainable management

10.1 Coastal Medium-Sized Cities in the Planning Challenge

The analysis and identification of the urban processes in medium-sized cities within a general framework involves exposing academic theories to the harsh reality highlighted by the data. In recent years, these cities have provided perfect scenarios for analysis and combine some of the most complex, most innovative processes of those identifiable in cities today (Bellet and Beltrao 2009; Benach and Walliser 2011; Camagni et al. 2014).

They have also become excellent observatories within which to assess some of the challenges involved in managing the most severe impacts of global capitalism at a local scale and for the rediscovery of unusual urban and functional profiles (Léo et al. 2010; INTELI 2011; Andrés López and González González 2019). In addition, from the point of view of urban planning, they enable us to review in detail the results of decades in which the practice of urbanism was not conceived as an instrument for promoting local development and was identified solely as a means of creating more land for building and an imbalance between public and private funding (Jurado Almonte 2010; Fernandes et al. 2011; Arellano Ramos and Roca Cladera 2012; Sevilla Buitrago 2015).

In Spain, these problems are accentuated in coastal areas. This is particularly evident on the coast of Andalusia, one of the most dynamic areas in the region together with the metropolitan areas of the major cities. This is especially true in the eastern provinces—Málaga, Granada, Almería—where the coastal strip is under great pressure from competing economic sectors that use the scarce available resources very intensively.

The main objective of this research is to characterize, systematize and critically evaluate, from a geographical perspective, the current situation of medium-sized cities in terms of planning, and in so doing assess the implications for the design of new more sustainable territorial planning models. Our initial hypothesis is that through the study and identification of complex territorial processes in representative cases; it is possible to obtain useful lessons and information for the subsequent design and improvement of the effectiveness of territorial planning at different scales. This results in a society with more and better tools for the implementation of territorial strategies with development models that can work towards sustainability (Herrschel 2009; Kunzmann 2009).

In this case, after describing the scientific debate about medium-sized/intermediate cities and the challenges they entail for local and regional territorial planning, we present the analysis of three case studies of medium-sized cities on the Mediterranean coast of Andalusia: Vélez-Málaga, Motril and Roquetas de Mar.

These case studies provide a representative catalogue of processes and conflicts within territorial models that have difficult unsolved problems and face serious challenges when it comes to becoming more sustainable. This phase of study and diagnosis is supported by statistical and documentary evidence and by the drawing up of synthetic maps of territorial processes. Various useful lessons can be learnt. These are presented and discussed in the last phase of the work.

The Mediterranean coast reflects the logic of location and concentration common in other countries on the European shore. These processes find their maximum expression on the Spanish coast—especially in Catalonia, Balearic Islands, Valencia and Andalusia—due to the strength of tourist activity. This economic sector found in local urban planning its main accomplice for the massive construction of the coastline (Giannakourou 2005; Benach and Walliser 2011; Andrés López and González González 2019). This hyper-urbanization is substantiated during the first decade of the twenty-first century, in which until 35% of the Spanish Mediterranean coastline was built on, with even higher percentages in the coastal provinces of Malaga (49%) and Alicante (45%) (Rullán 2011; Górgolas Martín 2020).

Particularizing in the Andalusian case, during the last decades, the tourist and urban pressure has focused on the coastline—the first 500 m from the sea. A phenomenon that increases its impact on the eastern coast of the region, where the intense and profitable agricultural activity of crops under plastic or subtropical crops must coexist with tourist activity (Arellano Ramos and Roca Cladera 2012; Ruiz Moya and Cejudo García 2022).

Within this context of current regional planning, in this paper we use the concept of "medium-sized city" to refer to an urban space with what could be considered a medium-sized population for Andalusia. In statistical terms, this means a population with a minimum of 10,000 or 20,000 inhabitants (according to the dynamism and importance of the city in question in terms of basic urban functions), and a maximum could be considered a threshold for the top tier of urban settlements represented, according to the particular case, by their status as a "regional centre" or by a total population of over 100,000 inhabitants (Sánchez Escolano 2016).

The medium-sized cities in Andalusia have traditionally not been confined within a statistical classification in relation to their demographic densities and instead have performed an important functional role, by concentrating the supramunicipal services and facilities, so becoming functional nodes in the network of towns and villages at a subregional level. As a result, many of the medium-sized cities inland and even on the coast—as in the case studies discussed here—are heirs to the historic "agrocities" of Andalusia, which developed as supramunicipal service hubs, while maintaining their traditional role as agricultural centres (Ruiz Moya and Cejudo García 2022).

The fundamental value of this research, therefore, lies in providing a more detailed, better informed understanding of the situation in the Mediterranean in general and more specifically in Andalusia. In this way, we will obtain information about the urban system and medium-sized cities in particular and the role they play in these territories and especially about the planning procedures and tools required to build scenarios that are favourable for sustainability, quality of life and development.

10.2 Conflicts Among Sub-Regional Planning and Local Governments in Coastal Andalusia

Cities, long established as the most dynamic spaces in socioeconomic terms, are the scenarios that best represent the incapacity of territorial planning to solve the spatial tensions arising from their productive development model.

In Andalusia, the medium-sized cities on the coast are, together with the major urban agglomerations, the spaces that concentrate the greatest demographic, economic, social and urban growth in the region. This is based on the development of a production system based on two main sectors, namely intensive agriculture and tourism, which compete for the same resources (land, water and by extension landscape) to continue growing, so generating numerous spatial conflicts that the different authorities try to solve through territorial planning (Sánchez Escolano 2016; Ruiz Moya and Sánchez Escolano 2022). In this paper, we will be looking at the medium-sized cities of Vélez-Málaga, Motril and Roquetas de Mar, as suitable case studies for illustrating all these processes.

Under the Spanish Constitution of 1978, the regions are responsible for territorial and town planning. In the specific case of Andalusia, these were acquired by the Regional Government in 1981 when the Statute of Autonomy for Andalusia was passed. Since then, Andalusia has developed its own legislative, administrative and instrumental corpus in both territorial and town planning, and these affect the spatial planning of the region at different scales (Hildenbrand Scheid 2011; De Córdoba 2019).

In both territorial and town planning, the medium-sized cities play a leading role in the subregions in which they are located (Vélez-Málaga in the eastern Costa del Sol, Motril on the Costa Tropical of Granada and Roquetas de Mar in the south coast of Almería) by hosting the infrastructure, facilities and amenities at a supramunicipal level. In recent decades, they have also seen the greatest urban growth. For this reason, in the coastal spaces under most pressure (south coast of Almería in 2002; Doñana and the Bay of Cádiz in 2004 or the western end of the coast of Huelva and the eastern Costa del Sol in 2006…), the Regional Government of Andalusia has opted for specific territorial planning instruments at a subregional scale (hereinafter referred to its Spanish acronym, PSOT), even before it introduced a specific planning instrument for the whole region (Plan for the Organization of the Territory of Andalucía, December 2006) (Górgolas Martín 2020; Fig. 10.1).

However, these subregional planning instruments proved incapable of containing the generalized hyper-development of the "monumental decade of Spanish urban development" at the beginning of the twenty-first century (Jurado Almonte 2010), which had an especially strong impact on the first few metres of the region's coastline. All of this was done with the blessing of the Town Plans approved by each Town Council, which provided a large "stockpile" of land for the development of tourism for residential purposes in these areas, on numerous occasions facilitating or even encouraging urban sprawl and the appearance of new built-up areas a long way from the main town, in megaprojects in areas which until then had been known for their

Fig. 10.1 Map of the territorial planning instruments at a subregional scale approved on the coast of Andalusia. *Source* The authors

attractive landscapes (González Sicilia 2004; Jurado Almonte 2010; Rullán 2011; Górgolas Martín 2018).

This situation was clearly unsustainable in social, environmental and spatial terms and the regional administration attempted to control it by introducing territorial planning. This coincided with the bursting of the property bubble in 2008 and the enormous recession into which regional and national urban development was pushed over the period 2009–2015. During this period, the plan for the Protection of the Coastal Corridor of Andalusia (hereinafter the PPCLA, 2015) was formulated and passed as a fundamental instrument for the preservation of the remaining building-free spaces on the coastal strip. This plan focused on the first 500 m inland from the line marking the Terrestrial Maritime Public Domain and sought to enhance its environmental, territorial and landscape protection. It was based on an integrated, sustainable conception of the entire coast of Andalusia and proposed the opening of biological corridors that linked the coast with the natural areas in its hinterland (Zamorano Wisnes 2016).

The lack of control of urban development on the coast of Andalusia prior to 2008 was so severe that, in areas in which building speculation had won the battle against territorial sustainability and rationality, the PPCLA was given the power to declassify the stockpiles of land that had earlier been classified as suitable for tourism or residential development. In fact, the PPCLA declassified eight such reserves of land approved for development (1,395,274 m^2) on the coastal strip of Vélez-Málaga—in

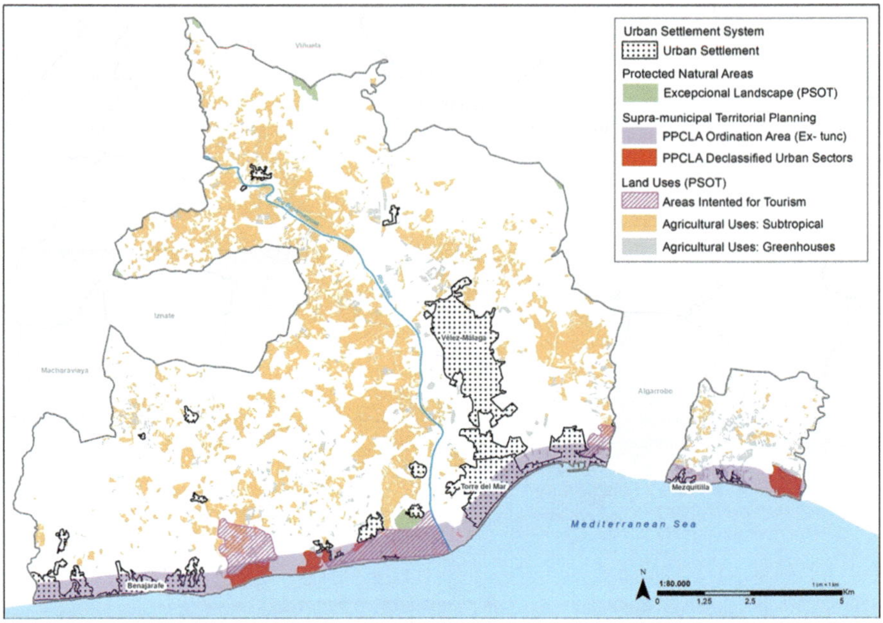

Fig. 10.2 Territorial tensions produced by the land use model in Vélez-Málaga (eastern Costa del Sol–Axarquía). *Source* The authors

the area around the Delta of Vélez and the Niza Valley—which represented 12% of all the area protected by the PPCLA; and three in Roquetas de Mar (351.443 m^2), which although less important in terms of area (2%), were situated on the northern edge of the Punta Entinas-Sabinar Protected Natural Space, which was strongly endangered by the expansion of tourism and agricultural business in the city (Figs. 10.2 and 10.4).

Although this planning instrument ensured the protection of almost 8% of the total surface area of Vélez-Málaga and Motril, and about 29% of that of Roquetas de Mar, it was annulled by the Supreme Court of Justice of Andalusia in 2018 due to a procedural error in its approval process. This happened just three years after it came into force. Under the best-case scenario, this left the power to manage and resolve territorial conflicts in the hands of the municipal urban plans within the PSOTs (in those areas with subregional planning) (Rando Burgos 2018).

In this sense, the main tensions observed in the case studies resulted from the incapability of the spatial planning instruments to tackle the complex spatial dynamics resulting from their production model, which are expressed in the different ways of occupying and using land.

Although the three cities experienced demographic and urban growth fuelled by the great dynamism of the agriculture sector—either through subtropical crops or intensive greenhouse cultivation—and residential sun and beach tourism, they did so with different degrees of intensity. While Vélez-Málaga showed a proportional increase in its population and its housing stock within 2000 and 2020 (both of which doubled), Roquetas experienced exponential growth, tripling its population and almost quadrupling the number of houses. For its part, Motril showed a clear dichotomy between strong real estate growth (as exemplified by the construction boom on the coast which had occurred later than in other areas due to its geographical relief and infrastructure problems), and a slight increase in population distorted by the separation from Motril in 2018 of the town of Torrenueva, until then its main coastal tourism district (Table 10.1).

However, in spite of the different rates of growth, the three cities share the same model of territorial occupation based on the use of scarce resources that are essential for both agriculture and tourism, so giving rise to numerous conflicts that town planning has so far proved incapable of managing (Ruiz Moya and Cejudo García 2022).

Although the three cities we studied share a similar level of development in terms of their territorial occupation model, Table 10.2 shows that there are significant differences in the distribution of land uses due to the particular characteristics of each city.

Vélez-Málaga is one of the most populated medium-sized cities on the coast of Andalusia and is also the functional capital of the Axarquía region. In spite of it having a strong tourist sector with substantial growth in the parts of the town nearest the coast (Torre del Mar, la Caleta de Vélez or Benajarafe) such that practically the whole shoreline has been developed (Tables 10.2 and 10.3), the closeness of the Málaga-Estepona-Torremolinos urban conurbation is a great handicap to further growth in tourism, which represents just 3% of land use in the municipal area. In addition, the combination of the typical tourist resort on the coast and the explosion of second homes scattered throughout the Axarquía (Mérida Rodríguez et al. 2021) offers a slightly confusing image of the type of tourism they want to attract. This has enabled the consolidation of a local economic system based on two main sectors, in which tourism is accompanied by a strong agricultural specialization in the fertile

Table 10.1 Changes in the housing stock of the medium-sized cities on the coast of eastern Andalusia

Year	Vélez-Málaga		Motril		Roquetas de Mar	
	N° houses	% growth rate	N° houses	% growth rate	N° houses	% growth rate
1991	29,306	0	22,438	0	19,481	0
2001	36,235	23.64	28,866	28.65	28,909	48.4
2011	47,230	30.34	35,760	23.88	57,285	98.6
% growth rate 1991–2011	61.16		59.37		194.06	

Sources Spanish National Statistics Institute. Census of Population and Housing 1991, 2001, 2011 and the authors

Table 10.2 Distribution of the main land uses in the medium-sized cities on the coast of eastern Andalusia

Distribution of the main land uses (area)	Vélez-Málaga		Motril		Roquetas de Mar	
	km^2	%	km^2	%	km^2	%
Residential use	11.99	7.59	5.83	5.66	14.31	23.98
Tourist use	4.14	2.62	Not delimited		13.18	22.08
Agricultural use	34.65	21.94	27.73	26.94	21.90	36.70
Natural protected areas	0.76	0.48	0.14	0.13	15.14	25.37
PSOT protected areas	0.00	0.00	20.53	19.94	8.58	14.39
PPCLA ordination area	11.89	7.53	8.35	8.11	17.27	28.94
Municipality area	157.94	100.00	102.94	100.00	59.67	100.00

Source Drawn up by the authors on the basis of data from the IECA, Rediam, SIOSE and photointerpretation

plain along the River Vélez. This area is devoted above all to the production of subtropical fruits (especially avocadoes), of which it is one of the main producers and distributors—together with its neighbours on the Costa Tropical of Granada—in the European Union (Fig. 10.2).

For its part, Motril exemplifies the tendencies in a medium-sized city that is consolidating its capacity as a hub connecting both the small villages around it and

Table 10.3 Demographic trends in the urban areas of Vélez-Málaga (2000–2020)

Districts	Inhabitants 2000	Inhabitants 2020	Growth Rate 2000–2020 (%)
Vélez-Málaga (main town)	30,547	41,453	35.70
Almayate Alto	447	1501	235.79
Almayate Bajo	1966	2948	49.95
Benajarafe	1775	3484	96.28
Cabrillas	252	213	−15.48
Cajiz	633	802	26.70
Caleta de Vélez	2160	3842	77.87
Chilches	939	3661	289.88
Lagos	651	834	28.11
Torre del Mar	14,679	22,575	53.79
Trapiche	366	377	3.01
Triana	741	675	−8.91
VÉLEZ-MÁLAGA (Municipality)	55,156	82,365	49.33

Source Spanish National Statistics Institute (INE Nomenclátor). Drawn up by the authors

large provincial capitals. Although it acts as the functional capital of the coastal region of the province of Granada in that it provides most supramunicipal services, it is a typical representative of the economic and territorial division seen throughout the Costa Tropical. Hemmed in by a very uneven mountain relief and many cliffs created by the closeness of the Penibética Mountain Range to the sea; the coastal strip in Motril is very narrow and densely occupied and has few beaches—11% of coastal urban saturation—(Górgolas Martín 2020). Its initial tourism development has been restricted by a number of factors such as the lack of land set aside for this purpose in territorial planning, the hiving off of its main holiday centre into a separate municipality—which has reduced even more the number of beaches suitable for holidaymakers—the limited tourism infrastructure in the town and the opting for and continuation of an urban development model based on second homes. As a result, tourist development has so far been restricted to the western end of the town in the area around Varadero, Puntalón and the Port, and more recently in the small outlying district of Calahonda (Tables 10.2 and 10.4).

The town also has a strong agricultural sector, covering 27% of its total surface area, which is concentrated in the areas in which the relief permits. The flattest areas—river plains and coastal depressions—at the eastern end are occupied by greenhouses. As these areas are filled, greenhouses are also being built on hillsides at higher altitudes (400 m or 500 m). This requires substantial levelling work and earth movements in sloping areas, in spite of this being prohibited under the subregional plan. The subtropical fruit production (mostly avocadoes and custard apples) is concentrated at the western end of the municipality—up to altitudes of 200-300 m—and on the fringes of the main town, from which it stretches out towards the greenhouse area of Carchuna (Fig. 10.3). Now that all the flat areas have been occupied, new areas on the hillsides are being colonized, and another trend enters into conflict with the legal, environmental and territorial protection granted to the main mountain reliefs in the

Table 10.4 Demographic trends in the urban areas of Motril (2000–2020)

Districts	Inhabitants 2000	Inhabitants 2020	Growth rate 2000–2020 (%)
Motril (main town)	42,342	49,501	16.91
Calahonda	1367	1721	25.90
La Garnatilla	171	174	1.75
Los Tablones	35	165	371.43
Torrenueva	1847	Independent	
El Varadero	2579	3866	49.90
Carchuna	1564	2270	45.14
Puntalón	206	599	190.78
Las Ventillas	61	164	168.85
MOTRIL (Municipality)	50,172	58,460	16.52

Source Spanish National Statistics Institute (INE Nomenclátor). Drawn up by the authors

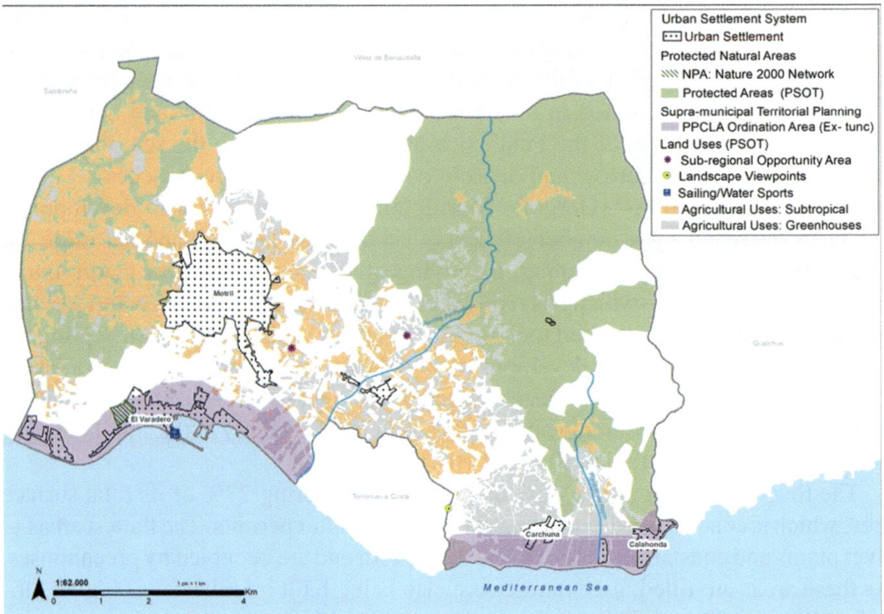

Fig. 10.3 Territorial tensions resulting from the land use model in Motril (Costa Tropical de Granada). *Source* Drawn up by the authors

Motril area by the PSOT for the Costa Tropical, under which 20% of its municipal area enjoys some form of protection (Table 10.2).

For its part, Roquetas de Mar is one of the most dynamic cities on the Mediterranean coast of Andalusia. Its intense demographic and residential growth in recent decades in the last two decades have converted it into the functional centre of the southwest coast of Almería—services, facilities and amenities—so displacing El Ejido, the historical capital of this subregion or *comarca*, as endorsed in subregional planning (Sánchez Escolano 2013).

Its impressive social and economic development is based on the importance of intensive greenhouse agriculture, which has practically used up all the land in the town with the exception of the coastal strip (37% of the municipal surface). In fact, the enormous profits obtained from greenhouse farming, which has produced what is popularly known as a "Sea of Plastic", explain the huge increase in population in the small inland districts of the municipality—El Parador de las Hortichuelas, Las Marinas and Campillo del Moro—and in all its more scattered, outlying areas (Table 10.5).

In addition, once Roquetas had become the functional hub of the southwest coast of Almería due to its geographical position, its tourism business began to flourish. The land set aside in the Town Plan for tourism use is now being developed, and the influence of tourism is spreading towards inland districts which were previously dominated by agriculture such as Campillo del Moro or Las Marinas. There is also the

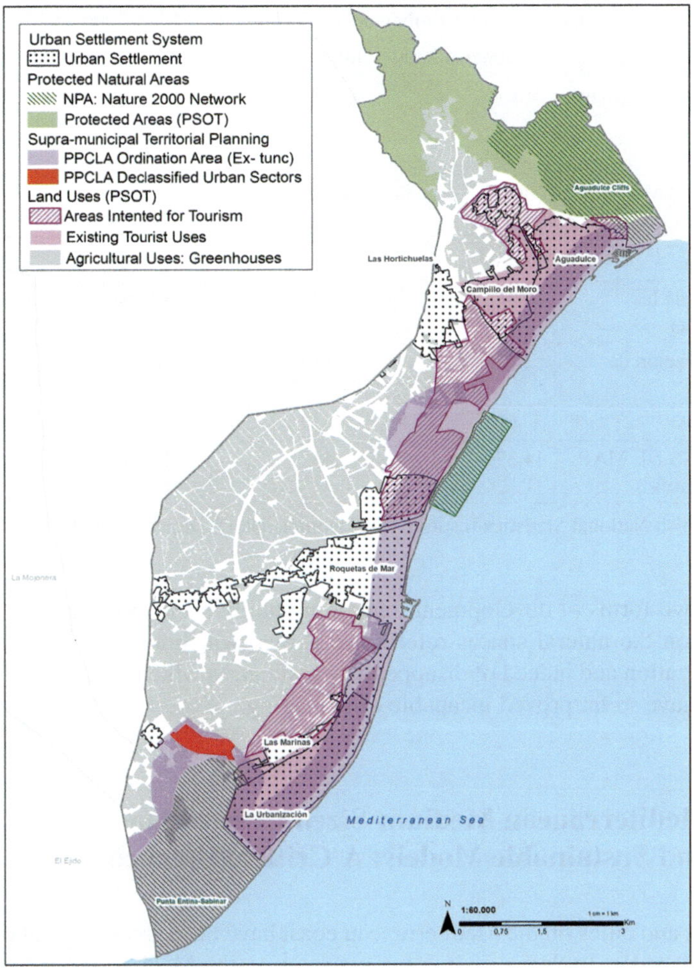

Fig. 10.4 Territorial tensions resulting from the land use model in Roquetas de Mar (southwest Almería). *Source* Drawn up by the authors

attraction of the two natural spaces at each end of the municipal area: the Natural Park of Punta Entinas-Sabinar in the West and the Cliffs of Aguadulce on the eastern edges. All of this has led to an emerging tourist cluster of second homes in the district known as La Urbanización de Roquetas and of more traditional tourist accommodation in Aguadulce—centred around the Marina—(22% of the municipal surface) (Table 10.2; Fig. 10.4).

However, this scenario of economic dynamism, the strengthening of its functional role and the related new business opportunities have also given rise to disputes and abuse due to the fact that very little land was available in the municipal area. The outskirts of the town have come under increasing pressure from greenhouses and

Table 10.5 Demographic trends in the urban areas of Roquetas de Mar (2000–2020)

Districts	Inhabitants 2000	Inhabitants 2020	Growth rate 2000–2020 (%)
Roquetas de Mar (main town)	21,616	47,309	118.86
Aguadulce	8994	16,772	86.48
Campillo del Moro	4444	12,828	188.66
Cortijos de Marín	1486	2529	70.19
Las Marinas	1451	4520	211.51
El Parador de las Hortichuelas	3578	9841	175.04
La Urbanización de Roquetas	2175	4117	89.29
El Solanillo	626	517	−17.41
ROQUETAS DE MAR (Municipality)	44,370	98,433	121.85

Source Spanish National Statistics Institute (INE Nomenclátor). Drawn up by the authors

tourism; two forms of development that are mutually exclusive and have tended to encroach on the natural spaces referred to above, which are in serious danger of denaturalization and indeed of disappearing, a situation which local and subregional planning have so far proved incapable of managing.

10.3 Mediterranean Medium-Sized Cities, Planning and Sustainable Models: A Critical Balance

The towns and cities on the Mediterranean coast have been areas of great dynamism in recent decades. In Spain, as in other countries in the Mediterranean Europe, this trend has also provoked some of the greatest territorial conflicts. In these areas, it is possible to foresee and detect the uneven, distorting impacts of global capitalism. Within these territories, planning in its different forms and scales has often become an administrative formula for covering up the looting of local resources (Kunzmann 2009; Herrschel 2009; Fernandes et al. 2011). It has also helped legitimize speculation within urbanism, which in many cases flouted the most basic principles of territorial planning. In fact, in recent years there have even been incentives to encourage the abusive consumption of strategic and scarce resources (above all water and land and by extension landscape), while the spatial tensions and impacts created by these processes have been largely ignored.

In a context of global warming, which in the Mediterranean region is manifested in extreme meteorological phenomena, such as longer, more intense droughts and a rise in the sea level, the massive building development on the coast has led to the irresponsible multiplication of risks for the population, so creating urban spaces that are not integrated into their environment and act as the main anthropic structure upsetting the environmental equilibrium. In this way, planning has acted as a guarantee of disequilibrium, creating a situation that is now very difficult to reverse (Giannakourou 2005; Olmedo Pérez 2017; Ruiz Moya and Sánchez Escolano 2022).

As a result, these spaces have been badly managed and neglected by outdated and uncoordinated planning methods, which have discarded the typical compactness of Mediterranean cities—as manifested in multifunctional towns made up of many different settlements linked by strong networks—which was always so highly praised in territorial planning textbooks. Once, this Mediterranean model of town planning had been distorted and replaced by urban development that devoured limited resources; a planning system has gradually emerged that is subject to and dependent on the local initiatives that have appeared in each town. In other words, instead of territorial planning strategies and values being integrated into planning at a local scale, the system has been distorted to such an extent that territorial planning is reduced to legitimizing the often disorganized, badly coordinated initiatives of Town Council planning.

This fact that the local scale has emerged as the main frame of reference when it comes to constructing territorial models has led to their definitive distancing from sustainability as a formula of reference for day-to-day work. In fact, it has given free rein to the massive occupation of land with urban sprawl all along the coastline. In this way, town planning, with a very reductionist view of the processes and a short-term approach using overly bureaucratized instruments, has tried to advance by making small occasional changes to outdated documents. The result in administrative terms has been a legislative mess that is difficult for local technicians and managers to understand and which has made the general public increasingly wary of planning and its objectives. In addition, on numerous occasions, Town Plans do not conform to territorial planning schemes at a regional or subregional scale—as has happened in cities such as Vélez-Málaga and Motril—so intensifying territorial tensions by not taking into account the supramunicipal territorial model within which they should operate.

All of this has been particularly clear on the coast, especially when the High Court of Justice of Andalusia declared the PPCLA null and void due to an administrative technicality, without analysing its content or its importance for the management of the coast as a whole. This, together with administrative laxity and the consent of local and regional territorial managers, is a prime example of the mistrust in territorial planning. This is recognized in the role attributed to territorial planning in the new regional legislative framework established in the Law for the Promotion of the Territorial Sustainability of Andalusia (2021), as the scenario of reference within which the planning proposals of each Town Council must fit. However, the lack of any coherent management of the coastline is particularly worrying, in that it enables oversimplified localist planning of the coastal strip by the Planning Department of each Town Council without even considering the possibility of an integrated management of these areas within broader territorial planning.

All the above only serves to reinforce the social, political and administrative discredit of territorial and town planning (as an entertainment for academics which does little to improve the quality of people's lives). Town planning in particular is seen as a murky business that centres on intensive, highly speculative processes.

In this research, we have analysed the highly complex, dynamic realities reflected in the case studies. These can help to bring about advances in the academic recognition of issues that must be addressed in the design of new territorial models. Finding out more about the interaction between planning, the sustainable management of natural resources and the democratic design of territorial models will be a key factor in achieving the objectives that enable us to take on the challenges of the future with some guarantee of success.

Acknowledgements This research has been conducted with the support of an FPU predoctoral contract from the Spanish Ministry of Science and Universities (FPU 18/03123) and from the project "The rural territories of Andalusia in the face of depopulation and aging: challenges and opportunities". Regional Ministry of Economic Transformation, Industry, Knowledge and Universities. Regional Government of Andalusia R+D+I FEDER (B-HUM-460-UGR20).

References

Andrés López G, González González MJ (2019) Urbanização difusa e crescimento urbano irregular processos e tendências nas cidades médias da região de Castilla y León (Espanha). Finisterra: Revista Portuguesa de Geografía 54(112):3–26

Arellano Ramos B, Roca Cladera J (2012) Sprawl in European cities. The metropolitan peripheries, the main stage of the dispersion of urbanization in Europe? Arch, City Environ 20:95–114

Bellet C, Beltrao ME (eds) (2009) Las ciudades medias o intermedias en un mundo globalizado. Universitat de Lleida, Lérida

Benach N, Walliser A (2011) Introduction to the special issue: urban problems and issues in contemporary Spanish and Portuguese cities. Urban Res Pract 4:227–231

Camagni R, Capello R, Caragliu A (2014) The rise of second-rank cities: what role for agglomeration economies? Eur Plan Stud 23(6):1069–1089

De Córdoba MBF (2019) La insoportable rigidez del PGOU. La necesidad de un cambio de modelo. Ciudad y Territorio. Estudios Territoriales 51:451–466

Fernandes E, Da Silva R, Alexandrino R (2011) As cidades médias como nódulos de equilíbrío da rede de cidades. Revista Paranaense De Desenvolvimento 121:41–63

Giannakourou G (2005) Transforming spatial planning policy in Mediterranean countries: Europeanization and domestic change. Eur Plan Stud 13:319–331

González Sicilia A (2004) El turismo residencial en la ley de ordenación urbanística de Andalucía. Revista De Derecho Urbanístico y Medio Ambiente 212:132

Górgolas Martín P (2018) Planeamiento urbanístico y suburbanización irregular en el litoral andaluz: directrices y recomendaciones para impulsar la integración urbano-territorial de asentamientos. Ciudad y Territorio: Estudios Territoriales 10(195):35–52

Górgolas Martín P (2020) El urbanismo en el litoral andaluz tras la última burbuja inmobiliaria. Cambio de ciclo o reincidencia. Tirant Humanidades, Valencia

Herrschel T (2009) City regions, polycentricity and the construction of peripheralities through governance. Urban Res Pract 2(3):240–250

Hildenbrand Scheid A (2011) Los planes de ordenación del territorio (POTA y POTS) de la Comunidad Autónoma de Andalucía y sus afecciones para el planeamiento urbanístico. En: Jurado Almonte JM (Coord) Ordenación del Territorio y Urbanismo: conflictos y oportunidades. Universidad Internacional de Andalucía, pp 163–201

INTELI (2011) Creative-based strategies in small and medium-sized cities guidelines for local authorities. Óbidos Municipality (Portugal). Urba- Act, INTELI, Creative Clusters

Jurado Almonte JM (2010) El "vilipendiado" urbanismo: una visión desde la experiencia en la gestión pública. Cuadernos Geográficos 47:275–295

Kunzmann KR (2009) Medium-sized towns, strategic planning, and creative governance in the South Baltic Arc. Germany. Meck-Pomm Goberment, pp 1–22

Léo P, Philippe J, Camacho Ballesta JA, Monnoyer M, Navarro Espigares JL, Stare M (2010) Services et hautes qualifications, un noveau défi pour le devéloppement des villes moyennes. Économies Et Sociétés (EGS) 2(11):139–172

Mérida Rodríguez M, Perles Roselló MJ, Sortino Barrionuevo JF, Carruana Herrera D (2021) Percepción social de las repercusiones paisajísticas de la urbanización dispersa en espacios rurales del entorno litoral mediterráneo. Investigaciones Geográficas 76:53–73

Olmedo Pérez S (2017) El desarrollo autonómico de la competencia de Ordenación del Territorio y sus relaciones con el urbanismo. Balance de la experiencia andaluza. Práctica Urbanística 147:1–8

Rando Burgos E (2018) La anulación del Plan de Protección del Corredor Litoral de Andalucía por el Tribunal Superior de Justicia de Andalucía: ¿nuevas dificultades en la planificación territorial de Andalucía? Revista De Estudios De La Administración Local y Autonómica: Nueva Época 10:109–131

Ruiz Moya N, Cejudo García E (2022) La desarticulación territorial del litoral mediterráneo oriental: el refuerzo del hábitat diseminado frente al debilitamiento de los núcleos principales. En: Tirado Ballesteros JG, Piñeiro Antelo MA, Paül V, Lois González R (eds) ¿Renacimiento rural? Los espacios rurales en época de pos-pandemia. pp 93–103

Ruiz Moya N, Sánchez Escolano LM (2022) Las ciudades medias en la planificación. Escalas, tendencias y conflictos entre urbanismo y ordenación del territorio en Andalucía. Revista de Estudios Regionales (aceptado, pendiente de publicación)

Rullán O (2011) La regulación del crecimiento urbanístico en el litoral mediterráneo español. Ciudad y Territorio. Estudios Territoriales 43(168):279–297

Sánchez Escolano LM (2013) Modelo territorial innovador y articulación urbana en el poniente almeriense. Investigaciones Geográficas 59:57–74

Sánchez Escolano LM (2016) Ciudades medias y articulación espacial en el litoral mediterráneo andaluz (1950–2014). Nuevas tipologías urbanas para el diagnóstico y el desarrollo territorial. Universidad de Granada, Granada

Sevilla Buitrago A (2015) Urbanismo, crisis y austeridad. Ciudades 18:31–48

Zamorano Wisnes J (2016) La ordenación del litoral en Andalucía. El Plan de Protección del Corredor Litoral de Andalucía. En: Núñez Lozano MC (dir) Estudios jurídicos sobre el litoral. pp 87–130

Chapter 11
Housing Change During the COVID-19 Pandemic in the Metropolitan Areas of the Canary Island Capitals (Spain)

Josefina Domínguez-Mujica, Juan M. Parreño-Castellano, and Víctor Jiménez Barrado

Abstract During the pandemic, the geographical study of housing has acquired a new dimension because of the relationship forged between residents and this living space in a situation of containment of mobility and impediment to use the public and community space. It is from this perspective that this research on the metropolitan areas of the capital cities of the Canary Islands (Spain) was designed. A statistically representative survey reveals the singularities in the way residents relate to housing in these areas, as well as their responses to the adequacy of their housing in the situation of confinement. This allows us to appreciate certain socio-demographic and spatial contrasts, which are even more evident when we analyse the changes in housing and the relationship they have with a second residence. This last dimension reveals a resilient adaptation, as a significant part of the population sought more adequate housing conditions during the pandemic, although this adaptation reflects notable contrasts according to the different economic capacity of the respondents and the territorial characteristics of housing in the metropolitan areas of the Canary Islands' capitals.

Keywords Housing · Pandemic · Housing change · Metropolitan areas · Canary Islands

J. Domínguez-Mujica (✉) · J. M. Parreño-Castellano · V. Jiménez Barrado
Department of Geography, University of Las Palmas de Gran Canaria, Pérez del Toro, 1, 35003 Las Palmas, Spain
e-mail: josefina.dominguezmujica@ulpgc.es

J. M. Parreño-Castellano
e-mail: juan.parreno@ulpgc.es

V. Jiménez Barrado
e-mail: victor.jimenez@ulpgc.es

E. Navarro-Jurado et al. (eds.), *Urban Dynamics in the Post-pandemic Period*, The Urban Book Series,
https://doi.org/10.1007/978-3-031-36017-6_11

11.1 Introduction: State of the Question and Research Hypotheses and Objectives

The measures to prevent the spread of COVID-19 were a Copernican shift in the way we conceive, move and use both communal and public as well as individual and private space. Geography has traversed all these scales, recording changes in behavioural patterns (Rowe et al. 2022; Lekić et al. 2022), envisioning a new and different paradigm (Donadio 2020; Ferrás 2020) or, finally, proposing adaptation and resilience measures during the conjuncture (Iracheta 2020; Bereitschaft and Scheller 2020).

One such space is housing, whose relationship with individuals seems to have undergone alterations since the onset of the health crisis. This paper addresses this issue from the first perspective, i.e. recording changes in behavioural patterns, focusing its interest on the resilient response of residents in relation to their housing. To this end, the paper takes the metropolitan areas of the Canary Islands' capital cities as the place of study, which allows for a multidimensional approach.

Although the analysis is developed primarily in the housing dimension, the territorial scale cannot be discarded and is fundamental to decipher the dynamics of a singular space. The Canary Islands is an insular and ultraperipheral region, highly dependent on its external connectivity, and has therefore been seriously affected by a situation in which the links of the global urban network have been interrupted (Fareed et al. 2022), isolating this type of territory to a greater extent, and offering more possibilities for the development of particular dynamics.

At a global level, changes in residential mobility have already been observed at this scale, due to the reactivation of rural–urban migration and other proximity relations such as tourism and trade (Callau-Berenguer et al. 2022). However, it is not yet clear whether this can be replicated in a small territory such as the Canary Islands. In any case, the rural-nature relationship is perceived as safe, while the dense city has been stigmatised (Åberg and Tondelli 2021).

In fact, within the urban scale, in this parenthesis of confinement, studies have shown that the diffuse city, zoning, certain economic activities and the massive use of the car have a detrimental effect on climate, pollution and human health (Martín-Vide et al. 2021; Aceves et al. 2021). In contrast, natural strongholds within cities, such as open and green spaces, have been shown to have positive effects beyond the material and to benefit mental health and well-being (Tendais and Ribeiro 2020). On the other hand, unequal social and economic conditions within cities have had disparate consequences. In this way, disease has once again manifested itself as an indicator of urban inequality, which augurs differentiated recovery according to areas of the city (Barata-Salgueiro 2020).

Therefore, the environmental, socio-economic, and material aspects of each neighbourhood have determined vulnerability to COVID-19 (Cohen et al. 2022; Romero-Aravena and Mendes 2021), including in terms of risk of infection (Esteban et al. 2021). This means that individual aspects such as age, employment status, tenure status or housing environment are important in determining and analysing changes in relation to housing in the Canary Islands.

However, despite its importance, geographical analyses such as this have gone as far as the housing scale, since the study of habitat without considering housing is incomplete (Valente and Jacinto 2021). Hence, other variables such as the type of dwelling, its surface area, the number of household members or the level of facilities are essential for gauging housing change with the pandemic in the metropolitan areas of the Canary Islands' capital cities.

The functions of housing during the pandemic have exceeded what is considered to be usual, acting not only as homes but also as places of work, leisure, forts in the face of contagious disease, etc. Consequently, the "geography of the house", i.e. the organisation of its spaces and its hierarchy, has been altered in a conjunctural way because it was previously designed for other conditions and for other purposes (Da Graça 2020; Boccagni et al. 2022), which, for example, relegated the importance of a window or a terrace. In fact, the latter has had an unusual and unprecedented prominence, diminishing their storage function or their simple lack of use, and increasing their usefulness as a place for entertainment, exercise or even privacy with respect to those living with them (phone calls and reading) (Khalil and Eissa 2022).

The literature indicates that this situation of "life compression" has implied the implementation of strategies within a limited space in order to combine the need to be alone and to develop a communal life, as well as to mitigate the lack of contact with the outside world (Porcelloni and Mazzanti 2020), which is not always possible due to the shortcomings of the dwelling. These inadequacies are commonly related to a low number of rooms, small surface area and lack of open or semi-open spaces (Ramos 2021), generating a more emphasised sense of seclusion.

Thus, some research suggests that confinement in houses has limited regular activities in the outside world (socialising, sightseeing or walking), but this same situation of impairment has generated benefits, such as increased productivity, mainly due to time savings (Ozçelik and Kayihan 2022). For other studies, these positive aspects depend to a greater extent on the material capacity to establish boundaries between uses within the properties themselves (Goodwin et al. 2021).

In this way, it can be observed that any alteration is mediated by the affectation of material to behavioural issues, although within a much more restricted context of change than in any other consumer good, as housing is a costly element of slow and/or difficult transformation, exchange and transaction.

In correspondence with the transformations and the new significance acquired by housing during the pandemic, the research we carried out has the following objectives. Firstly, to analyse the level of adequacy of housing during the pandemic, taking into account the differences that can be seen depending on the socio-demographic profile of the respondents and the characteristics of the dwelling itself. Secondly, to identify changes in the usual residence in relation to the location of the dwelling. And, finally, to recognise the role of the ownership of a second residence in the effects of housing change during the pandemic.

To achieve these objectives, the research we carried out revolves around the following two hypotheses. The first: given that confinement was perceived as a temporary situation, a considerable part of the population in the metropolitan areas of the Canarian capitals was able to adapt to this scenario. Second, the change of residence or the desire to change presents important disparities. These are related not only to the conditions of the properties themselves but also to the purchasing power of the population, factors that allow us to recognise the unequal capacity for resilience to the pandemic from a socio-territorial perspective.

11.2 Source and Methodology of Analysis

In order to find out the impact that the COVID-19 pandemic has had in the Canary Islands in relation to mobility and migration trends, work, housing, consumption patterns and leisure, a survey was carried out between 26 April and 21 June 2022, in the period concerning the transition from the sixth to the seventh waves of infection in Spain.

The universe of those surveyed was made up of persons aged 18 years old and over residing in the Canary Islands, according to data from the continuous register on 1 January 2021. A total of 1297 computer-assisted telephone surveys were carried out using a multiple stratified random sampling procedure according to geographical areas of residence (islands and, for the two capital islands, large districts), sex and age groups (18 to 34, 35 to 64, 55 to 69 and 70 and over). The confidence level for the total sample is 95% with a standard error of 2.92% for the worst-case scenario of $p = q = 0.50$.

The survey was structured in seven main blocks: socio-demographic profile of the respondent, mobility linked to work and study, migrations, work activity, housing, change of home and second home, purchasing and consumption habits and changes in leisure and tourism patterns, all of these related with the pandemic.

In the case we are concerned with in this paper, we are interested in the level of housing adequacy during the pandemic, the changes in the usual residence that occurred due to the health crisis situation and the relationship between pandemic and second residence. The analysis has focused on the two metropolitan areas or two large urban capital areas of the Canary Islands (Fig. 11.1): Las Palmas de Gran Canaria (LPGC), which includes besides the city of Las Palmas de Gran Canaria itself, the municipalities of Telde, Santa Brígida and Arucas; and Santa Cruz de Tenerife-La Laguna (SCT-Laguna), which includes the municipalities of Santa Cruz de Tenerife, San Cristóbal de La Laguna, Tegueste, Tacoronte, El Rosario and Candelaria. The former had 538,276 inhabitants in 2021, and the latter 438,298, which together accounted for 44.9% of the population of the Canary Islands (Ministry of Transport, Mobility and Urban Agenda 2021; National Statistics Institute 2022).

The sample in the two metropolitan areas was reduced to 399 surveys, which increased the sampling error to 4.91% (Table 11.1). The sample was subdivided 50.9% and 49.1% between men and women, respectively. 43.1% of respondents

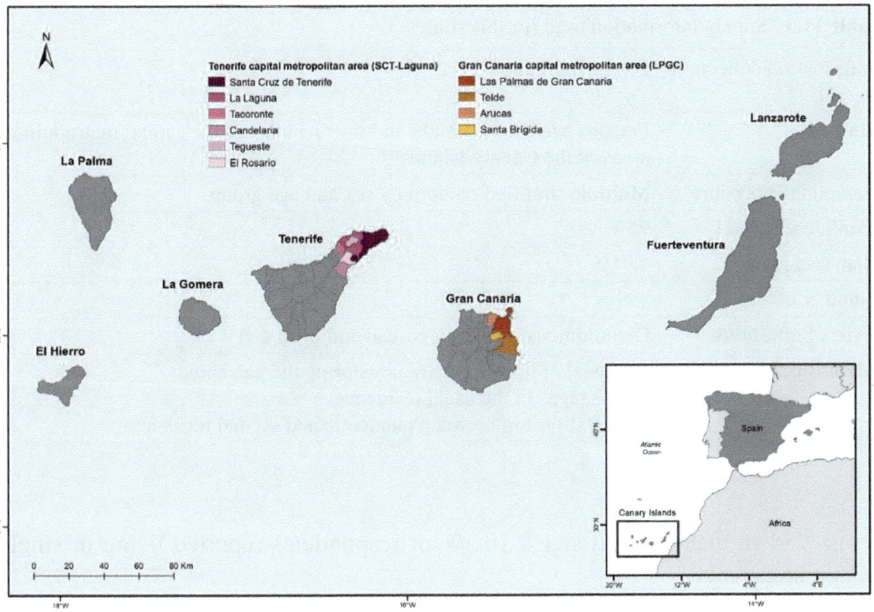

Fig. 11.1 Location map of the capital metropolitan areas of the Canary Islands. *Source* Authors

were aged 35–54; 25.8% were aged 55–69; 25.6% were aged 18–34 and 5.5% were aged 70 and over. These percentages are close to those officially recorded in the continuous population register. As for other profile variables that were not used for stratification, 88% of respondents were born in Spain, and most respondents were working during the pandemic (48.4%), while the rest were unemployed (13.0%), studying (11.5%), retired (11.0%) or affected by a temporary employment regulation (9%), a measure within the framework of the European Union's Temporary Support to mitigate Unemployment Risks in an Emergency (SURE). In relation to the latter, at the time of the survey, the number of workers had increased significantly, and conversely, the number of those affected by temporary employment regulation, and studying had decreased. The proportion of retired and unemployed people had also increased slightly.

The questions focused on the block analysed in this paper were formulated by means of dichotomous, multicategorical or multiple response choices. In the latter case, the option of open text was given, but in general, and in order to facilitate the statistical treatment of the information, an inductive methodology categorised in the questionnaire itself was chosen.

In relation to the housing profile of the respondents, five questions were introduced concerning household size, usable floor area, type of dwelling, tenure status and type of environment in which the dwelling was located. The most frequent household sizes were 2 and 3 people (35.1 and 28.3%, respectively), with the average household size

Table 11.1 Survey information used for this study

Information collection period	26 April–21 June 2022
Universe	Persons aged 18 years old and over residing in the capital metropolitan areas of the Canary Islands
Sampling Procedure	Multiple stratified random by sex and age group
Confidence Level	95%
Standard Error	4.91%
Number of surveys	399
Type of questions	Dichotomous, multicategorical and open text
Main Blocks	(1) Level of housing adequacy during the pandemic (2) Changes in the usual residence (3) Relationship between pandemic and second residence

Source Authors

being 2.84 in metropolitan areas. 10.3% of respondents reported living in single-person households.

The predominant type of dwelling was the flat or apartment with horizontal division of the property. 65.7% lived in this type of dwelling, while detached dwellings accounted for 22.6% and semi-detached dwellings for 11.8%. 84.2% of the dwellings were described by the respondents as dwellings in an urban environment and were mostly owned (61.4% of the respondents), with rental accommodation accounting for 23.6%. In terms of living area, more than half of the properties were between 60 and 90 square metres (50.5%). A total of 31.4% lived in a regular dwelling with a floor area between 90 and 120 square metres, while smaller dwellings (<60 m^2) were only seen in 8.2% of respondents and larger dwellings (>120 m^2) accounted for 18%.

In summary, the housing profile of the survey is similar to that provided by the census data, so we can consider some of these characteristics as crossover variables to better understand the housing changes and modifications that the pandemic may have provoked in relation to housing (Table 11.2).

The rest of the questionnaire referring to the object of analysis of this work is made up of questions focused on the adaptation, change and modification of the usual dwelling, as well as on secondary housing. It is these questions that will be analysed below, and which will allow us to detect the level of housing change that has been induced by the pandemic. For this purpose, the methodology of analysis is based on the creation of frequency tables and counting categories and contingency matrices that allow us to detect trends and causal relationships.

Table 11.2 Crossover and housing variables of the survey

Selected crossover variables	
Socio-demographic profile	
Gender	Men: 50.9%; women: 49.1%
Age group	18–34: 25.6%; 35–54; 43.1%; 55–69: 25.8%; 70 and over: 5.5%
Relationship with economic activity	Employed: 48.4%, unemployed: 13.0%, students: 11.5%, retired: 11.0%, affected by a TER: 9%, others: 7.1%
Territorial profile	
Place of birth	88% were born in Spain
Municipality of residence	Municipalities of metropolitan area of Las Palmas de Gran Canaria and Santa Cruz de Tenerife-La Laguna

Housing variables	
Number of household members	1 person: 10.3% 2 persons: 35.1% 3 persons: 28.3% 4 and over: 26.3%
Type of dwelling	Flat or apartment: 65.7% Detached dwelling: 22.6% Semi-detached dwelling: 11.8%
Usable area	<60 m^2: 8.2% 60–90 m^2: 50.5% 90–120 m^2: 31.4% >120 m^2: 18%
Tenure regime	Ownership: 61.4% Rent: 23.6% Others: 15%
Living environment	Urban: 84.2% Non-urban: 15.8%

Source Authors

11.3 Housing and Confinement: The Peculiarities of Metropolitan Areas in the Canary Islands

A first analysis of the housing conditions in the metropolitan areas of the Canary Islands reveals the predominance of flats or apartments, as we have already pointed out, although in these metropolitan areas, there are areas where there is a proliferation of detached and semi-detached dwellings, which represent 34% of all dwellings, and which the interviewees themselves define as 16% within a rural environment. Moreover, given the typological diversity in terms of the surface area of the dwellings and the fact that many of them have walkable roofs, terraces and balconies, it cannot be interpreted that confinement represented a generalised enclosure for the whole population, or an absence of possibilities of contact with open spaces; in fact, only just under a quarter of those surveyed indicated that they were not in contact with such spaces. This phenomenon should be considered as typical of the housing conditions in the metropolitan areas of the Canary Islands (Fig. 11.2), where traditional forms of construction coexist with more modern ones and, especially those related to a favourable climate, which allows for the enjoyment of open spaces on roofs and overhangs (Arroyo Pérez 2022).

On the other hand, it is also striking that the average occupancy of the dwelling is slightly lower in the metropolitan areas than in the Canary Islands as a whole (2.84 people per dwelling compared with 2.95 in the survey as a whole), which indicates a more reduced composition of households, with a greater number of single-

Fig. 11.2 Typologies of housing in Las Palmas de Gran Canaria. Top left, subsidised private housing development; middle left, multi-family tower buildings; bottom left, subsidised public housing development; top right, garden city houses; middle right, multi-family buildings in a closed block; bottom right, traditional self-built housing on hillside (*Riscos*). *Source* Authors

or double-person households, corresponding to forms of cohabitation typical of an ageing society and with a higher proportion of people living alone. For this reason, almost 70% of single-person households in the study areas correspond to dwellings of less than 90 square metres. The lower density of occupancy also has implications for the perception that the dwelling was comfortable at the time of confinement, in contrast to what occurred in other geographical areas of higher density.

Given that the level of satisfaction in relation to the characteristics of the housing and its use during confinement is very high (83% of respondents report this), it is worth looking in more detail at the factors or indicators of comfort.

As shown in Fig. 11.3, the greatest adaptation of the dwelling to confinement was that of Internet access. It also stands out that the dwelling preserved privacy in more than 86% of the cases, with the only data showing lower values for the practice of sport and contact with open spaces. The conditions of suitability of the dwelling for confinement do not show notable differences according to gender. From the point of view of age, the biggest gap occurs in the responses about the suitability of the dwelling to preserve privacy and to have contact with open spaces, which young people consider to be lower. These circumstances probably derive from a greater desire for independence at a young age, especially if they are still living with their parents. In relation to the suitability for entertainment or sport, adults consider it to be lower compared to the young and old, possibly because the former are able to find alternative forms of sport in unfavourable circumstances (Jaenes et al. 2021). Finally, in relation to the suitability of the dwelling for Internet connection, young people achieve the highest values for satisfaction.

When the comfort of the dwelling is combined with whether it is a flat or apartment or a detached and/or semi-detached house, the former is logically the least suitable for being in contact with open spaces and practising sport. However, satisfaction with the suitability of these dwellings is high, at almost 70%. Finally, it is worth noting the very high level of convenience of Internet connection in the case of flats or apartments compared to the other typologies, reaching more than 93% of the responses, with

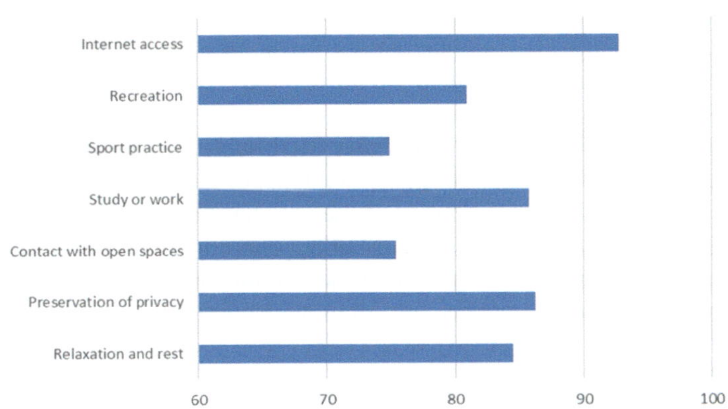

Fig. 11.3 Adequacy of housing to containment according to survey respondents. *Source* Authors

this fact being more highly valued by those who worked (94%). This factor is of considerable importance if we bear in mind that almost 24% of the total population of the Canary Islands surveyed had been teleworking for at least three months after the official declaration of the pandemic.

Without questioning the fact that the majority of the population perceives that they had adequate housing for the period of confinement, it should be borne in mind, firstly, that, on average, almost 18% considered that the property in which they lived had some characteristic that made it unsuitable. Secondly, the pandemic, and especially confinement, is perceived as an occasional element. The idea of suffering more regular confinement could provoke changes in a perception based on temporary resilience.

In line with these two statements, recent studies on the level of happiness in the home have highlighted the fact that the population of the Canary Islands between 25 and 65 years of age is the one with the lowest level of happiness in Spain, with an average value of 7.1 out of 10, with young people between 25 and 35 years of age being particularly unhappy (6.8). The low quality of housing and urbanisation are two of the causes cited, and there is an inverse relationship between the size of the dwelling, the number of household members and the level of happiness (Aedas Home 2021).

11.4 Changing House During the Pandemic: Resilience to an Enduring Asset

This research demonstrates the link between the pandemic and the search for alternative housing at the time of the health crisis (Table 11.3). Thus, a not insignificant proportion of respondents (15.5%) changed their place of residence during the pandemic, and in half of the cases, this change also involved moving to another municipality. However, the number of respondents who say they intend to move house in the near future is drastically reduced (4.7%), and of these, even fewer say they want to change their municipality of residence. Those who would do so indicate in greater numbers that they would move to another municipality on the island. The large difference in the percentages between those who changed their primary residence during the pandemic and those who plan to do so in the near future is the better manifestation of the large impact of the pandemic on the change of primary residence. On the other hand, those who owned their own home changed their dwelling during the pandemic in a smaller proportion than those who resided in rented or other tenancies, demonstrating the retaining power of home ownership.

The main reason given by respondents for moving house during the pandemic and/or the intention to move is mainly to live in a larger property. Living in a cheaper property and a change of job or studies also play a minor role. Only a quarter mentioned the pandemic as a reason for moving. In other words, we can consider that the pandemic acted as a direct cause in a quarter of the changes, while for much of the remainder,

Table 11.3 Changes in place of residence in % during the pandemic (based on survey responses)

Changes in residence			
People changing primary residence	People changing municipality of residence	People with intention to change primary residence soon	
15.5	7.3	4.7	
Reasons for moving house or intentions for moving house			
To live in a larger property	To live in a cheaper property	For jobs reasons	For other reasons
16.0	11.7	11.7	39.4
Owning a second home or intentions to buy a second home soon			
People with a second home	Intentions to buy a second home in the near future		
22.8	11.4		

Source Authors

the health situation acted as an indirect or underlying cause. In fact, about a third or more of those who changed housing during confinement say that the main dwelling did not allow them to be in contact with open spaces, or to study or work or to play sports or entertain themselves.

The change of home during the pandemic has also been detected by other surveys. Along these lines, the survey conducted by Aedas Home in the Canary Islands found that, at the beginning of 2021, 34% of the population between 25 and 65 years of age wished to change their home in the next two years, the highest percentage in Spain. In a complementary manner, among the main reasons given for buying a home were the intention to change neighbourhood or the need for a larger home, but others such as the search for quieter areas, the search for homes with a terrace or garden or the desire to live in a home with higher material quality also stood out (Aedas Home 2021).

On the other hand, it should be noted that the proportion of people who had a second home during confinement amounted to 22.8% of the respondents. In terms of age, the majority of those who owned a second home were over 55 years of age. However, the link between owning a second home and changing one's usual residence is not high, as only a quarter of those who owned a second home made it their usual residence. Prohibitions on mobility and the fact that many second homeowners may also be owners of a usual residence that is comfortable to live in during confinement are possible reasons for this.

Although the level of use of second homes with the pandemic was not very high according to the results of the survey we conducted, it does appear that the pandemic has directly or indirectly increased the demand for second homes among the population living in metropolitan areas. 11.4% of those surveyed stated that they would like to buy a second home in the near future, in line with other studies carried out in 2021 for Spain as a whole, in which 11% of the demand also stated this, a figure that represented an increase of more than two points compared to 2020 (Fotocasa Research 2022a).

Finally, the information at the municipal level of the metropolitan areas allows us to appreciate some aspects of spatial behaviour during the pandemic, one of the aspects of particular interest to geographers (Sancho Comíns and Olcina 2021). From this perspective, the least housing change in the study areas occurred in the metropolitan areas. Lower incomes and the predominance of housing better adapted to the requirements generated by the pandemic seem to be the reasons for this. The environmental conditions typical of the rural environment that can be seen in some of the nuclei of the metropolitan areas favoured a lower willingness to move on the part of the population residing there, as many of these residents consider that they live in an environment like that of the rural environment.

In contrast, residents in the more urban municipalities of S/C de Tenerife, La Laguna and Las Palmas de G.C. were those who changed residence the most. From a geographical perspective, there is a certain correspondence between this fact and the ownership of a second home. There is no doubt that the higher income level of a segment of the population residing in the capital centres enables them to have a second home which they appreciate and value during the pandemic and confinement (Mayordomo Maya et al. 2022). But the main reason for this is the fact that inner-city residential areas abound in small dwellings in buildings with high levels of buildability. Given that the social groups living in these housing conditions in central areas have a sufficient level of income to live in other larger properties in peripheral areas, the pandemic seems to have favoured the mobility of part of these social groups towards these spaces.

11.5 Conclusions

Housing has multiple functions. The most common are those of shelter and preservation of personal integrity, habitation and reproduction of the home and the workforce, safeguarding of heritage, socialisation and generation of leisure, production, identity formation and preservation of intimacy or social representation. All this defines housing as a multifaceted concept in which material, cognitive, affective and social aspects coexist (Amérigo and Pérez-López 2010).

Moreover, when the pandemic begins in March 2020, some of the functions of production and leisure have to be transferred to the housing environment, while others, which were already inherent to the dwelling itself, become more important, such as shelter and the preservation of personal integrity. In this context, housing satisfies these functions in an unequal manner depending on parameters such as surface area and its relation to household size, housing quality, the typology of the building or the presence of open spaces (González and Piñeira 2020).

The first objective of this work was to assess the level of suitability of the dwellings for their functions during confinement, concluding that this level was sufficient in

83% of cases, on average. Only when it was intended to carry out leisure functions was the level of suitability somewhat lower.

We conclude, moreover, that this result can only be explained by the existence of transversal reasons. Some of them refer to the building singularity of the Canarian metropolitan areas, with a high percentage of suburban and self-built properties and a widespread presence of multi-family buildings with walkable roofs. Others are related to equipment and household size factors, such as the high level of Internet connection, which facilitated production functions, or the low average household size, which favoured personal intimacy. Finally, other reasons relate to the resilience of the population, sustained by a temporary perception of their own confinement.

Based on these results, we must ask ourselves whether, in the medium term, the situation of pandemic and confinement has come to strengthen the private space of the home to the detriment of other spaces for community and public use. This would imply a strengthening of multifunctional housing, which would have consequences for the model of coexistence.

The second objective of this study is to find out the impact of the pandemic on the change of usual residence. As we pointed out in the previous section, the health crisis led to an increase in the process of moving house, both for direct reasons (a search for more suitable housing conditions) and indirect ones (a change in purchasing power or in the price of property). Moreover, it affected more those living in rented accommodation. Those who owned their homes remained in their homes to a greater extent, either because these were better suited to the diversification and functional redefinition of housing, or because ownership produces emotional and economic attachment to the property owned. Given that ownership, on average, is associated with higher incomes (Fernández-Rodríguez 2019), it can be inferred that these social groups showed less vulnerability (Caravantes and Romero 2021).

It does not seem that the urgency to change housing has taken hold in society, but what it has favoured are changes in the housing market and in the urban structure itself. Based on the motivations for change that were argued in the survey, we have concluded that there is a demand for larger, more affordable housing with more open spaces. The combination of these factors may be suggesting, in addition to a change in housing typology (Alves and San Juan 2021), implications in the urban growth model of Canarian cities given that housing in low-density areas is that which responds to them. If we bear in mind, moreover, that housing will be increasingly multifunctional, and residential location appears to be more independent of the distribution of classic residential location factors such as work. Therefore, it is likely that the dynamics of housing change among residents in urban or central spaces will lead to a process of greater metropolisation, contradicting a trend that has been registered of a recovery of central spaces. On the other hand, the pandemic has encouraged ownership rather than renting. Undoubtedly, other factors explain the balance in demand for these two tenure regimes, but market analyses seem to suggest an association between ownership, housing multifunctionality and shelter.

The third objective of the study was to find out the relationship between pandemic and secondary housing. In the previous section, we said that only a quarter of those who had secondary housing used it regularly in the confinement. Surely the suitability of the main dwelling was greater than the secondary dwelling for the requirements introduced by the pandemic. However, once this period of forced confinement is over, the pandemic and post-pandemic are characterised by an increase in the demand for second homes.

Demand studies point in the same direction. In the years prior to the pandemic, demand for second homes was on a downward trend. However, in 2021, there has been a significant change, as this demand has grown by 44% in the last two years (Fotocasa Research 2022b). In this sense, the second home responds to the function of leisure and refuge, not only in terms of health, as it is related to less dense spaces and pseudo-urban morphologies, but also in the economic dimension. The pandemic seems, therefore, to have stimulated the trend towards the ownership or use of more than one property, which may also have territorial consequences, insofar as it may favour a second process of metropolisation and the occupation of other areas.

In summary, the results suggest that the pandemic has boosted private space and its multifunctional dimension, bringing the functions of shelter and leisure to the forefront; it has stimulated the change of habitual residence and modified the requirements of demand by promoting new typologies and property as a tenure regime, and it has boosted the purchase of second homes. All this allows us to say that the health crisis will foreseeably have its consequences on the urban dynamics and territorial structure of Canarian cities, as can be seen in other territorial areas (Liu and Su 2021). Its greater or lesser impact will depend on the interference of other market factors and cyclical situations and, above all, on the extent to which this new housing conceptualisation is consolidated.

Acknowledgements Grant "The post-COVID-19 territorial balance in the Canary Islands. New strategies for new times (ProID2021010005)" funded by the Canary Islands Government (Smart Specialization Strategy of the Canary Islands RIS-3 and co-funded by the ERDF Operational Program).

Grant "Housing and international mobility in cities of Canary Islands: the emergence of new forms of urban inequality" (RTI2018-093296-B-C21) funded by MCIN/AEI/10.13039/501100011033 "ERDF A way of making Europe".

References

Åberg HE, Tondelli S (2021) Escape to the country: a reaction-driven rural renaissance on a Swedish Island post COVID-19. Sustainability 13:12895. https://doi.org/10.3390/su132212895

Aceves H, Mercado S, López O, Arévalo J (2021) Documento: Procesos de construcción, emisión de dióxido de carbono y resultados socio-económicos durante la pandemia del covid-19 en México. Telos: revista de Estudios Interdisciplinarios en Ciencias Sociales 23(2):485–502. https://doi.org/10.36390/telos232.17

Aedas Home (2021) Observatorio Aedas Home. Julio 2021. https://www.aedashomes.com/

Alves P, San Juan L (2021) El impacto de la crisis sanitaria del Covid-19 sobre el mercado de la vivienda en España. Banco de España. https://www.bde.es/f/webbde/SES/Secciones/Publicaciones/InformesBoletinesRevistas/ArticulosAnaliticos/21/T2/Fich/be2102-art16.pdf

Amérigo M, Pérez-López R (2010) Ambientes residenciales. In: Aragonés JI, Amérigo M (eds) Psicología Ambiental. Pirámide, pp 59–75

Arroyo Pérez CM (2022) Impacto del confinamiento domiciliario durante la pandemia COVID-19 en seis parejas andaluzas: estresores, oportunidades, pareja como apoyo y post-confinamiento. Un estudio cualitativo a través de grupos focales. Degree Thesis, Universidad de Córdoba. http://hdl.handle.net/10396/22568

Barata-Salgueiro T (2020) Viver na cidade sob a pandemia da Covid-19. Finisterra 55(115):113–119. https://doi.org/10.18055/Finis20375

Bereitschaft B, Scheller D (2020) How might the COVID-19 pandemic affect 21st century urban design, planning, and development? Urban Sci 4(4):Article 0056. https://doi.org/10.3390/urbansci4040056

Boccagni P, Brodesco A, La Bruna F (2022) 'Stayhome' as a YouTube performance: representing and reshaping domestic space under the 2020 COVID lockdown in Italy. Eur Soc 24(2):129–153. https://doi.org/10.1080/14616696.2022.2043411

Callau-Berenguer S, Roca-Torrent A, Montasell-Dorda J, Ricart S (2022) How to guarantee food supply during pandemics? Rethinking local food systems from peri-urban strategic agents' behaviour: The case study of the Barcelona Metropolitan Region. Investigaciones Geográficas 77:363–379. https://doi.org/10.14198/INGEO.19554

Caravantes GM, Romero J (2021) Vivienda pública y estado de bienestar en España: balance y estado de la cuestión en la época del COVID-19. Boletín de la Asociación de Geógrafos Españoles 91. https://doi.org/10.21138/bage.3152

Cohen M, Gutman M, Díaz MR, Fodde MB, Cabrera C, Mora B, Díaz C (2022) COVID-19 in real time: comparing the struggle of three low-income neighbourhoods in Buenos Aires. Environ Urban 34(1):209–228. https://doi.org/10.1177/09562478211060687

Da Graça M (2020) Rever a geografia do quarteirão e da casa: vários usos no mesmo espaço. Finisterra 55(115):133–138. https://doi.org/10.18055/Finis20360

Donadio T (2020) Repensar a cidade inteligente ou voltar ao "antigo normal"? Uma reflexão sobre o caso de Lisboa no contexto da Covid-19. Finisterra 55(115):121–126. https://doi.org/10.18055/Finis20214

Esteban MM, Gil JL, Fernández E, Fernández R, Jiménez R (2021) Covid-19 en la ciudad de Madrid y vulnerabilidad. Análisis de las dos primeras olas. Revista Española de Salud Pública 95:Article e1–10. https://www.sanidad.gob.es/biblioPublic/publicaciones/recursos_propios/resp/revista_cdrom/VOL95/ORIGINALES/RS95C_202110119.pdf

Fareed Z, Ghaemi Asl M, Irfan M, Rashidi MM, Wang H (2022) Exploring the co-movements between COVID-19 pandemic and international air traffic: a global perspective based on wavelet analysis. Int Migr 61(3):116–131. https://doi.org/10.1111/imig.13026

Fernández-Rodríguez JF (2019) Gestión híbrida de la vivienda social en España. Hacia modelos colaborativos de provisión pública al margen del lucro. Fundación Alternativas. https://bit.ly/3eRBhsV

Ferrás C (2020) Ciudades dispersas y aldeas virtuales en la postpandemia del Covid-19. Finisterra 55(115):243–248. https://doi.org/10.18055/Finis20279

Fotocasa Research (2022a) Experiencia en compra y venta en 2021. Fotocasa. https://s36360.pcdn.co/wp-content/uploads/2022a/02/Experiencia-en-compra-y-venta-en-2021.pdf

Fotocasa Research (2022b) Perfil del comprador de segunda residencia. Fotocasa. https://s36216.pcdn.co/wp-content/uploads/2021/08/NdP-perfil-comprador-segunda-residencia-2021.pdf

González JM, Piñeira MJ (2020) The unequal city in Palma (Majorca): geography of confinement during the COVID-19 pandemic. Boletín de la Asociación de Geógrafos Españoles 87. https://doi.org/10.21138/bage.2998

Goodwin B, Webber N, Baker T, Bartos AE (2021) Working from home: negotiations of domestic functionality and aesthetics. Int J Hous Policy. https://doi.org/10.1080/19491247.2021.1983245

Iracheta AX (2020) La ciudad que quisiéramos después de COVID-19. ACE: Arch, City Environ 15(43):Article 9512. https://doi.org/10.5821/ace.15.43.9512

Jaenes JC, Peñaloza R, García P, Costa M, Jaenes-Amarillo PI (2021) La vida en confinamiento de deportistas de alto rendimiento. In: del Campo Tejedor A (ed) Pensar la pandemia: más allá de la sanidad y la economía. Dykinson, pp 129–143. https://doi.org/10.2307/j.ctv1ks0d5x.10

Khalil M, Eissa D (2022) Balconies during COVID-19 lockdown: exploring the change in patterns of use in Cairo. Open House Int 47(2):235–253. https://doi.org/10.1108/OHI-08-2021-0178

Lekić O, Nikolić N, Folić B, Vitošević B, Mitrović A, Kosanović S (2022) COVID-19 and city space: impact and perspectives. Sustainability 14:1885. https://doi.org/10.3390/su14031885

Liu S, Su Y (2021) The impact of the COVID-19 pandemic on the demand for density: evidence from the US housing market. Economics Letters 207:Article 110010. https://doi.org/10.1016/j.econlet.2021.110010

Martín-Vide J, Zayas A, Salvador F, Moreno-García Mª-C. (2021). Air quality in Barcelona during the COVID-19 lockdown and the global effect on CO_2 emissions. Boletín de la Asociación de Geógrafos Españoles 91. https://doi.org/10.21138/bage.3139

Mayordomo Maya S, Hermosilla J, Fernández M (2022) Pandemia, gobernanza y municipalismo en las coronas metropolitanas a partir de su estudio en Riba-Roja de Túria (València). Investigaciones Geográficas 77:381–405. https://doi.org/10.14198/INGEO.20100

Ministry of Transport, Mobility and Urban Agenda (2021) Áreas urbanas en España. 2021. https://apps.fomento.gob.es/CVP/handlers/pdfhandler.ashx?idpub=BAW087

National Statistics Institute (2022) Continuous register statistics. https://www.ine.es/dyngs/INEbase/es/operacion.htm?c=Estadistica_C&cid=1254736177012&menu=resultados&idp=1254734710990

Ozçelik S, Kayihan KS (2022) Compressing urban living in the dwelling: pandemic living praxis. Open House Int 47(2):296–315. https://doi.org/10.1108/OHI-06-2021-0115

Porcelloni L, Mazzanti C (2020) Spazio sicuro e non-sicuro: un'indagine sulle nuove strategie dell'abitare nel contesto della pandemia di Covid-19. Documenti Geografici 1:633–646. https://doi.org/10.19246/DOCUGEO2281-7549/202001_40

Ramos AI (2021) Satisfacción de habitabilidad durante el confinamiento por Covid-19. Estudio comparativo de dos tipologías de vivienda en Argentina. Contexto 15(23):28–44. https://doi.org/10.29105/contexto15.23-281

Romero-Aravena H, Mendes FH (2021) La construcción social de climas urbanos y su relación con la pandemia de Covid-19 en Santiago de Chile. Cuadernos de Geografía: Revista Colombiana de Geografía 30(2):376–395. https://doi.org/10.15446/rcdg.v30n2.88701

Rowe F, Robinson C, Patias N (2022) Sensing global changes in local patterns of energy consumption in cities during the early stages of the COVID-19 pandemic. Cities 129:Article 103808. https://doi.org/10.1016/j.cities.2022.103808

Sancho Comíns J, Olcina J (2021) La cartografía temática como recurso idóneo para el conocimiento de la pandemia de COVID-19: ejemplo de aplicación en España. Boletín de la Asociación de Geógrafos Españoles 91. https://doi.org/10.21138/bage.3141

Tendais I, Ribeiro AI (2020) Espaços verdes urbanos e saúde mental durante o confinamento causado pela Covid-19. Finisterra 55(115):183–188. https://doi.org/10.18055/Finis20184

Valente D, Jacinto G (2021) Habitar en pandemia. Aislamiento social y desigualdades urbanas en asentamientos y barrios populares de la ciudad de Tandil (provincia de Buenos Aires, Argentina). Revista Universitaria de Geografía 30(2):145–172. https://doi.org/10.52292/j.rug.2021.30.2.0029

Chapter 12
Two Megaprojects—One City. Learning between Large-Scale Urban Development Projects on Tenerife, Spain

Marcus Hübscher

Abstract Inner-city megaprojects have become a typical element in the neoliberal city and are associated with multiple conflicts such as undemocratic planning, cost explosions and (green) gentrification. This chapter analyzes two neighboring megaprojects (Cabo-Llanos Plan and Santa Cruz Verde 2030) in Santa Cruz de Tenerife (Spain) and explores if there are learning processes of primary and secondary stakeholders between the two projects. I conducted 18 qualitative interviews with the town hall, architects, neighborhood associations and real estate experts. The results indicate how each stakeholder group learned particularly in the area that affects them most, for example with regard to urbanism, housing, and public spaces. The project's initiators show a clear learning process, as the new megaproject (Santa Cruz Verde 2030) takes up several deficits revealed by the former megaproject (Cabo-Llanos Plan). However, this chapter questions the motivation behind this learning, as the apparently sustainable mask of the project rather conceals the entrepreneurial logics. Hence, this study reveals the potential of integrating stakeholders into planning processes, but it also shows how spatial injustices are reproduced.

Keywords Megaprojects · Planning · Local stakeholders · Learning process · Santa Cruz de Tenerife

12.1 Introduction

Megaprojects are one of the most visible elements in today's neoliberal urbanism (Del Cerro Santamaría 2013a). They are expected to solve the city's most urgent problems at a stroke, increase inter-urban competitiveness (Kennedy 2014) and reposition the city within the global hierarchy (Swyngedouw et al. 2002). On the other hand, the

M. Hübscher (✉)
Institute of Urban Development and Construction Management, University of Leipig, Grimmaische Straße 10, 04109 Leipzig, Germany
e-mail: huebscher@wifa.uni-leipzig.de

E. Navarro-Jurado et al. (eds.), *Urban Dynamics in the Post-pandemic Period*, The Urban Book Series,
https://doi.org/10.1007/978-3-031-36017-6_12

list of megaproject-induced conflicts in the urban arena is long and contains aspects such as the underestimation of costs (Flyvbjerg 2005), undemocratic governance (Tarazona Vento 2017) and (green) gentrification (Argüelles et al. 2021; Gellert and Lynch 2003). Hence, Flyvbjerg rightly observes how "no one is learning from past mistakes, or no one wants to learn" (Flyvbjerg 2005, p. 20), addressing the inability of project's initiators to improve their practices (Hübscher 2022, p. 86).

This particular need to improve megaprojects' sustainability becomes even more evident in light of COVID-19. Not only has the pandemic reinforced existing spatial inequalities in the city and has left different social groups more vulnerable. This current crisis also gives reason to critically reflect on the underlying logics of how urban space is produced (Haase 2020; Honey-Rosés et al. 2020, p. 14). With regard to megaprojects, this implies how to enhance learning processes and thus decrease spatial injustices (Fainstein 2009b).

Within the academic discussion, this process takes place by learning from "successful megaprojects" (Brady and Davies 2013; Brookes et al. 2015). Apart from that, this chapter argues that also secondary stakeholders (who are affected indirectly by megaprojects) show learning processes. I will investigate the case of Santa Cruz de Tenerife, a Spanish city with about 200,000 inhabitants (ISTAC Instituto Canario de Estadística 2022) located on the Canary Islands. In the island's urbanism, megaprojects are an essential element and contribute to the "new image" that decision-makers are trying to establish (Hübscher 2019). There is a long list of large-scale projects, such as the new megaport in the south of the island (Armas-Díaz and Sabaté-Bel 2020) and the planned Formula 1 circuit, which has just been approved by the Island's government (Vargas 2022). Simultaneously, there is an increasing consciousness of citizens concerning the impacts of these projects, which other scholars frame as "the right to the island" (Sabaté Bel and Armas Díaz 2022), based on Lefebvre's idea of the "right to the city" (Purcell 2002, p. 99).

In Santa Cruz de Tenerife, I will explore learning processes between two local megaprojects, namely the Cabo-Llanos Plan and Santa Cruz Verde 2030 (Fig. 12.1 and Table 12.1). Both megaprojects are comparable due to their neighboring location and their similar trajectory. The Cabo-Llanos Plan converted one part of the inner-city oil refinery during the 1990s (Arencibia de Torres 2005). Santa Cruz Verde 2030 is a new megaproject east of Cabo-Llanos (Fig. 12.1) that was announced in 2018, as the remaining refinery (managed by CEPSA, Compañía Española de Petróleos), decided to leave the city. While the Cabo-Llanos Plan provoked polarization, displacement and demonstrated serious urban deficits (García Herrera 2003; García Herrera et al. 2007; González Chávez 2018), Santa Cruz Verde is expected to cause (green) gentrification (Hübscher 2019) and will continue the observed touristification (Armas Díaz 2016; García Hernández et al. 2020) and transformation of Santa Cruz' waterfront toward service-oriented economies (Díaz Rodríguez et al. 2008). Both megaprojects are located in a highly polarized setting, with high and low socio-economic vulnerabilities adjacent to both projects' sites (Hübscher 2020, 2022, p. 88).

Hence, my main research question is to ask whether there has been a learning process between the two megaprojects in Santa Cruz de Tenerife, seen from different stakeholders' perspectives. I do not only aim to find out who is learning, but also

Fig. 12.1 The Cabo-Llanos neighborhood and Santa Cruz Verde 2030. *Source* Own elaboration based on Hübscher (2022)

which aspects form part of the process. Hence, this text contributes to the understanding of why megaprojects continue to produce well-known shortcomings and fail to integrate lessons learnt from predecessor projects.

This chapter is structured as follows. Section two presents the theoretical framework of this study. Section three will discuss the methods applied. Section four analyzes the material, while section five summarizes the main findings and draws a conclusion.

12.2 Pursuing Megaprojects: An Incorrigible Habit?

Megaprojects are defined as large-scale development projects that are complex in every aspect (Brookes 2014; Flyvbjerg 2005). Rather than focusing on a quantitative or scale-based definition (Stoddart-Stones 1988), I understand megaprojects as entities that impact the surrounding social, spatial and functional systems in a decisive way (Heng Li 2011). Transforming not only the project site itself, but also adjacent areas is one of the immanent characteristics of megaprojects (Gellert and Lynch 2003, p. 19).

Spain is the country with the highest number of urban megaprojects debated, at least in the English-speaking discussion (Hübscher 2021). This comes as no surprise, considering projects such as Guggenheim (Bilbao) or Ciudad de las Artes y las Ciencias (Valencia, Fig. 12.2). These projects reached an impressive global visibility (Del Cerro Santamaría 2013b; Tarazona Vento 2017), copied by other city administrations

Table 12.1 Comparison of the two urban megaprojects in Santa Cruz de Tenerife

	Cabo-Llanos plan	Santa Cruz Verde 2030
Primary stakeholders	• CEPSA • Town Hall of Santa Cruz • Government of the Canary Islands • Local Port Authority	• CEPSA (Mubadala Company, Carlyle Group) • Town hall of Santa Cruz
Surface	• 1,036,000 m^2	• 573,000 m^2
Time frame	• Since the 1950s: displacement of neighborhoods • During the 1990s: dismantlement of refinery's infrastructure • Public land still lies idle	• June 2018: announcement of Santa Cruz Verde 2030 • 2022: dismantlement of the refinery starts
Former land use	• One third of the local oil refinery • The neighborhoods El Cabo and Los Llanos were evicted and demolished (during the 1950s)	• Remaining area of the oil refinery (two thirds)
Objectives	• Build a new representative city center with housing, commercial, administrative and recreational functions • Produce a new, modern image of the city • Promote urban tourism	• Reshape the relation between sea and city • Provide green public spaces • Reduce shortage on housing market • Promote urban tourism
Concept	• Three shopping malls • High-rise buildings with housing and offices • "Starchitecture": Opera house (Auditorio), Twin skyscrapers (Las Torres) • Large avenues, focus on individual transport	• Extensive green and blue infrastructure (parks and new urban beach) • Mix of housing and urban hotels • Sustainable mobility concepts, focus on public transport

Source Own elaboration based on Ayuntamiento de Santa Cruz de Tenerife and CEPSA (2018), García Herrera et al. (2007), and Hübscher (2022)

around the globe in the hope of profiting from a similar "Cinderella transformation" (Del Cerro Santamaría 2020). In this sense, megaprojects are recognized as a means to foster growth (Altshuler and Luberoff 2003, p. 2), increase a city's visibility and reach a higher position in the global urban hierarchy (Harvey 1989; Moulaert et al. 2001).

Spain's largest urban project is currently Santa Cruz Verde 2030 on Tenerife (Millet 2022). And yet, the project has not caused controversial public debate to date, which could be due to two different reasons. Firstly, one might suppose that this new megaproject provides a considerably enhanced concept (compared to former local megaprojects) and meets the needs of local stakeholder groups. This is a hypothesis that supports the idea of a learning process.

Contrary to that, a second interpretation refers to Lehrer and Laidley's (2008) observation of project initiators actively trying to avoid public protest. This can be

Fig. 12.2 Inner-city megaprojects with global visibility. From left to right: Parque del Fórum (Barcelona), Ciudad de las Artes y las Ciencias (Valencia) and Elbphilharmonie (Hamburg). *Source* The author

achieved by presenting an attractive mix of functions and extensive green or public spaces. In addition, constructing the projects on derelict brownfields where no direct displacement happens is a further means to reduce conflicts. However, the final output of such megaprojects is often less sustainable than promised and must be analyzed critically (Lehrer and Laidley 2008, p. 787).

Hence, learning processes across megaprojects are desirable, but difficult to achieve, which is due to several reasons (Brookes et al. 2015, p. 5; Flyvbjerg 2005, p. 20). Firstly, the so-called principle of the "hiding hand" euphemizes cost–benefit analysis of megaprojects, so that their realization becomes more probable (Flyvbjerg 2014, p. 13). This phenomenon actively opposes any learning from past mistakes. Secondly, the high complexity and individuality of megaprojects reduce the stakeholders' opportunity to compare and learn (Brookes et al. 2015, p. 5).

However, despite these observations, my hypothesis is that the immediate spatial and temporal proximity between both megaprojects in Santa Cruz indeed enabled a certain learning process. I understand such a learning process as a change of the stakeholders' behavior or opinions based on certain experiences (De Houwer et al. 2013). Unlike other studies that worked with quantitative data to explore such learning processes (Brookes et al. 2015), I propose a qualitative approach based on expert interviews to understand possible learning processes of local stakeholders. This stakeholder approach is rather underestimated in academia (Di Maddaloni and Davis 2017), but becomes more and more established in research on megaprojects (Doucet et al. 2010; Shatkin 2008, p. 383).

12.3 Materials and Method

I conducted 18 interviews with both primary stakeholders (such as the representatives of the town hall of Santa Cruz and CEPSA), but also a broad range of secondary stakeholders (neighborhood organizations, real estate market, scientists, architects, relevant political parties etc., see Table 12.2). Firstly, I chose these groups based on the CABERNET stakeholder model, which summarizes relevant actors in large-scale redevelopment projects (Ferber et al. 2006). Secondly, I applied snowball sampling.

I accessed the networks of several key stakeholders in the city, in order to reach otherwise inapproachable contacts. Hence, my selection contains a certain bias and does not claim to be representative, but still gives insights into the logics of each of the interviewees.

Most of the interviews were conducted face to face, three were done online via video-telephony and only one interviewee (I18, CEPSA) preferred to answer the questions in written form, which of course limits the informative value considerably. The interviews were structured by means of a guideline that entailed different aspects such as planning, politics, the relationship between the refinery and the city and the Cabo-Llanos neighborhood. The transcribed material was analyzed with MAXQDA, applying the rules of qualitative contents analysis (Rädiker and Kuckartz 2020; Valles 2007). With regard to analyzing the interviews, I developed the categories based on the material and structured the codes according to the aspect they refer to (public space, housing market, etc.), the type of stakeholder and the planning process.

12.4 Learning Between Two Megaprojects?

This section will present the results of the case study and analyzes the "who" and "what" of the learning processes between the two megaprojects in Santa Cruz. While some of the interviewees actively emphasize that Cabo-Llanos should serve as experience for Santa Cruz Verde 2030, even without me asking (I17: 97, Art historian), others avoid answering the question if learning should take place. For example, CEPSA responds that the projects "do not have much in common at all" (I18: 54) and are thus not comparable. However, the learning processes become obvious in several thematic areas, such as housing market and green public spaces. The most intensive discussion takes place about the urbanism produced by the two megaprojects (Hübscher 2022, p. 90).

12.4.1 Urbanism

With regard to urbanism, the learning processes are expressed by the fact that 17 out of 18 interviewees criticized Cabo-Llanos: "[it] is a poor neighborhood, seen from the urbanism's point of view" (I15: 120, Left-wing Party). The aspects that are criticized most are Cabo-Llanos' character as a dormitory town (I10: 254, Chamber of Architects) and the sterile atmosphere (I8: 104, Geographer). Some stakeholders argue that the planning model of Cabo-Llanos was not bad per se, but "the town hall lost control about what was going to be built there" (I7: 93, Architect).

Primary stakeholders such as the representative of the town hall confirm that they have to learn from the mistakes in Cabo-Llanos, but he associates these mistakes rather with the general urban model of the 70s and 80s that was applied there (I13: 50, Town hall). Also, the chief officer in the urban development office confirms that

Table 12.2 Overview of the interviews. Primary stakeholders are highlighted in bold emphasis

No.	Organization/field	Function	Date	Place
I1.	Professional Association of Real Estate Experts (APEI)	Regional delegate	23.08.2019	Office, Santa Cruz
I2.	**Urban Planning Office, Town Hall Santa Cruz de Tenerife**	**Representative (Liberal Party, Ciudadanos)**	**30.08.2019**	**Office, Santa Cruz**
I3.	Real estate	Agent, self-employed	02.09.2019	Office, Santa Cruz
I4.	Local association of industrial monument preservation	President	04.09.2019	Public café, Santa Cruz
I5.	Local environmental association (Ecologistas en Acción)	Representative	12.09.2019	Public café, San Cristóbal de la Laguna
I6.	**Conservative Party (Partido Popular)**	**Member of the city parliament, former chief officer in the Urban Planning Office**	**09.03.2020**	**Public café, Santa Cruz**
I7.	Architecture	former president of the Chamber of Architects of the Canary Islands	12.03.2020	Office, Santa Cruz
I8.	University of La Laguna, Department of Geography	Geographer, research associate	03.09.2020	Office, San Cristóbal de la Laguna
I9.	Journalism	Former head of a local newspaper	03.09.2020	Public café, Santa Cruz
I10.	Chamber of Architects of the Canary Islands	Three employees/ members	07.09.2020	Public café, Santa Cruz
I11.	Labor Party (Partido Socialista Obrero Español, PSOE)	Member of the city parliament, former chief officer in the Urban Planning Office	09.09.2020	Online
I12.	Buenos Aires Neighborhood	Neighbor	10.09.2020	Public café, Santa Cruz
I13.	**Local government of Santa Cruz**	**Representative of city government**	**11.09.2020**	**Town hall, Santa Cruz**
I14.	Local neighborhood association	President and vice president	11.09.2020	Online
I15.	Left Party (Unidas Podemos)	Member of the city parliament	11.09.2020	Town hall, Santa Cruz
I16.	Local neighborhood association of homeowners	President	11.09.2020	Public café, Santa Cruz
I17.	Institute for History of Art, University of La Laguna	Art historian, research associate	17.09.2020	Online

(continued)

Table 12.2 (continued)

No.	Organization/field	Function	Date	Place
I18.	**CEPSA (oil refinery)**	**Representative**	**26.01.2021**	**Written document**

Source Own elaboration, based on Hübscher (2022)

the Cabo-Llanos model should not be applied for the remaining refinery (I6: 82, Conservative Party).

Despite that, secondary stakeholders do not believe that politicians in charge have learnt from past mistakes. There is a notable fear that Santa Cruz Verde 2030 will have the same urbanism as Cabo-Llanos (I5: 112, Environmental association). It is particularly the imbalance of functions in Cabo-Llanos that draws criticism. There is no cultural (I17: 24, Art historian) or public infrastructure (such as green spaces, health centers or schools), but a strong focus on commercial space (I11: 44, Socialist Party; I10: 257, Chamber of Architects).

These findings add to what is observed in other studies. While Lehrer and Laidley speak of an "increasing commodification and circumscription of urban public space" (2008, p. 786), for Fainstein, it is rather the inaccessibility of these spaces that is problematic within megaprojects (Fainstein 2009a, p. 772). This case study reveals another—and even more extreme—form of withdrawing the public space in megaprojects: In Cabo-Llanos, public spaces have indeed been planned (Ayuntamiento de Santa Cruz de Tenerife 2017) and thus stimulated private investment (I16: 61, Owner association), which is a typical rationale in the neoliberal city (Harvey 1989; Kennedy 2014). However, none of these promises have been delivered during the last 30 years by the public authorities (I6: 82, Conservative Party), which underlines the poor integration between megaproject and the city (Swyngedouw et al. 2002, p. 547).

This lack of public space in Cabo-Llanos has led to the rise of neighbors' protest (Torres 2021) and the foundation of the local association "Los Llanos for living" (Spanish: Los Llanos Para Vivir), which fights for a higher quality of life in the neighborhood. The association criticizes that the infrastructure in Cabo-Llanos was designed exclusively for a regional or even international target group (shopping centers, trade fair and the opera El Auditorio, see Fig. 12.3), while local needs of the inhabitants were neglected (I16: 10, Owner association)—an often observed imbalance in megaprojects (Warrack 1993). As a result, the association demands that Santa Cruz Verde 2030 should not limit the development of the public space in Cabo-Llanos (I16: 49, Owner association).

Considering the multiple deficits in Cabo-Llanos and the large spatial scale of the new megaproject, the architects interviewed firstly call for a less bureaucratic and more flexible urban planning (I7: 97, Architect). Secondly, they question whether the formal instruments of municipal planning are the appropriate instrument here (I10: 34, Chamber of Architects). Contrary to that, this is what many scholars criticize,

Fig. 12.3 Two megaprojects in Santa Cruz de Tenerife. Left: the current oil refinery (and future megaproject Santa Cruz Verde 2030); Right: the Cabo-Llanos neighbourhood. *Source* The author

namely how for megaprojects, exceptional planning mechanisms are established (Moulaert et al. 2001, p. 80).

12.4.2 Green (and Blue) Public Space

Santa Cruz Verde 2030 aims to convert 41% of the current oil refinery into green spaces and thus takes up one of the most obvious deficits in Cabo-Llanos (Hübscher 2022, p. 91): "I live here, and there is no park where I can take my kids" (I10: 99, Chamber of Architects). Although the planned amount of green infrastructure is evaluated positively by several interviewees (I1: 56, Real estate association; I12: 75, Neighbor), experiencing the above-mentioned nonfulfillment of public promises decreases the credibility of public stakeholders drastically. Instead, it is expected that in the end, the share of green spaces will be replaced according to the economic interests, "because it always happens like that" (I1: 56, Real estate association).

Interestingly, only some of the interviewees saw the probability of green gentrification, although this is an immanent feature in many megaprojects (Argüelles et al. 2021). Currently deprived neighborhoods adjacent to the refinery will then be located near the city's largest green space, while the new urban beach will be reachable within a walking distance of five minutes. This future revaluation of space is only seen by some primary (I6: 80, Conservative Party) and secondary stakeholders (I5: 120, Environmental association; I14: 88, Neighborhood association).

One can argue that the megaprojects' initiators indeed have learned from past mistakes, as they have put the "green" into the focus of Santa Cruz Verde 2030 ("verde" means green in Spanish) and thus make up for the lack of such spaces in Cabo-Llanos. However, this entails a hidden entrepreneurial logic, namely the revaluation of adjacent neighborhoods. Accordingly, Ballard et al. observe how projects' initiators apply a strategy of framing their megaprojects "in terms of objectives such as livability, sustainability and equity" (Ballard et al. 2017, p. 117), although this often rather remains a rhetorical promise. The announced construction of the city's first urban beach within the project of Santa Cruz Verde 2030 is identified as such a strategy because it responds to a strong public desire: "the beach for the people, the beach for Santa Cruz that we never had" (I8: 74, Geographer).

12.4.3 Housing Market

The paragraph above insinuates the interplay between green and blue infrastructures and the dynamics on the housing market (Kronenberg et al. 2021). There is a very high level of skepticism with regard to the intended residential function in Santa Cruz Verde 2030 (21% of the whole area) among secondary stakeholders, based on the Cabo-Llanos experience (Hübscher 2022, p. 91).

Similar to other Spanish coastal cities, Santa Cruz has a tense housing market with high growth rates in land and rental prices. There is a particular shortage of affordable flats on the rental market (I3: 16, Real estate agent). The announced objective to include social housing in Santa Cruz Verde 2030 (Ayuntamiento de Santa Cruz de Tenerife and CEPSA 2018, p. 4) is hence an obvious intend to soothe the tense situation on the city's housing market. However, the question if there is indeed "more commitment to affordable housing" (Diaz Orueta and Fainstein 2009, p. 762) than in past megaprojects, is still an open one as for Santa Cruz Verde 2030—and needs to be evaluated ex-post.

From the real estate perspective, there are certainly high expectations of economic profit: "There will be speculation" (I3: 42, Real estate agent). This is also based on the speculative dynamics in Cabo-Llanos, which was labeled the "golden mile" of Santa Cruz (Pérez 2019). Here, exclusively high-standard housing has been developed (García Herrera et al. 2007), which worries the neighbors due to the lack of affordability (I12: 91, Neighbor). Some interviewees even associate the primary stakeholders of Santa Cruz Verde 2030 (namely the parties governing in the town hall, Coalición Canaria and Partido Popular) with real estate interests (I15: 51, Left-wing Party; I14: 84, Neighborhood association). The neighborhood association concludes that "for these politicians [Cabo-Llanos] was not a mistake" (I14: 100, Neighborhood association). This is also confirmed by one politician in charge, who does not show willingness to avoid speculation and displacement (I2: 21, Liberal Party). Thus, local stakeholders doubt that the new megaproject will be less elitist than Cabo-Llanos and provide a spatially just access to housing as "the speculators are the same that sit in the town hall today" (I14: 51, Neighborhood association).

In addition, the landowner (CEPSA) is assigned a strong interest in real estate development, too, which is due to the new international owner structure (I9: 45, Journalist; I7: 85, Architect). CEPSA, originally a Spanish power company, is currently owned by Mubadala Company, an investment fund from Abu Dhabi and the Carlyle Group (I18: 35, CEPSA). Carlyle is specialized in private equity and is associated with the ongoing financialization processes on housing markets (Metzger 2020). It is particularly this new owner structure, symbolizing how global logics have local impacts in the field of megaprojects (Swyngedouw et al. 2002, p. 550).

12.5 Conclusion: Different Stakeholders—Different Learning Processes

Santa Cruz' latest megaproject differs from its adjacent predecessor (the Cabo-Llanos Plan) significantly because Santa Cruz Verde 2030 proposes (1) a much larger importance of public (green and blue) spaces, (2) a visible focus on sustainability and (3) an apparently less elitist residential use. Hence, this new project contains many of the characteristics observed in the "new generation of megaprojects" (Diaz Orueta and Fainstein 2009, p. 761; Hersh 2017). It also shows the learning process of primary stakeholders in Santa Cruz, who incorporated the most obvious deficits of Cabo-Llanos into the new megaproject.

Despite that, this case study also reveals significant learning processes in the secondary stakeholders' argumentation. Here, I want to emphasize the plurality of these processes because based on past experiences, each stakeholder group has developed individual lessons learnt according to the aspects that affect them most (Hübscher 2022, p. 92):

- Inhabitants call for more affordable housing, and public spaces.
- Representatives of real estate demand to accelerate the whole process in order not to disappoint private investors.
- Architects and urbanists propose more flexible instruments that avoid the rigid corset of municipal planning.
- Political parties in the opposition wish for political consent in planning Santa Cruz Verde 2030.

The mere existence of such lessons learnt does not necessarily go hand in hand with an increased sustainability with regard to the megaproject; it rather shows each stakeholder's ambition to achieve their goals in a more efficient way. In the case of Santa Cruz Verde 2030, I even observe how "old" spatial injustices (such as those in Cabo-Llanos) are reproduced, although with different logics. In the Cabo-Llanos Plan, the lack of green spaces is simply associated with maximizing the profit.

Contrary to that, the extensive provision of green and blue infrastructure in Santa Cruz Verde 2030 is a more subtle way of revaluation, which is perceived only by a handful of interviewees, and which is framed by green gentrification (Argüelles et al. 2021). For many other interviewees (both primary and secondary), greening the city is unconditionally desirable. This is why understanding the underlying profit-driven logics poses a challenge for the already achieved learning processes for secondary stakeholders. Here, I see green gentrification as a stealthy process that contributes to a future environmental injustice (Anguelovski 2015).

In this sense, I see a strong methodological advantage in applying the stakeholder approach, as it enables to comprehend the social process behind the megaprojects (Di Maddaloni and Davis 2017). In order to overcome the limits of this case study, further research might now shift the focus to a quantitative design and prove for each stakeholder group learning processes with representative samples in different cities.

This chapter shows how stakeholders have learned from past mistakes, but still fail to integrate their perspectives into a broader understanding (Brookes et al. 2015, p. 5; Hübscher 2022, p. 93). There is an erosion of democratic understandings (Tarazona Vento 2017, p. 68), which becomes visible in the planning of Santa Cruz Verde 2030, as secondary stakeholders are not asked to participate, and primary stakeholders limit the participation to some expert groups (Swyngedouw et al. 2002, p. 579).

On that basis, I see at least two decisive questions that need to be addressed. Firstly, it remains unclear how to transfer the lessons learnt between the stakeholders and facilitate the integration into planning and practice. Secondly, despite the observed learning process between two megaprojects in this particular local setting, we need to find means to transfer this knowledge to a broader (international) audience.

This new megaproject on Tenerife opens up old wounds in the society's memory of how urban space is produced—namely as "elite playing fields" (Moulaert et al. 2001, p. 81), where only the most influential stakeholders decide. Primary stakeholders are well advised to incorporate the existing knowledge of local stakeholders, to overcome the elitist character of the megaproject and avoid reproducing spatial injustices. With the expectations that high, it will be challenging to fulfill the objective of Santa Cruz Verde 2030 as "panacea" of the city.

References

Altshuler A, Luberoff D (2003) Mega-projects: the changing politics of urban public investment. Brookings Institution Press, Washington, DC

Anguelovski I (2015) From toxic sites to parks as (green) LULUs? New challenges of inequity, privilege, gentrification, and exclusion for urban environmental justice. J Plan Lit 31(1):23–36. https://doi.org/10.1177/0885412215610491

Arencibia de Torres J (2005) Refinería de Tenerife, 1930–2005: 75 años de historia. CEPSA, Santa Cruz de Tenerife

Argüelles L, Cole H, Anguelovski I (2021) Rail-to-park transformations in 21st century modern cities: green gentrification on track. EPE: Nat Space 5(2):810–834. https://doi.org/10.1177/25148486211010064

Armas Díaz A (2016) Reestructuración urbana y producción de imagen: los espacios públicos en Santa Cruz de Tenerife. Doctoral thesis, Universidad de la Laguna

Armas-Díaz A, Sabaté-Bel F (2020) Struggles on the port of granadilla: defending the right to nature. Territ, Polit, Governance 10(2):256–276. https://doi.org/10.1080/21622671.2020.1773918

Ayuntamiento de Santa Cruz de Tenerife (2017) Modificación del PGOU-92 y adaptación básica al DL—1/2000 (PGOU-05). https://www.urbanismosantacruz.es/es/modificacion-del-pgou-92-y-adaptacion-basica-al-dl-12000-pgou-05. Accessed on 15 Feb 2022

Ayuntamiento de Santa Cruz de Tenerife, CEPSA (2018) Acuerdo de Colaboración Público-Privada para el Plan Santa Cruz Verde 2030. https://www.santacruzdetenerife.es/scverde2030/fileadmin/user_upload/web/SCverde2030/NotadePrensa26062018.pdf. Accessed on 10 Jan 2019

Ballard R, Dittgen R, Harrison P, Todes A (2017) Megaprojects and urban visions: Johannesburg's Corridor of Freedom and Modderfontein. Transform: Critical Perspect South Afr 95:111–139

Brady T, Davies A (2013) Governing complex infrastructure developments: learning from successful megaprojects. International symposium for next generation infrastructure, pp 1–8

Brookes N (2014) Mankind and mega-projects. Front Eng Manag 1(3):241–245. https://doi.org/10.15302/J-FEM-2014033

Brookes N, Locatelli G, Mikic M (2015) Learning across megaprojects. The INNOMET working group report. University of Leeds, Leeds

De Houwer J, Barnes-Holmes D, Moors A (2013) What is learning? On the nature and merits of a functional definition of learning. Psychon Bull Rev 20:631–642

Del Cerro Santamaría G (2013a) Urban megaprojects: a worldwide view. Bingley, Emerald

Del Cerro Santamaría, G. (2013b). The alleged Bilbao miracle and its discontents. In: Cerro Santamaría GD (ed) Urban megaprojects: a worldwide view. Emerald, New York, pp 27–60

Del Cerro Santamaría G (2020) Complexity and transdisciplinarity: the case of iconic urban megaprojects. Transdiscipl J Eng Sci 11:17–31. https://doi.org/10.22545/2020/0131

Diaz Orueta F, Fainstein S (2009) The new mega-projects: genesis and impacts. Int J Urban Reg Res 32:759–767. https://doi.org/10.1111/j.1468-2427.2008.00829.x

Díaz Rodríguez MdC, García Herrera LM, Armas Díaz A (2008) Puertos y espacios públicos renovados: el puerto de Santa Cruz de Tenerife. Paper presented at the XVIII Coloquio de Historia Canario-Americana (2008), Las Palmas de Gran Canaria

Di Maddaloni F, Davis K (2017) The influence of local community stakeholders in megaprojects: rethinking their inclusiveness to improve project performance. Int J Project Manage 35:1537–1556. https://doi.org/10.1016/j.ijproman.2017.08.011

Doucet B, Van Kempen R, Van Weesep J (2010) Resident perceptions of flagship waterfront regeneration: the case of the Kop Van Zuid in Rotterdam (article). Tijdschr Econ Soc Geogr 102(2):125–145. https://doi.org/10.1111/j.1467-9663.2010.00611.x

Fainstein S (2009a) Mega-projects in New York, London and Amsterdam. Int J Urban Reg Res 32(4):768–785. https://doi.org/10.1111/j.1468-2427.2008.00826.x

Fainstein S (2009b) Spatial justice and planning. Justice Spatiale/Spatial Justice 1:1–13

Ferber U, Grimski D, Millar K, Nathanail P (2006) Sustainable brownfield regeneration. University of Nottingham, Nottingham. Available at: https://issuu.com/guspin/docs/nameaa6734. Accessed on 20 Nov 2020

Flyvbjerg B (2005) Machiavellian megaprojects. Antipode 37(1):18–22

Flyvbjerg B (2014) What you should know about megaprojects, and why: an overview. Proj Manag J 45(2):6–19. https://doi.org/10.1002/pmj.21409

García Hernández JS, Armas Díaz A, Díaz Rodríguez MdC (2020) Desposesión de vivienda y turistificación en Santa Cruz de Tenerife (Canarias-España): los desahucios a inquilinos en el barrio de El Toscal. Boletín de Asociación de Geógrafos Españoles 87:1–38. https://doi.org/10.21138/bage.2982

García Herrera LM (2003) ¿De Cabañas a Palacios? Vivienda y Proceso de Elitización en el Cabo-los Llanos (Santa Cruz de Tenerife, Canarias). Scripta Nova. Revista Electrónica de Geografía y Ciencias Sociales 7(146):1–15. http://www.ub.edu/geocrit/sn/sn-146(075).htm

García Herrera LM, Smith N, Mejías Vera MÁ (2007) Gentrification, displacement, and tourism in Santa Cruz de Tenerife. Urban Geogr 28(3):276–298. https://doi.org/10.2747/0272-3638.28.3.276

Gellert P, Lynch B (2003) Mega-projects as displacements. Int Soc Sci J 55(175):15–25. https://doi.org/10.1111/1468-2451.5501002

González Chávez CM (2018) Nuevas propuestas arquitectónicas y de equipamiento urbano en el siglo XXI. El futuro de Cabo-Llanos en Santa Cruz de Tenerife (Canarias). Arte y Ciudad - Revista De Investigación 14:33–64

Haase A (2020) Covid-19 as a social crisis and justice challenge for cities. Front Sociol 5(583638):1–7. https://doi.org/10.3389/fsoc.2020.583638

Harvey D (1989) From managerialism to entrepreneurialism: the transformation in urban governance in late capitalism. Geografiska Annaler. Series B, Hum Geogr 71(1):3–17

Heng Li G (2011) Editorial. International journal of project management special issue on "Complexities in managing mega construction projects". Int J Proj Manag 29(7):795–796

Hersh B (2017) Megaprojects. In: Hersh B (ed) Urban redevelopment. Routledge, London, pp 136–148. https://doi.org/10.4324/9781315767314

Honey-Rosés J, Anguelovski I, Bohigas J, Chireh V, Daher C, Konijnendijk C et al (2020) The impact of COVID-19 on public space: a review of the emerging questions. Available at: https://doi.org/10.31219/osf.io/rf7xa. Accessed on 20 Dec 2020

Hübscher M (2019) From megaprojects to tourism gentrification? The case of Santa Cruz Verde 2030 (Canary Islands, Spain). Boletín de la Asociación de Geógrafos Españoles 83(2813):1–47. https://doi.org/10.21138/bage.2813

Hübscher M (2020) Zwischen Deindustrialisierung und Gentrification: Innerstädtische Großprojekte in Santa Cruz de Tenerife (Spanien). Mitteilungen Der Österreichischen Geographischen Gesellschaft 161:63–90. https://doi.org/10.1553/moegg161s63

Hübscher M (2021) Megaprojects, gentrification, and tourism: a systematic review on intertwined phenomena. Sustainability 13(22):1–22

Hübscher M (2022) One city—two megaprojects. Learning Processes between Large-scale urban development Project son Tenerife, Spain. In: Gallegas Reina A (ed) Libro de trabajos aportados al XVI Coloquio de Geografía Urbana. Universidad de Malaga, Malaga, pp 86–95. https://geourbana.age-geografia.es/wp-content/uploads/2022/09/XVIColoquioGeografiaUrbanaMalaga_LibroActas.pdf

ISTAC Instituto Canario de Estadística (2022) Población. Municipios por islas de Canarias y años. http://www.gobiernodecanarias.org/istac/estadisticas/demografia/. Accessed on 02 Feb 2022

Kennedy L (2014) The politics and changing paradigm of megaproject development in metropolitan cities. Habitat Int 45:163–168

Kronenberg J, Andersson E, Barton D, Borgström S (2021) The thorny path toward greening: unintended consequences, trade-offs, and constraints in green and blue infrastructure planning, implementation, and management. Ecol Soc 2(36):1–23. https://doi.org/10.5751/ES-12445-260236

Lehrer U, Laidley J (2008) Old mega-projects newly packaged? Waterfront redevelopment in Toronto. Int J Urban Reg Res 32:786–803. https://doi.org/10.1111/j.1468-2427.2008.00830.x

Metzger P (2020) Die Finanzialisierung der deutschen Ökonomie am Beispiel des Wohnungsmarktes. Westfälisches Dampfboot, Münster

Millet D (2022) El Gobierno local allana el camino a la «mayor operación urbanística del país». El Día. Retrieved from https://www.eldia.es/santa-cruz-de-tenerife/2022/01/31/gobierno-local-allana-camino-mayor-62129684.html?pimec-source=www.eldia.es&pimec-widget=1&pimec-config=2&pimec-mod=0&pimec-pos=2

Moulaert F, Swyngedouw E, Rodriguez A (2001) Large scale urban development projects and local governance: from democratic urban planning to besieged local governance. Geogr Z 89(2):71–84

Open Street Map, Geofabrik GmbH (2022) OpenStreetMap data extracts. https://download.geofabrik.de. Accessed on 19 July 2022

Pérez V (2019) Con las 800 familias expulsadas de Cabo Llanos se incumplió realojarlas en el centro. Planeta Canario. Retrieved from https://planetacanario.com/con-las-800-familias-expulsadas-de-cabo-llanos-se-incumplio-realojarlas-en-el-centro/

Purcell M (2002) Excavating Lefebvre: the right to the city and its urban politics of the inhabitant. GeoJournal 58:99–108

Rädiker S, Kuckartz U (2020) Focused analysis of qualitative interviews with MAXQDA. Step by step. MAXQDA Press, Berlin

Sabaté Bel F, Armas Díaz A (2022) Commodification or the right to the island: the struggle against the construction of a hotel in La Tejita (Tenerife). Island Stud J 17(2):214-234. https://doi.org/10.24043/isj.386

Shatkin G (2008) The city and the bottom line: urban megaprojects and the privatization of planning in Southeast Asia. Environ Plan A 40:383–401. https://doi.org/10.1068/a38439

Stoddart-Stones R (1988) Development of project management systems for major projects. Int J Project Manage 6(1):34–38. https://doi.org/10.1016/0263-7863(88)90056-7

Swyngedouw E, Moulaert F, Rodriguez A (2002) Neoliberal urbanization in Europe: large-scale urban development projects and the new urban policy. Antipode 34(3):542–577. https://doi.org/10.1111/1467-8330.00254

Tarazona Vento A (2017) Mega-project meltdown: post-politics, neoliberal urban regeneration and Valencia's fiscal crisis. Urban Studies 54(1):68–84

Torres N (2021) Los vecinos de Cabo Llanos denuncian que llevan 20 años sufriendo el olvido municipal. Diario de Avisos. Retrieved from https://diariodeavisos.elespanol.com/2021/05/los-vecinos-de-cabo-llanos-denuncian-que-llevan-20-anos-sufriendo-el-olvido-municipal/

Valles M (2007) Entrevistas cualitativas. CIS Centro de Investigaciones Sociológicas, Madrid

Vargas N (2022). Del puerto de Fonsalía al hotel de La Tejita, los macroproyectos de lujo paralizados en Tenerife por la movilización ciudadana. El Diario. Retrieved from https://www.eldiario.es/canariasahora/ciencia_y_medio_ambiente/puerto-fonsalia-hotel-tejita-macroproyectos-lujo-paralizados-tenerife-movilizacion-ciudadana_1_9073013.html

Warrack A (1993) Megaproject decision making: lessons and strategies. West Cent Econ Research—Information Bull 16:1–14. https://era.library.ualberta.ca/items/cda92a3d-1b92-4572-91eb-878443ee78ec/view/c73c7a57-59c6-4b99-a376-37c2882d77ef/16.pdf

Chapter 13
Touristification Process in Seaside Destination Inland: The Case of Mallorca Island, Spain

Miquel Àngel Coll-Ramis, Fernando Almeida-García, Rafael Cortés-Macías, and Seyedasaad Hosseini

Abstract This study contributes to analyse the impacts due to touristification in rural and inland areas in the context of touristic intensification. The research of touristification in rural areas is a little-studied field of analysis. The intensification of the touristification impacts in the seaside destination surrounding needs a specific study. Additionally, it serves as comparison material for the research on local community views towards tourism's effects in tourist destinations of Spain. Hence, it can provide valuable insights by contributing to our understanding of touristification. In order to be able to analyse the impact of houses for tourist accommodations on the processes of touristification of inland areas, this analysis focuses on the case of the island of Majorca (Balearic Islands, Spain). The study of touristification in rural areas is a recent phenomenon and, in the case under study, must be understood as part of the expansion of mature coastal destinations. This expansion towards rural areas is a response to the need to find complementary products to those of sun and sand. The results obtained highlight that the process of spreading tourist housing in the rural areas of Mallorca has caused a strong growth in the price of general housing and has contributed decisively to tourism in many small towns in the interior of the island.

Keywords Touristification · Rural Spaces · Tourist Accommodation · Impact of Housing for Tourism · Mallorca

---

M. À. Coll-Ramis
Department of Geography, University of the Balearic Island, Palma, Spain
e-mail: miquelangel.coll@uib.es

F. Almeida-García (✉) · R. Cortés-Macías · S. Hosseini
Department of Geography, University of Málaga, Málaga, Spain
e-mail: falmeida@uma.es

R. Cortés-Macías
e-mail: rcortes@uma.es

E. Navarro-Jurado et al. (eds.), *Urban Dynamics in the Post-pandemic Period*,
The Urban Book Series,
https://doi.org/10.1007/978-3-031-36017-6_13

13.1 Introduction

In general terms, touristification is the transformation of an ordinary area into another touristic (Jover and Díaz-Parra 2019; Leite 2008; Rio Fernandes 2011). Touristification is a procedure in which a destination's decisions are made with the intention of intensifying tourism activity (Rio Fernandes 2011). Nonetheless, this transformation caused by touristification will have effects on local communities. The touristification of certain areas of destinations usually leads to various impacts perceived as negative by residents (De la Calle 2019). One of the most recognised elements in the touristification process of an area or city is gentrification, especially in areas with intense tourist flows and cultural attractions (Seng 2005).

In recent decades, backcountry areas have undergone economic restructuring, particularly in Spain, as a result of market economy internationalisation, which has led to a decline in income from traditional rural economies based on agriculture. The tourism industry in rural areas has been intensively developed under a logic capable of reducing the economic backwardness suffered by these geographical areas, which are frequently geographically isolated (Öter 2010). Thus, tourism has been considered by several authors as a viable treatment for the falling economics of rural areas and associated rural migration, as well as a vital component in the regional development of rural areas (Cawley and Gilmor 2008; Freytag and Bauder 2018). The subsequent intensification of tourism and touristification that occurs in some cases has played a significant role in reinventing and revitalising deteriorating rural villages around the world (Mansilla and Milano 2019).

The tourism expansion in rural and inland areas little attention has been paid (Milano 2013; Lorenzen 2021; Brança and Haller 2021), where the phenomenon is more diffuse and less perceptible than in urban areas. Tourism must be viewed as a multifaceted and worldwide phenomenon that can have social, cultural, and environmental repercussions (Hall 2019). Tourism provides certain positive effects such as job creation, the protection of some cultural and environmental resources, a greater cultural offer, and the revitalisation of economic activity in general (Andereck et al. 2005). Therefore, despite the beneficial characteristics of tourism, this expansion is also associated with a number of negative side consequences. A small rural population is more likely to be overwhelmed by tourists, and preventing over-tourism in those areas shall be challenging (Butler 2019). Typically, rural areas have limited access to resources, transportation, and infrastructure, which are affected by a deep seasonality. Many rural areas are more fragile than urban areas in terms of tourism impacts (Boháč and Drápela 2022). A large number of visitors can cause a rural area's residential space to become a tourist attraction, creating disturbances to the residents (Herrera et al. 2007). In many rural areas, tourism increases housing prices, replaces residential housing with visitor accommodations (such as hotels and apartments), supplants local stores with tourist businesses, and disrupts the daily lives and social networks of long-term residents (Cócola-Gant 2018). Touristification can generate negative consequences, such as changes in the economic, political, and social composition (Lorenzen 2021).

Mallorca is not only a world-renowned seaside destination, but it also has a rural interior space that is growing intensely. The insular tourist spaces register in a more intense way the transformations due to tourism. The touristification of the rural spaces of the island of Mallorca heralds the future transformations that the intensely touristic spaces of the Mediterranean or the Caribbean will experience. In order to know the process of touristification of the case study, we carry out a review of the literature, as well as a documentary analysis around touristification and the evolution of tourist housing in rural areas of Mallorca. Likewise, we elaborate maps related to the topics studied. Observation and detailed description of the rural areas yielded insights that allowed for a debate on this rural issue and the design of solutions. This leads to the research's main question: What is the impact of housing for tourism process in the rural areas of Mallorca due to tourist pressure?

This study contributes to analyse the impacts due to touristification in rural areas and peripheries of mature tourism spaces in the context of deeply touristic island. The study of touristification in rural areas is a little-studied field of analysis. The intensification of the touristification impacts in the seaside destination surrounding needs a specific study. Additionally, it serves as comparison material for the research on local community views towards tourism's effects in tourist destinations of Spain. Hence, it can provide valuable insights by contributing to our understanding of touristification.

13.2 Literature Review

In recent years, studies on touristification have focused on urban tourist destinations, especially in city centres (Cócola-Grant 2018). However, the impact of temporary tourist rentals and collaborative platforms has changed production and consumption patterns in the tourist accommodation sector (Ortuño and Jiménez 2019); at the same time, they have multiplied the possibilities for tourists to visit other territories and access a wide range of experiences, boosting the number of tourists in spaces that had traditionally remained on the fringes of major tourist flows (Parralejo and Diaz-Parra 2021). Consequently, in a post-productivism context, many rural areas have been enhanced by tourism in such a way that they are experiencing a socio-economic restructuring based on the promotion of tourism activity (Cócola-Grant 2018), which is leading to an increase in the arrival of tourists to these areas.

A combination of several factors has contributed to increased tourist inflows to rural areas, including improvements in accessibility, increased promotion at both local and national levels, and increased visibility in social media and other forms of communication, have dramatically changed the level and nature of tourism in rural areas in a relatively short time (Butler 2020). Furthermore, COVID-19 has boosted the interest of tourism demand for rural areas (Zoğal et al. 2022; Miguel et al. 2022), as rural areas are perceived as quiet spaces with an important weight of natural resources, where travellers can escape from overcrowding and be in contact with authentic spaces that have not been transformed by tourism; these areas are perceived as healthier and safer spaces than overcrowded spaces such as coastal or urban destinations.

However, despite the intensification of tourism in rural areas, most studies have focused on the potential of tourism for the revitalisation of rural areas. On the other hand, there are still few publications focusing on the impacts of rural tourism (Blázquez 2013; Drápela 2021; Milano et al. 2019; Boháč and Drápela 2022). In this sense, it is worth highlighting that many rural areas are highly sensitive natural and cultural spaces that can be significantly altered by the arrival of visitors and show different signs of touristification than large mass destinations. In such areas, a small number of tourists can generate significant impacts due to their low carrying capacity and the small size of local communities, making it urgent to advance in the study and planning of measures aimed at avoiding the touristification of rural areas (Butler 2020).

Touristification involves the transformation of a territory into a place of tourist consumption and the changes that this entails in the landscape, the environment, and the social, cultural, and economic dynamics (Jover and Díaz-Parra 2019; Ojeda and Kieffer 2020). One of the main effects of touristification is the expansion of consumer activities aimed at tourists, highlighting those related to tourist accommodation (De la Calle 2019). In rural areas, one of the main accommodation modalities is constituted by single-family houses or dwellings, a fact that differentiates rural areas from other tourist areas; likewise, hotels are not a very abundant offer and are usually small in size. Therefore, the most characteristic feature of the touristification of rural areas is the expansion of dwellings for tourist accommodation (Hof and Blázquez 2013; Blanco et al. 2019), a fact that has intensified in recent years due to the emergence of marketing platforms such as Airbnb that have facilitated the expansion of their offer in rural areas (Cors-Iglesias et al. 2020). In many cases, this process has been made possible by the existence of very permissive urban planning and tourism regulations regarding the commercialisation of housing for tourist use, which has led to great urban and territorial disorder in rural areas (Yus and Delgado 2010; Müller et al. 2021).

The develop of tourist infrastructures and tourist housing that occupy traditional dwellings, farms or that have been built on plots of cultivated or abandoned land and in areas of landscape value, has given rise to what some authors have called the gentrification of rural space (Smith and Phillips 2001; Stockdale 2010; Clark and Pissin 2020). This process is primarily reflected in the change of land use and the replacement of former residents. In this sense, the search for potential rents in the rural space through the commodification of tourist housing has resulted in a series of both positive and negative impacts. On the one hand, it allows owners to obtain a revaluation of their properties. It has also created several economic and employment opportunities in rural areas through construction, housing maintenance (gardening, cleaning, etc.), commerce, and a wide range of services. On the other hand, however, it has led to price rises that make it difficult for local people to access housing, forcing them to move elsewhere (Lorenzen 2021).

Therefore, this scenario makes it necessary to rethink tourism development policies in rural areas, which have focused primarily on promoting projects for the creation of tourism facilities and services as a strategic factor for economic development, but rarely have control measures been considered to prevent touristification and the negative impacts that tourism can generate in these areas (Butler 2020).

13.3 Case Study and Results

In order to be able to analyse the impact of houses for tourist accommodations on the processes of touristification of rural areas, this analysis focuses on the case of the island of Majorca (Balearic Islands, Spain). It is a mature tourist destination, with a high summer seasonality, given its specialisation in sun and beach tourism. The island received 8 million tourists staying in hotels alone and offered 266,391 beds in 2019 (INE 2019). This productive particularity has led to an intense concentration of tourist accommodation supply in the centres located on the coast (Pons et al. 2014). As coastal areas enter a cycle of maturity and show signs of exhaustion and reduced profitability, destinations have been forced to improve the main product, diversify it, or create it anew in adjacent areas (Coll 2016). In this way, new tourism products and areas have been emerging in inland areas of the island and contributing to the diversification of mature coastal destinations.

Murray et al. (2017) point out that the recent economic crises have been accompanied by political responses that have favoured the incorporation of new tourist areas such as rural areas in Mallorca. To this end, new accommodation formulas have been created, with the massive tourist exploitation of dwellings being the one that has contributed most to the expansion of tourist activity. As shown in Fig. 13.1, Mallorca has been experiencing an increase in the number of tourist bed places associated with a growing supply of tourism homes. In Mallorca, as other destinations, the number of tourism homes in rural areas has grown every year since the economic crisis of 2008, and especially since the emergence of peer-to-peer platforms in 2010 (Sánchez-Aguilera and Font 2016; Cors-Iglesias et al. 2020).

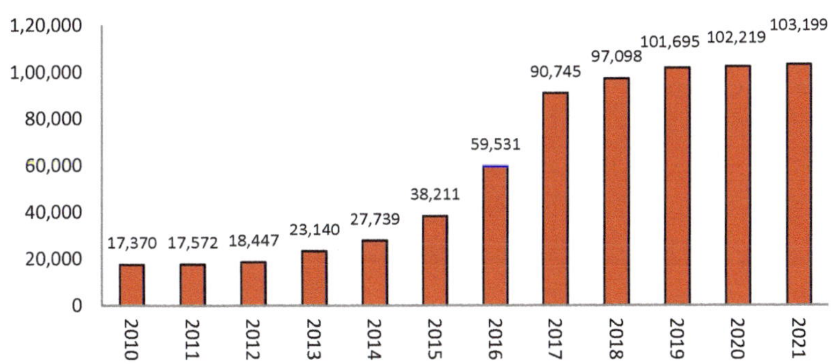

Fig. 13.1 Evolution of bed places in houses for tourist accommodations in Mallorca Island (*Source* Register of companies, activities and tourist establishments, Consell Insular de Mallorca)

The magnitude that the phenomenon of tourism homes has reached in Mallorca has sparked the interest of researchers who have analysed their socio-territorial impacts in Mallorca (Benítez-Aurioles 2020) or in certain towns such as Palma (Vives and Rullán 2017; Yrigoy 2019) or Pollença (Mateu 2003; Müller et al. 2021), where the proliferation of tourisim housing has been very intense. The case study of Mallorca helps to reflect the processes of touristification associated with the productive evolution of mature sun and beach destinations, which have seen in rural environments an opportunity for tourism expansion through the tourism homes business. In the current post-COVID-19 scenario, Murray and Cañada (2021) suggest that, in mass tourism destinations, there may be a deepening of the tourist commodification of housing in rural areas.

In this context, the tourist commodification of housing has been identified as a relevant factor in the processes of territorial touristification (Mansilla and Milano 2018). The tourism homes have radically changed the processes of generating space for tourist accommodation (De la Calle 2019), in such a way that it has gone from a coastal model where the hotel establishment plays a prominent role, to an accelerated process of continuous expansion of the accommodation offer that tends to occupy the entire interior of Mallorca (Fig. 13.2). This process can be seen more clearly in those territories whose economy is heavily dependent on tourism, as is the case of Mallorca, which forces it to constantly search for new productive spaces that satisfy a greater diversity of demand (Benur and Bramwell 2015).

If we analyse the number of tourist places in absolute figures, most of the supply in Mallorca is concentrated in the coastal municipalities, which have the tourist attraction of beaches, and which saw how, from the second half of the twentieth century, their coastline was subject to intense tourist urbanisation in order to meet the strong demand for mass sun and beach tourism. However, since the 1990s, this trend has been changing as a result of territorial and tourism policies to adapt from

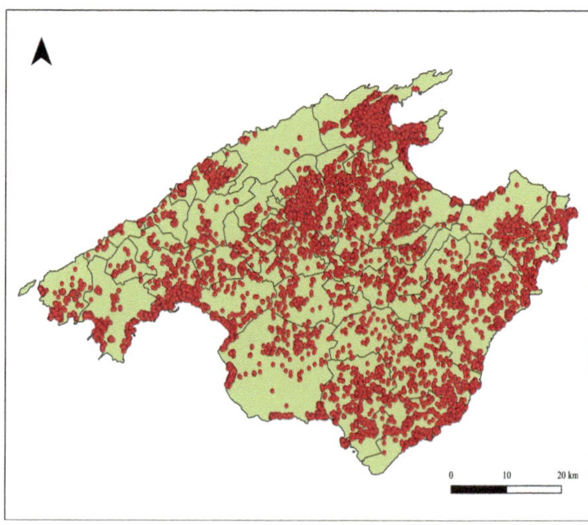

Fig. 13.2 Distribution of houses for tourist accommodations in Mallorca island (2021) (*Source* Register of companies, activities and tourist establishments, Consell Insular de Mallorca)

the Fordist cycle to a new post-Fordist model from 1990 onwards. This change can be seen from the analysis of the percentage growth in the supply of tourist places between 1990 and 2021 (Fig. 13.3), which clearly shows how the greatest relative growth has been concentrated in inland municipalities, which have developed post-Fordist formulas that value attractions such as tranquility, rurality, gastronomy, heritage, and culture, which attract a tourist profile that seeks to disconnect, relax, and visit new and less crowded places.

The development of these rural areas has been accompanied by the creation of a significant supply of accommodation, especially in the form of tourist accommodation (Table 13.1). In the inland municipalities of the island, this type of accommodation accounts for 83% of the total number of places on offer, a figure well above the 25% for the island as a whole.

In the process of growth of the supply, two stages must be distinguished, marked by the approval of the zoning provided for in the Law for the Tourist Commercialisation of Tourism Second Homes, approved by the Regional Government of the Balearic Islands in 2017. Prior to this date, there was a large increase in the number of tourist

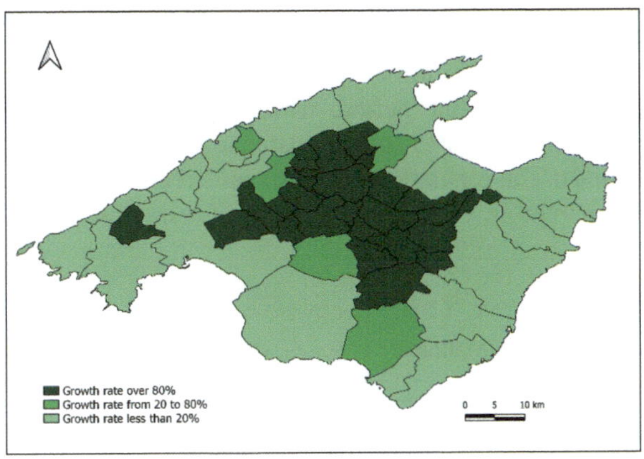

Fig. 13.3 Percentage growth of tourist bed places by municipality in Mallorca (1990–2021) (*Source* Register of companies, activities and tourist establishments, Consell Insular de Mallorca)

Table 13.1 Percentage of bed places in homes for tourist accommodations over the total market supply

	Total Mallorca	Municipalities of inland		Total Mallorca	Municipalities of inland
2010	6%	43%	2016	17%	79%
2011	6%	44%	2017	23%	83%
2012	6%	56%	2018	24%	83%
2013	7%	57%	2019	25%	83%
2014	9%	61%	2020	25%	83%
2015	12%	69%	2021	25%	83%

Source Register of companies, activities and tourist establishments, Consell Insular de Mallorca

accommodation places due to the rapid return on investment and the relative ease with which this type of accommodation could be registered. However, with the entry into force of the new law, the purchase of the rights to existing tourist accommodation places is compulsory in order to be able to register a dwelling for tourist use, and the areas of the island in which this type of tourist home accommodation is permitted are also restricted. Consequently, as can be seen in Table 13.1, from 2017 onwards, the growth in the number of tourism houses places has been contained.

The irruption of the tourism homes has generated changes in the socio-territorial dynamics of tourism on the island, demonstrating its high capacity for touristification due to its ability to settle the territory. Consequently, it has become an important source of income for rural communities that have progressively abandoned traditional agricultural activities (Cànoves et al. 2012; Coll et al. 2016). However, as in other territories (Gascón and Cañada 2016; Blanco et al. 2019; Moreno et al. 2021), the process of touristification has been promoted in an uncritical and poorly regulated way, which has generated negative impacts such as landscape transformation, water consumption, waste production, intensification of private vehicles, real estate speculation, gentrification, among others. The growing tourist pressure experienced by the population of inland municipalities (Table 13.2), and the negative effects it entails, is generating a growing social unrest which is manifested in a resistance to the touristification of their territory. This social debate has been reactivated with the post-pandemic period, which has led to the approval of Decree Law 3/2022, on urgent measures for the sustainability and circularity of tourism in the Balearic Islands. The regulation prohibits the creation of new tourist places for the next four years to improve the quality of life of residents and the tourist experience of visitors.

Table 13.2 Rate of houses for tourist accommodations tourist bed places per 100 inhabitants

	Total Mallorca	Municipalities of inland		Total Mallorca	Municipalities of inland
2010	1.9	1.3	2016	10.4	19.0
2011	2.0	1.5	2017	11.0	19.1
2012	2.1	2.5	2018	11.3	19.0
2013	2.7	3.0	2019	11.2	18.8
2014	3.2	3.8	2020	11.3	18.6
2015	4.4	6.5	2021	10.4	19.0

Source Register of companies, activities and tourist establishments, Consell Insular de Mallorca and Institute of Statistical of Balearic Islands

13.4 Discussion and Conclusions

Touristification and the processes related to tourist intensification have traditionally been analysed in urban areas (De la Calle 2019). The study of touristification in rural areas is a recent phenomenon and, in the case under study, must be understood as part of the expansion of mature coastal destinations. This expansion towards rural areas is a response to the need to find complementary products to those of sun and sand and is also a strategy for tourism financialisation (Morell 2020). Rural areas show themselves as a great business opportunity for investment funds.

What has been described in the case of Mallorca should be analysed as a forerunner of what is going to happen with rural areas close to mature tourist destinations, specialising in sun and beach offers, and especially in the tourist islands of the Mediterranean and the Caribbean Sea (Briguglio and Avelino 2019). Until a few years ago, the difficulty of access to the atomised supply of rural housing and the maintenance of agricultural activity limited the expansion of the supply of tourist housing in rural areas. Now that the problem of marketing in rural areas has been solved, tourist housing in rural areas has expanded rapidly, just as it did in urban destinations. Moreover, economic agents find a higher percentage of profit than in traditional coastal destinations (Sidali et al. 2015).

In relation to the initial research question, referring the impact of housing for tourism process in the rural areas of Mallorca, we can highlight that tourist housing has caused a strong growth in the price of general housing and has decisively contributed to the tourism of many small towns in the interior of the island.

Local communities in rural areas receive some positive impacts, especially in the case of areas that are in a process of economic stagnation (Andereck 2005). In such cases, tourism provides a source of employment, a way of maintaining the population. But also, as mentioned above, tourism has negative effects, generating serious problems related to access to housing and cultural changes that do not favour local communities (Cócola-Gant 2018; Gascón and Cañada 2016; Lorenzen 2021). Attitudes that seemed to be exclusive to urban destinations, such as tourism-phobia, are beginning to develop. In some rural destinations, the behaviour of tourists who complain about the customs or activities of the local community provokes irritation in the community "they want rural tourism, but they want to turn the villages into cities" (La Información 2022).

Touristification and its effects must be considered by policy of the local, regional, and national governments, especially those aimed at access to housing, territorial planning, and the protection of historical and ethnological heritage. Although tourism opens up other economic initiatives, it also hinders the development of traditional economic activities, both in the primary and secondary sectors. For this reason, aid policies for rural areas must ensure economic diversification (Esparcia, 2014). The economic stagnation of many rural and peripheral areas reduces their capacity to organise and react to tourism expansion.

Acknowledgements The research was funded by the projects: "New strategies for new tourist conflicts in Andalusian historic centers" (UMA20-FEDERJA-005), FEDER Andalusian Operational Program 2014-2020, Spain and "Residents versus tourists in Andalusian historic centres? Conflicts, strategies and new post-Covid scenarios" (P20_01198), Andalusian Plan for Research, Development and Innovation (PAIDI 2020), Spain.

References

Andereck KL, Valentine KM, Knopf RC, Vogt CA (2005) Residents' perceptions of community tourism impacts. Ann Tour Res 32(4):1056–1076. https://doi.org/10.1016/j.annals.2005.03.001

Benítez-Aurioles B (2020) Impacts of the peer-to-peer market on tourist accommodation on the Balearic Islands of Mallorca and Menorca. Isl Stud J 15(2):353–370. https://doi.org/10.24043/isj.108

Benur AM, Bramwell B (2015) Tourism product development and product diversification in destinations. Tour Manage 50:213–224. https://doi.org/10.1016/j.tourman.2015.02.005

Blanco-Romero A, Blázquez-Salom M, Morell M, Fletcher R (2019) Not tourism-phobia but urban-philia: understanding stakeholders' perceptions of urban touristification. Boletín de la Asociación de Geógrafos Españoles 83(3). https://doi.org/10.21138/bage.2834

Blázquez Salom M (2013) More villas and more barriers: gentrification and the enclosure of rural land on Majorca. Méditerranée. Revue Géographique Des Pays Méditerranéens/journal of Mediterranean Geography 120:25–36. https://doi.org/10.4000/mediterranee.6638

Branca D, Haller A (2021) Urbanization, touristification and verticality in the andes: a Profile of Huaraz. Peru. Sustainability 13(11):6438. https://doi.org/10.3390/su13116438

Briguglio L, Avellino M (2019) Has overtourism reached the Maltese Islands? University of Malta, Occasional papers on Islands and small states

Boháč A, Drápela E (2022) Overtourism Hotspots: both a threat and opportunity for rural tourism. Eur CtrySide 14(1):157–179

Butler RW (2019) Tourism carrying capacity research: a perspective article. Tour Rev 75(1):207–211. https://doi.org/10.1108/TR-05-2019-0194

Butler RW (2020) Overtourism in rural areas. In Overtourism (pp. 27–43). Palgrave Macmillan, Cham.

Cànoves G, Garay L, Duro JA (2012) Turismo rural en España: avances y retrocesos en los últimos veinte años. Papers de Turisme (51):7–21.

Cawley M, Gillmor DA (2008) Integrated rural tourism: concepts and practice. Ann Tour Res 35(2):316–337. https://doi.org/10.1016/j.annals.2007.07.011

Clark E, Pissin A (2020) Potential rents vs. potential lives. Environment and Planning A: Economy and Space 0(0). https://doi.org/10.1177/0308518X20971308

Cócola-Gant A (2018) Tourism gentrification. Edward Elgar Publishing, In Handbook of gentrification studies

Coll Ramis MA, Obrador Izara M, Seguí Llinàs M (2016) Consideraciones socio-territoriales en torno al negocio del alquiler de viviendas de uso turístico. El caso del municipio de Lloret de Vistalegre (Mallorca). J Tour Anal 22:28–38

Coll Ramis MA (2016) Análisis socio-espacial de la estacionalidad turística en Mallorca. [Tesis de doctorado]. Universidad de las Illes Balears.

Consell Insular de Mallorca (2021) Registro Oficial de empresas, actividades y establecimientos turísticos. https://www.caib.es/cathosfront/cens

Cors-Iglesias M, Gómez-Martín MB, Armesto-López XA (2020) Peer-to-peer accommodation in rural areas of Catalonia: defining typologies of rural municipalities. Sustainability 12(15):6145. https://doi.org/10.3390/su12156145

De la Calle Vaquero M (2019) Turistificación de centros urbanos: clarificando el debate. Boletín de la Asociación de Geógrafos Españoles 83:2. https://doi.org/10.21138/bage.2829

Drápela E (2021) Prevention of damage to sandstone rocks in protected areas of nature in northern Bohemia. AIMS Geosciences 7(1):56–73. https://doi.org/10.3934/geosci.2021003

Esparcia J (2014) Innovation and networks in rural areas. an analysis from European innovative projects. J Rural Stud 34:1–14. https://doi.org/10.1016/j.jrurstud.2013.12.004

Freytag T, Bauder M (2018) Bottom-up touristification and urban transformations in Paris. Tour Geogr 20(3):443–460. https://doi.org/10.1080/14616688.2018.1454504

Gascón J, Cañada E (2016) Turismo Residencial y Gentrificación Rural. PASOS, Revista de Turismo y Patrimonio Cultural, p 16.

Hall CM (2019) Constructing sustainable tourism development: the 2030 agenda and the managerial ecology of sustainable tourism. J Sustain Tour 27(7):1044–1060. https://doi.org/10.1080/09669582.2018.1560456

Herrera LMG, Smith N, Vera MÁM (2007) Gentrification, displacement, and tourism in Santa Cruz de Tenerife. Urban Geogr 28(3):276–298. https://doi.org/10.2747/0272-3638.28.3.276

Hof A, Blázquez-Salom M (2013) The linkages between real estate tourism and urban sprawl in Majorca (Balearic Islands, Spain). Land 2(2):252–277. https://doi.org/10.3390/land2020252

Instituto Nacional de Estadística (INE) (2019) Encuesta de Ocupación Hotelera, 2019, INE, Ministerio de Asuntos Económicos, Gobierno de España.

Jover J, Díaz-Parra I (2019) Gentrification, transnational gentrification and touristification in Seville. Spain. Urban Stud. 57(15):3044–3059. https://doi.org/10.1177/0042098019857585

La Información (2022) “La afluencia masiva de visitantes a los pueblos genera ya “turismofobia” https://www.lainformacion.com/economia-negocios-y-finanzas/turismofobia-mal-sufren-pueblos-verano/2872781/. Accessed on 29 Aug 2022

Leite NK (2008) Turismo e território: um estudo sobre a turistificação de Portimão (Algarve/Portugal) a partir da geografia do turismo (Doctoral dissertation).

Lorenzen M (2021) Rural gentrification, touristification, and displacement: Analysing evidence from Mexico. J Rural Stud 86:62–75. https://doi.org/10.1016/j.jrurstud.2021.05.015

Mansilla JA, Milano C (2018) Introducción a la ciudad de vacaciones. Apuntes sobre turismo y malestar social en Barcelona. In Ciudad de vacaciones: conflictos urbanos en espacios turísticos (pp. 19–80).

Mansilla JA, Milano C (2019) Becoming centre: tourism placemaking and space production in two neighborhoods in Barcelona. Tour Geogr 24. https://doi.org/10.1080/14616688.2019.1571097

Miguel C, Pechurina A, Kirkulak-Uludag B, Drotarova MH, Dumančić K, Braje IN, Giglio C (2022) Short-term rental market crisis management during the COVID-19 pandemic: stakeholders' perspectives. Int J Hosp Manag 102:103147. https://doi.org/10.1016/j.ijhm.2022.103147

Morell M (2020) Housing financialization in Majorcan holiday rentals. In Hann C, & Kalb D (eds) Financialization: relational Approaches, pp 302–320

Mateu Lladó J (2003) Turismo residencial y dispersión urbana en Mallorca (Illes Balears). Un ensayo metodológico en el municipio de Pollensa. Estudios Turísticos 197–218.

Milano C (2013) De espacio rural a espacio turístico: Una etnografía del desarrollo en el Delta de Parnaíba (Brasil). Quaderns-e De L'institut Català D'antropologia 18(1):128–143

Milano C, Cheer JM, Novelli M (2019) Overtourism: excesses, discontents and measures in travel and tourism. CABI, Wallingford

Moreno IL, Zarza MF, Pozo JIH (2021) Los conflictos de la turistificación y su estudio basado en el enfoque centrado en el actor. Revista Confluências Culturais 10(3):75–91

Müller N, Murray I, Blázquez-Salom M (2021) Short-term rentals and the rentier growth coalition in Pollença (Majorca). Environ Plan A: Econ Space 53(7):1609–1629. https://doi.org/10.1177/0308518X211033307

Murray I, Yrigoy I, Blazquez-Salom M (2017) The role of crises in the production, destruction and restructuring of tourist spaces: the case of the Balearic Islands. Investigaciones Turísticas 13:1–29. https://doi.org/10.14198/INTURI2017.13.01

Murray I, Cañada E, (2021) Turistificación confinada. In Cañada E, & Murray I (eds) #Tourism-PostCOVID19: Turistificación confinada. Alba Sud, Barcelona

Ojeda A, Kieffer M (2020) Touristification. Empty concept or element of analysis in tourism geography? Geoforum 115:143–145. https://doi.org/10.1016/j.geoforum.2020.06.021

Ortuño A, Jiménez JL (2019) Las viviendas turísticas ofertadas por plataformas on-line: estado de la cuestión (No. 2019-04). FEDEA, Madrid.

Öter Z (2010) Touristification of cultural heritage, identities tested by marketing of handcrafts to tourists in Turkey: the case of Görece. Eur J Tour Res 3(1):77–79

Parralejo JJ, Díaz-Parra I (2021) Gentrification and touristification in the central urban areas of Seville and Cádiz. Urban Science 5(2):40. https://doi.org/10.3390/urbansci5020040

Pons Esteva A, Rullán Salamanca O, Murray Mas I (2014) Tourism capitalism and island urbanization: tourist accommodation diffusion in the Balearics, 1936–2010. Isl Stud J 9(2):239–258

Ramos RY, Delgado MAT (2010) Urbanismo difuso en suelo rústico. Deterioro ambiental y corrupción en la provincia de Málaga (El caso de la Axarquía). Sociedad: Boletín de la Sociedad de Amigos de la Cultura de Vélez-Málaga (9):114–115

Rio Fernandes JA. (2011) Area-based initiatives and urban dynamics. The case of the Porto city centre. Urban Res & Pract 4(3):285–307. https://doi.org/10.1080/17535069.2011.616747

Sánchez-Aguilera D, Font Garolera J (2016) Nuevas formas de alojamiento en el medio rural: la difusión de la economía colaborativa en tres destinos de la montaña española. In *Turismo y crisis, turismo colaborativo y ecoturismo* (pp. 412–413). Societat d'Història Natural de les Balears, Palma, Spain.

Seng OC (2005) State-civil society relations and tourism: singaporeanizing tourists, touristifying Singapore. Sojourn: J Soc Issues Southeast Asia 20(2):249–272

Sidali KL, Kastenholz E, Bianchi R (2015) Food tourism, niche markets and products in rural tourism: combining the intimacy model and the experience economy as a rural development strategy. J Sustain Tour 23(8–9):1179–1197. https://doi.org/10.1080/09669582.2013.836210

Smith D, Phillips D (2001) Socio-cultural representations of gentrified Pennine rurality. J Rural Stud 17(4):457–469. https://doi.org/10.1016/S0743-0167(01)00014-6

Stockdale A (2010) The diverse geographies of rural gentrification in Scotland. J Rural Stud 26:31–40. https://doi.org/10.1016/j.jrurstud.2009.04.001

Vives-Miró S, Rullan O (2017) Desposesión de vivienda por turistización?: Revalorización y desplazamientos en el Centro Histórico de Palma (Mallorca). Revista De Geografía Norte Grande 67:53–71. https://doi.org/10.4067/S0718-34022017000200004

Yrigoy I (2019) Rent gap reloaded: airbnb and the shift from residential to touristic rental housing in the Palma Old Quarter in Mallorca. Spain. Urban Stud 56(13):2709–2726. https://doi.org/10.1177/0042098018803261

Yus Ramos R, Torres Delgado MA (2010). *Urbanismo difuso en suelo rústico. Málaga.* Gabinete de Estudios de la Naturaleza de la Axarquía-Ecologistas en Acción, Vélez-Málaga, Spain.

Zoğal V, Domènech A, Emekli G (2022) Stay at (which) home: second homes during and after the COVID-19 pandemic. J Tour Futur 8(1):125–133. https://doi.org/10.1108/JTF-06-2020-0090

Part III
Urban Transformations: New Dynamics, Conflicts and Challenges in a Context of Uncertainty

Chapter 14
Towards Urban Degrowth? Urban Planning as a Common Thread of Contradictory Dynamics in Cities

Ibai de Juan Ayuso, Carmen Hidalgo Giralt, and Antonio Palacios García

Abstract Across recent decades, the numerous social, economic, and environmental changes occurring within urban spaces have moved the discipline of Urban Studies to generate proposals and solutions wherever necessary. One crucial challenge within that context involves the contradictions inherent in urban dynamics and processes which operate under a premise of continuous growth based on natural, fragile, and finite resources. The recent COVID-19 pandemic has exacerbated those contradictions, spotlighting the weaknesses of urban systems deriving from their growing complexity. In response, the theory of degrowth (currently on the rise, given the socio-economic and environmental convulsions of the twenty-first century) advocates for reductions in the consumption of natural resources through a socio-economic reorganization of space, and here, urban planning will prove fundamental. The prevailing paradigm of urban growth is now being questioned through varied initiatives that seek to limit the expansion of urban centers. The objective of this study is to analyze certain urban actions comparable with degrowth (zero growth, limited growth, sustainable growth, etc.) from a social, ecological, and urban perspective. Reflections on urban degrowth will help to situate this concept in a post-pandemic space and time and to identify the potential emergence of a new paradigm: the urbanism of degrowth.

Keywords Urban degrowth · Urban planning · Urban reorganization · Urban crisis · Urban sustainability · Urban performance

I. de Juan Ayuso
Universidad Politécnica de Madrid, Madrid, Spain
e-mail: ibai.dejuan123@gmail.com

C. Hidalgo Giralt (✉) · A. Palacios García
Universidad Autónoma de Madrid, Madrid, Spain
e-mail: carmen.hidalgog@uam.es

A. Palacios García
e-mail: antonio.palacios@uam.es

E. Navarro-Jurado et al. (eds.), *Urban Dynamics in the Post-pandemic Period*, The Urban Book Series,
https://doi.org/10.1007/978-3-031-36017-6_14

14.1 Introduction

To a great extent, contemporary cities owe their success to the functional role they play in the expansion and reproduction of capital (Durán and Reyes 2018: 458). Currently, 55% of the world population lives in urban areas, which are responsible for 80% of global GDP (ONU HABITAT 2020); thus as David Harvey (2013: 22) warns, "it cannot surprise us [...] that the logistic curve of the growth over time of capitalist production is practically identical to that of the urbanization of the world population." This apparently inseparable relationship between the reproduction of the city and of capital can be studied from a metabolic perspective. The city becomes the centrality of accumulation, absorbing matter and energy from natural resources and discharging matter as waste and energy as heat in increasing volume and at increasing speed. This generates an unequal exchange between the city and its surroundings that, in a globalized context, raises certain contradictions to planetary dimensions.

Urban processes operate under the premise of continuous growth based on fragile and finite natural resources, generating contradictions that have not been satisfactorily resolved by urban practice. The COVID-19 crisis has further exacerbated these contradictions, revealing the weakness of increasingly complex urban systems. Supply disruptions within logistical networks and supply deficits of raw materials produced by pandemic effects and by the energy crisis derived from the war in Ukraine are just a few examples of this new reality.

In this context, the proposal of "degrowth" has recently emerged, addressing the urban sphere and advocating a reduction in the socio-economic metabolism of the planet's most opulent societies in accordance with its biophysical limits (Latouche 2009). This implies a socio-economic reorganization of the space in which urbanism plays a fundamental role.

Given recent initiatives arising in cities that have begun to question the axiom of growth, it is worth asking what these initiatives linked to degrowth consist of, from an urban perspective. Are they truly advocating degrowth, or are they linked to the frameworks of limited growth and sustainability? If they can be considered in line with genuine degrowth, will it be possible to convert theory into practice? Indeed, will degrowth become the future urban paradigm? And how is degrowth integrated into reflections on post-pandemic urbanism? In response to these research questions, the objective of this paper is to detect and analyze urban actions that can be assimilated into degrowth (zero growth, limited growth, sustainable growth, etc.) with the aim of determining whether this currently constitutes an urban practice or a mere theoretical approach.

14.2 Methodology

To address our objective, in terms of methodology, a non-systematic inventory of urban actions with an impact on planning has been undertaken on a national scale, in order to analyze practices directly or indirectly related with the theory of degrowth from a social, ecological, and urban planning perspective. This inventory has been assembled from three principal sources: urban planning itself; the academic bibliography specialized in urban degrowth; related newspaper articles from 2007 to the present; and the news referenced by Google with the keywords "urban degrowth." Thus, we have compiled a record of actions from the institutional sphere and pertinent to urban degrowth that can help to determine the current stage of this "new paradigm," to evaluate the implementation in urban centers of this theoretical model and to define its social, ecological, and urban scope. This exercise in reflection around the issue of degrowth will also help position urban practice in post-pandemic space and time and to identify the potential emergence of "degrowth urbanism."

14.3 Degrowth, in an Urban Key

Since the 1970s, an increasingly intense debate has developed around the biophysical limits of the planet, especially since the report by the Club of Rome. Nicholas Georgescu-Roegen (through his work "The law of entropy and the economic process") and André Gorz (through the concept of "*décroissance*") are both considered pioneers in this current of thought (Kallis et al. 2015), although they offer different perspectives.

According to Paul Ariès, degrowth is a "shell-word" that points to the language of productivism and that seeks a reconsideration of the very concept of quality of life (cited in Cantero 2013). Carlos Taibo (2021: 58) breaks this proposal down into seven verbs: "decrease, de-urbanize, de-technologize, de-patriarchalize, decolonize, decommodify and de-complexize." This necessarily involves a deceleration in flows of both matter and energy within the socio-economic metabolism and a deep restructuring, conducted in an equitable manner. This process will further entail social reorganization and a new distribution of work, preferably articulated through governance models that are outside both the market and the State, such as the communal management of resources by a population (Helfrich and Bollier 2015: 208–212). What is sought is the recovery of a model that (although employed for centuries in the management of finite resources) has been gradually dismantled since the industrial revolution. In short, degrowth serves as an umbrella concept under which are grouped a set of currents critical of the prevailing development model based on unlimited consumption of resources. It proposes a quantitative and selective reduction of the socio-economic metabolism of the most opulent societies, adapting them to the planet's biophysical limits through cultural change that will ultimately lead to a political change, as the basis of a new development model (Latouche 2009; Riechmann 2019).

Definitions of degrowth have been criticized for their great generality, characterized as "substantive meta-ideas of leftism" (Mora 2011) destined to remain in the philosophical and moral fields (Torres 2011) and not appreciative of perspectives of class and gender (Harribey 2004; Soriano 2021: 194–197; Treillet 2009). However, analysis of the implications of degrowth on the urban environment can provide a greater degree of specificity to the proposal while responding to a certain extent to such criticism.

Although debate remains incipient, urban degrowth has been studied from a critical perspective in cities such as Milan, Seoul, Vancouver, Amsterdam or Copenhagen (Echabarren et al. 2020; Savini 2021; Xue 2016). The literature review shows that the main objectives of degrowth can be viewed through three dimensions: urban, social, and ecological. The urban dimension concerns the morphological changes required in the urban fabric to reduce the city's socio-economic metabolism. This implies definition of an optimal size and density for the city (Marcellesi 2013; Xue 2014), applying measures of structuring, containment, and extinction upon any planning already undertaken (Sanz and Navazo 2012) while at the same time avoiding the construction of new high-capacity infrastructures and large commercial zones in urban peripheries (Xue 2020). One proposal from the field of planning that contemplates the entire life-cycle of a territory (from its occupation to its potential vacancy) is "circular urban planning" (del Valle 2019), which seeks to integrate into the urban discipline a new conception of limits to the urbanization process (Lehtinen 2018).

Such morphological changes have clear implications for the mobility model that underlies every urban structure. Urban degrowth promotes the reduction of urban road traffic capacity while promoting sustainable transport alternatives other than private motorized vehicles (Marcellesi 2013; Reyes 2010; Savini 2021). This is opposed to a mono-functional urban model based on isolated single-family housing and the separation of uses imposed by the private vehicle as the only option (and thereby accentuating unsustainability).

Second, as regards the social dimension, an emphasis on relative proximity (so that journeys by private transport are unnecessary, while journeys on foot are encouraged) and greater habitability of city streets can encourage greater use of alternatives, which translates to an increase in both social interaction and social cohesion (Hiernaux-Nicolas, 2014; Valera, 2014). In this context, degrowth contemplates the development of communal alternatives around social reproduction that are outside the State and the market (Ruiz-Alejos and Prats 2021; Savini 2021). Ultimately, urban degrowth seeks social equity and the satisfaction of all citizens' needs.

Finally, the ecological dimension of urban degrowth comprises a broad set of proposals aimed at achieving greater urban sustainability. This demands a relocation as well as a quantitative and selective reduction of both production and consumption (Krähmer 2022; Marcellesi 2013; Xue 2014), implying a decrease in all the factors that today comprised the socio-economic system. Often, such proposals are related to the introduction of agriculture and renewable energies into the urban environment (Reyes 2010; Wächter 2013), and these can now be seen as responding to a context of difficult access to raw materials and energy sources, as experienced throughout the pandemic and the war in Ukraine, given the current impossibility of moving to a 100% renewable model (Turiel 2020). Consequently, reduction in the consumption of natural resources has become a main objective of urban degrowth.

14.4 Main Results

This research examines activities undertaken in Spanish municipalities from the field of urban planning and possibly associated with degrowth; to a variable extent, these activities respond to the urban, social, and ecological contingencies upon which actions for degrowth are conceived. It should be noted that, at the time of compiling this research inventory, no actions were found that were entirely consistent with all areas of action that urban degrowth implies. Thus, the inventory was opened to include initiatives that address only certain aspects (see the summarized results in Table 14.1). Initiatives were then classified into three groups based on their degree of agreement with the dimensions of degrowth (urban, social, and/or ecological).

14.4.1 Actions with Application to the Three Degrowth Dimensions: Málaga, Tudela, and Mallorca

In 2019, the *Observatorio de Medio Ambiente Urbano* (Urban Environment Observatory) of Málaga commissioned the consultancy firm n'UNDO—a promoter of degrowth—to study the empty spaces of the city. A set of urban environments of diverse morphology (expansions, garden cities, agricultural lands) present in the city of Málaga were analyzed, and proposals were made to re-articulate the territory in order to raise the city's degree of sustainability. In this case, the perception of degrowth was integrated without an initial predisposition towards urban sprawl and unlimited economic growth; all proposals featured dimensions aimed at improving social cohesion and the ecological sustainability of the urban fabric.

The other action in which all three degrowth dimensions applied—the General Municipal Plan (or PGM) of Tudela (Navarra), approved in 2019—recognized that the growth and development expectations foreseen in prior planning had not been met. This plan considers that population movements (both natural and migratory) do not require further expansion of the city but are compatible with the built space, without effects on the citizens' quality of life. The plan's objectives include the rehabilitation of the built-up city and the promotion of sustainable mobility, prioritizing pedestrian, and cycle travel. This reflects its commitment to a socially cohesive city. Not only does this document limit future growth, but it seeks to preserve agricultural lands on the urban periphery (Fig. 14.1), bringing this case study in line with a degrowth perspective, albeit in a tentative way.

Fig. 14.1 Tudela (*Source* ©The Spanish Traveler [This file is licensed under the Creative Commons Attribution-Share Alike 2.0 Generic-CC BY-SA 2.0] https://upload.wikimedia.org/wikipedia/commons/d/d3/Panor%C3%A1mica_de_Tudela.jpg)

Finally, in the municipality of Mallorca, the government, through the new General Plan, wants to reduce the available land in the municipality initially contemplated as land for development by 46%, from 496.85 to 268.29 hectares. This will entail an increase in protected rural land, as well as parks and green corridors (Re-dacción MallorcaDiario.com 2021). At the same time, it aims to limit changes of commercial use in dwellings and to set a minimum distance of 500 m between residential uses, schools, and social amenities.

14.4.2 Actions with Application in Two Degrowth Dimensions: Sarriguren, Carcaboso, and Errekaleor

Among urban centers that can serve as examples of application in two dimensions, the eco-city of Sarriguren (Fig. 14.2) presents a case of urban sustainability related to the ecological dimension of degrowth. Sarriguren arose conceptually from a "Sectoral Plan for Supramunicipal Incidence" (or PSIS) projected on the southeastern outskirts of Pamplona, although it was implemented in the municipality of Valle de Egüés, the planning of which is currently being revised. This is a neighborhood of over 10,000 inhabitants comprised of multi-family homes in open blocks with ample green zones and ground-floor shops on the main avenues, some of which permit pedestrian traffic only. The project was regarded as "Good Practice" by the United Nations Center for Human Settlements in 2000, and its bio-climatic homes earned it the European Urbanism Award in 2008.

The municipality of Carcaboso (Cáceres) started from a distinct context. After the bursting of a significant housing bubble, the City Council launched a set of projects aimed at converting this town into a "municipality in transition." This translated into cessions of land, to be re-allocated for purpose of cultivation by the local population. Initially, the sale of agricultural products was prohibited, but eventually trade was

Fig. 14.2 Eco-city of Sarriguren (*Source* ©Zariquiegui [This file is licensed under the Creative Commons Attribution-Share Alike 2.0 Generic- CC BY-SA 2.0] https://upload.wikimedia.org/wikipedia/commons/e/e7/Sarriguren_eraikitzen.jpg)

allowed as this was an economic activity within the province that benefited families. Currently, the village's farmland and chicken-raising facility are collectively managed. Initiatives promoted in Carcaboso earned the town the 5th Sustainable Community Award, given in 2016 by the UNESCO center in Extremadura (Table 14.1).

Elsewhere, the neighborhood of Errekaleor (Fig. 14.3), located on the south-eastern outskirts of Vitoria, was initially developed to accommodate immigration in the mid-twentieth century; this was slated for demolition and relocation of residents in the early twenty-first century, while new peripheral neighborhoods (Salburua and Zabalgana) were projected. In 2013, the Errekaleor Bizirik ("Errekaleor Lives") self-management project launched with the occupation of the neighborhood, in contravention to approved planning. Constant attempts at eviction and abandonment by the local administration left the neighborhood without basic services such as electricity or water, driving the community of Errekaleor to become the largest sustainable island in southern Europe. Currently, the neighborhood produces its own energy and hot water through thermal and photovoltaic panels; it possesses its own farmland, which serves the consumption of the community in part. Although this initiative does not derive from institutional planning, it does constitute an example of counter-planning in line with the proposals of urban degrowth.

Table 14.1 Inventory of degressivity actions in the plans

Initiative	Locality type		Degrowth dimension			
	Autonomous Community	Province	Municipality	Urban	Social	Ecological
1. n´UNDO VACÍOS Plan for Málaga	Andalucía	Málaga	Málaga	x	x	x
2. PGM Tudela	Navarra	Navarra	Tudela	x	x	x
3. PGOU Mallorca	Balearic Islands	Mallorca	Mallorca	x	x	x
4. Eco-city of Sarriguren	Navarra	Navarra	Valle de Egüés		x	x
5. Agro-ecological projects at Carcaboso	Extremadura	Cáceres	Carcaboso		x	x
6. Errekaleor neighborhood	Basque Country	Álava	Vitoria		x	x
7. Sustainable Strategic Plan for Gotarrendura	Castilla y León	Ávila	Gotarrendura			x
8. Law 7/2014 on Measures for Rehabilitation, Regeneration, and Urban Renewal and on Sustainability, Coordination, and Simplification in Urban Planning	Castilla y León	Segovia	Sepúlveda	x		
9. Citizen consultation	Madrid	Madrid	Rivas-Vaciamadrid	x		
10. La Pinada eco-district	Valencian Community	Valencia	Paterna	–	–	–

Source Own elaboration

14.4.3 Actions with Application in a Single Degrowth Dimension: Sepúlveda, Gotarrendura, Rivas-Vaciamadrid, and La Pinada

Sepúlveda (Segovia) (Fig. 14.5) is among the municipalities in the Autonomous Community of Castilla y León affected by Law 7/2014 on Measures for Rehabilitation, Regeneration, and Urban Renewal and on Sustainability, Coordination, and

Fig. 14.3 Errekaleor neighborhood (*Source* ©Zariquiegui [This file is licensed under the Creative Commons Attribution-Share Alike 2.0 Generic- CC BY-SA 2.0] https://upload.wikimedia.org/wikipedia/commons/e/e7/Sarriguren_eraikitzen.jpg)

Simplification in Urban Planning. This Law resulted from the aftermath of a real estate boom characterized by a large surplus and unconsolidated developable land, and it requires that lands not consolidated within a certain period be returned to common rural use. The municipality of Sepúlveda has since seen its growth expectations reduced by 146,069 m^2, on land that had been previously slated for industrial, residential, and public uses.

Meanwhile, the municipality of Gotarrendura (Ávila) (Fig. 14.4) has only 160 inhabitants, permitting fairly easy implementation of actions for urban sustainability. In 2004, the Sustainable Strategic Plan was launched, fostering the installation of solar and thermal energy equipment in undeveloped plots near the urban center. These solar farms produce an average of 6,706,620 kWh/year, meeting 60% of the demand for hot water previously obtained from fossil fuels. Buildings dependent on the municipal administration now employ boilers fueled by biomass. Given this progress, the town in 2011 received the distinction of "City of Science and Innovation" from the Spanish Ministry of Science and Innovation. Similarly, the urban regulations of the municipality (also approved in 2004) recognize a negative demographic dynamic, so that future growth of the urban fabric is not foreseen.

More recently, on October 25, 2021, the City Council of Rivas-Vaciamadrid (Fig. 14.6), located in the southeastern metropolitan area of Madrid, launched a

Fig. 14.4 Gotarrendura (*Source* ©David Pérez (DPC) [This file is licensed under the Creative Commons At-tribution-ShareAlike 4.0 International- CC BY-SA 4.0] https://upload.wikimedia.org/wikipedia/commons/thumb/3/32/Gotarrendura_01_by-dpc.jpg/1024px-Gotarrendura_01_by-dpc.jpg)

Fig. 14.5 Sepúlveda (*Source* ©Pelayo2 assumed [based on copyright claims) This file is licensed un-der the Creative Commons Attribution-ShareAlike 4.0 Inter-national- CC BY-SA 3.0]. https://upload.wikimedia.org/wikipedia/commons/thumb/3/3d/Sepulveda_02_by-dpc.jpg/1200px-Sepulveda_02_by-dpc.jpg)

citizen consultation on a temporary halt to residential urban growth in order to open a participatory process known as "Stop to Think." With 18% participation, the inhabitants of Rivas voted in favor of this process for presenting proposals that could be integrated into the General Plan for Urban Ordering (PGOU). This action has been accompanied by a controversial one-year standstill in the issuance of residential building licenses to match the growth in population with public service capabilities (Ezquiaga 2022).

Lastly, the La Pinada eco-district, situated within the municipality of Paterna in the Valencia metropolitan area, and developed through the Loma Larga Partial Plan, comprises 25 hectares designed for about 1,000 families. This initiative has been included in the "Smart Sustainable Districts" network within the Climate KIC program of the European Institute of Innovation and Technology. The district features semi-detached single-family homes as well as multi-family homes in open fenced blocks with privatized facilities. There is at present little ground-floor retail, given strong competition from the leisure and services on offer at the Valterna Plaza shopping center in the center of the neighborhood. Notable here are the private nature of the neighborhood's educational centers (French Lyceum, Montessori School) and

Fig. 14.6 Rivas-Vaciamadrid (*Source* ©David Daguerro de Madrid [This file is licensed under the Creative Commons Attribution-Share Alike 2.0 Generic- CC BY-SA 4.0] https://upload.wikimedia.org/wikipedia/commons/4/46/Rivas_Vaciamadrid-nevado-DavidDaguerro.JPG)

residences for the elderly (Ballesol). The municipal sports center is the only public facility. The district is delimited by high-capacity infrastructures to the south and east and a green barrier to the west and north.

14.5 Some General Reflections After Analysis of the Initiatives

Overall, two of the ten actions analyzed in relation to urban degrowth are located in Navarra (in the municipalities of Tudela and Valle de Egüés), and another two are in Castilla y León (in the municipalities of Sepúlveda and Gotarrendura). The rest are located in the Autonomous Communities of the Basque Country (Vitoria, Álava), Extremadura (Carcaboso, Cáceres), Andalucía (Málaga), the Valencian Community (Barrio de la Pineda in Paterna, Valencia), the Community of Madrid (Rivas-Vaciamadrid), and the Balearic Islands (Mallorca). No apparent geographic pattern indicates that particular regional or local administrations have been favoring urban measures focused on degrowth; this implies a future line of research, given that territorial and urban planning are responsibilities of the Autonomous Communities and individual municipalities, respectively. Territorial strategies have the capacity to articulate regions by establishing general land uses, which should have a significant influence on urban development models. However, at present, spatial planning instruments at regional or sub-regional level have not been developed in all territories, sometimes due to a lack of political will. On the other hand, even when there are strategies, sometimes the just don't have the expected influence on the territory due to their generalist nature.

In light of these results and given the characteristics of proposals for degrowth, a greater degree of applicability is found in small urban centers (Carcaboso and Gotarrendura) or in specific neighborhoods of medium-sized cities (Errekaleor and Sarriguren). Degrowth does not seem to be a current trend in larger cities, despite the cases of Rivas-Vaciamadrid or Málaga, perhaps due to the unpopularity of the term; and yet, limitations on growth have been conceived in coherence with the unique circumstances of a municipality (Tudela).

Beyond the limited number of initiatives considered for this study, it should further be noted that very few meet the three dimensions observed in the inventory (urban, social, and ecological), with the only exceptions being the cases of Málaga, Tudela, and Mallorca. The rest are oriented towards two dimensions (Carcaboso, Errekaleor, and Sarriguren), one dimension (Sepúlveda, Gotarrendura, and Rivas-Vaciamadrid) or even none (La Pinada). This underlines the difficulty in applying effective measures for degrowth (that is, which jointly contemplate all its dimensions). In the absence of such a confluence, actions promoted from the institutional sphere run the risk of passing through the "growth filter" and thus promoting an apparently sustainable urbanism that may nevertheless be based on the externalization of environmental and social costs (as in the case of La Pinada).

Actions related to degrowth are not exempt from a certain degree of conflict. Initiatives promoted by social groups can emerge in response to urban planning itself, as has occurred in Errekaleor, leading to a confrontation of perceptions on what is truly sustainable. Such conflicts can also arise from the urban administration, generally derived from political parties that (as in the case of Rivas-Vaciamadrid) have openly declared themselves in opposition to degrowth.

14.6 Conclusions

Analysis of the results makes possible an affirmation that actions related to urban degrowth remain very limited in Spain's urban areas. Indeed, they tend to be presented as measures for limiting growth or increasing urban sustainability rather than authentic initiatives for degrowth. Clearly, degrowth has yet to significantly permeate the social discourse or the institutional sphere.

Moreover, little synchronicity is observed between the theory and practice of urban degrowth. Debate is beginning to emerge, and degrowth has a fairly established theoretical body already behind it, but operational practices remain largely untried, at least in Spain's urban settings. Multiple reasons may be cited; among these, it is conceivable that despite the social, economic, and environmental problems caused by the capitalist model; the public administrations responsible for urban development continue to view growth as a way of sustaining a positive economic trend. Neither the business world nor society has openly come out in favor of adopting degrowth theories as a model for development, pointing to a lack of awareness of the socio-economic and environmental contradictions upon which the reproduction of the capitalist urban environment is presently based.

In such a situation, degrowth cannot be confirmed as the future of urbanism. Nor have the sharp impacts from COVID-19 caused a paradigm shift in urban planning. Even so, geopolitical phenomena like the current war in Ukraine or events related to the worsening ecological crisis may yet alter (significantly and in short order) the rules of an apparently inexorable hegemonic system. The limitations of the present work suggest that new research be proposed to enrich the inventory of initiatives considered as well as the analytical categories pertaining to urban degrowth.

References

Cantero J (2013) El movimiento por el "decrecimiento", ¿Una alternativa a la crisis sistémica? Primeros pasos en Francia y España. Clivatge 2:10–55

del Valle G (2019) Nuevos mecanismos de intervención en la ciudad existente: la recualificación a través del decrecimiento: objetivo: una ciudad resiliente a través de un urbanismo circular [Universidad Politécnica de Madrid]. https://doi.org/10.20868/UPM.THESIS.56876

Durán R, Reyes L (2018) En la espiral de la energía (Libros en acción, Vol. 2)

Echevarren RF, Mosera BM, y Nardo JFN (2020) Decrecimiento urbano: una alternativa complementaria. [Trabajo de fin de grado, Universidad ORT Uruguay]

Ezquiaga M (2022, January 5) Frenazo en seco al crecimiento de Rivas Vaciamadrid. El País. https://elpais.com/espana/madrid/2022-01-05/frenazo-en-seco-al-crecimiento-de-rivas-vaciamadrid.html

Font P (2022) La batalla por el colapso. Crisis ecosocial y élites contra el pueblo. Ecorama

Harribey JM (2004) Alternativas: Por una sociedad ahorrativa y solidaria—Rebelion. https://rebelion.org/alternativas-por-una-sociedad-ahorrativa-y-solidaria/

Harvey D (2013) Ciudades rebeldes. Del derecho de la ciudad a la revolución urbana. Akal

Helfrich S, Bollier D (2015) Procomunes. In D'Alisa G, Demaria F, & Kallis G (eds), Decrecimiento. Vocabulario para una nueva era (pp. 208–213). Icaria

Hiernaux-Nicolas D (2014) Identidades cosmopolitas y territorialidades en las sociedades posmodernas. In Sánchez-González D, & Domínguez L.Á (eds), Identidad y Espacio Público (Gedisa, pp. 41–53)

Kallis G, Demaria F, D'Alisa G (2015) Decrecimiento. In D'Alisa G, Demaria F, & Kallis G (eds), Decrecimiento. Vocabulario para una nueva era (Icaria, pp. 36–59)

Krähmer K (2022) Degrowth and the city. Multiscalar strategies for the socio-ecological transformation of space and place. City 1–30. https://doi.org/10.1080/13604813.2022.2035969

Latouche S (2009) La apuesta por el decrecimiento. ¿Cómo salir del imaginario dominante? Icaria

Lehtinen AA (2018) Degrowth in city planning. Fennia 196(1): 43–57. https://doi.org/10.11143/FENNIA.65443

Marcellesi F (2013) El decrecimiento: ¿Utopía o necesidad para las ciudades del siglo XXI? Pasaia Protuktiboak, 8–18.

Mora FR (2011, October 27) Controversia con Serge Latouche: ¿Revolución integral o decrecimiento? Decrecimiento.Info. http://www.decrecimiento.info/2011/10/controversia-con-serge-latouche.html

ONU HABITAT (2020) World cities report 2020. The value of sustainable urbanization. http://onuhabitat.org.mx/images/WCR2020/

Redacción MallorcaDiario.com. (2021, October 4). Cort apuesta por frenar el crecimiento urbanístico en el nuevo Plan general. MallorcaDiario.Com. https://www.mallorcadiario.com/presentacion-plan-general-urbanismo-palma-limitacion-suelo-disponible

Reyes E (2010) Decrecimiento y medio urbano. Tres discursos sobre el decrecimiento en las ciudades. In Decrecimientos. Sobre lo que hay que cambiar en la vida cotidiana (pp. 51–64). Catarata

Riechmann J (2019) Otro fin del mundo es posible, decían los compañeros. MRA

Ruiz-Alejos C, Prats V (2021) In quest of implementing degrowth in local urban planning policies. Local Environ 27(4):1–17. https://doi.org/10.1080/13549839.2021.1983789

Sanz A, Navazo M (2012) Metabolismo urbano, energía y movilidad: los retos del urbanismo en el declive de la era del petróleo. Ciudad y Territorio Estudios Territoriales 44(171):87–95

Savini F (2021) Towards an urban degrowth: habitability, finity and polycentric autonomism. Environ Plan A 53(5):1076–1095. https://doi.org/10.1177/0308518X20981391

Soriano C (2021) Antropoceno, reproducción de capital y comunismo. Maia

Taibo C (2021) Iberia vaciada. Despoblación, decrecimiento, colapso. Catarata

Tejero H, Santiago E (2019) ¿Qué hacer en caso de incendio? Manifiesto por el Green New Deal. Capitán Swing

Torres J (2011, October 25) Sobre el concepto de decrecimiento. Decrecimiento.Info. http://www.decrecimiento.info/2011/10/sobre-el-concepto-de-decrecimiento.html

Treillet S (2009, March 26) Un proyecto de sociedad reaccionaria. Viento Sur. https://vientosur.info/un-proyecto-de-sociedad-reaccionaria/

Turiel A (2020) Petrocalipsis. Crisis energética global y cómo (no) la vamos a solucionar. Alfabeto.

Turiel A (2021) Antonio Turiel, comparecencia en el Senado: Comisión de Transición Ecológica - YouTube. https://www.youtube.com/watch?v=QZ-OE90YW6Q&ab_channel=Algemesiporunaeconom%C3%ADabasadaenrecursos

Valera S (2014) La identidad social urbana como instrumento para mejorar el bienestar humano. In Sánchez-González D, & Domínguez LÁ (eds), Identidad y Espacio Público (Gedisa, pp. 97–119)
Wächter P (2013) The impacts of spatial planning on degrowth. Sustainability 5(3):1067–1079. https://doi.org/10.3390/su5031067
Xue J (2014) Is eco-village/urban village the future of a degrowth society? An urban planner's perspective. Ecol Econ 105:130–138. https://doi.org/10.1016/J.ECOLECON.2014.06.003
Xue J (2016) Eco-metropolis planning conditiones by the growth ideology: the case of Grater Copenhagen. Urban Des Plannin 171(3):133–142. https://doi.org/10.1680/jurdp.16.00037
Xue J (2020) Urban planning and degrowth: a missing dialogue. Local Environ 27(4):404–422. https://doi.org/10.1080/13549839.2020.1867840

Chapter 15
Shrinking Cities in Spain: Shrinking Medium-Sized Cities in the Twenty-First Century. Depopulation and Employment

José María Martínez-Navarro, Luis Alfonso Escudero-Gómez, and Juan Antonio García-González

Abstract Many countries have experienced a phenomenon of hollowing out of their rural areas and small cities, which has led to a territorial imbalance in favour of large metropolitan areas. Meanwhile, in the first decade of the twenty-first century, medium-sized cities underwent intense expansion demographically and, above all, in terms of urbanisation. However, they were then affected by the Great Recession and are now in decline. The phenomenon of shrinking cities is common in North America and central and northern European countries, such as France and Germany. Indeed, urban shrinkage is a subject of academic interest but is still an incipient field of study in the literature. This research aims to encourage the analysis of shrinking cities by studying two critical aspects of urban decline: depopulation and the evolution of employment. The method analyses the evolution of the population of medium-sized cities and workers in the first two decades of the twenty-first century. Results show no direct correlation between depopulation and the evolution of employment in Spanish medium-sized cities as a general phenomenon, but certainly in some specific cases.

J. M. Martínez-Navarro · L. A. Escudero-Gómez (✉) · J. A. García-González
Department of Geography, University of Castilla-La Mancha, Ciudad Real, Spain
e-mail: luisalfonso.escudero@uclm.es

J. M. Martínez-Navarro
e-mail: josemaria.martinez@uclm.es

J. A. García-González
e-mail: juanantonio.garcia@uclm.es

J. M. Martínez-Navarro
Facultad de Educación de Cuenca, Edificio Fray Luis de León, Camino Pozuelo, s/n, 16071 Cuenca, Spain

L. A. Escudero-Gómez
Facultad de Humanidades de Toledo, Plaza de Padilla 4, 45071 Toledo, Spain

J. A. García-González
Facultad de Humanidades de Albacete, Edificio Benjamín Palencia, Campus Universitario, s/n, 02071 Albacete, Spain

E. Navarro-Jurado et al. (eds.), *Urban Dynamics in the Post-pandemic Period*,
The Urban Book Series,
https://doi.org/10.1007/978-3-031-36017-6_15

Keywords Shrinking cities · Urban shrinkage · Depopulation · Medium-sized cities · Spain

15.1 Introduction

An increasing number of cities are today undergoing urban shrinkage processes. From a historical perspective, urban shrinkage is not new (Nelle et al. 2017) and was characteristic of many cities in the United States in the second half of the twentieth century (Bradbury et al. 1982). In Europe, however, it is a recent but significant phenomenon (Haase et al. 2016). The process started to develop in the early twenty-first century (Oswalt 2005) and has now spread, to differing degrees, across all of Europe (Nijman and Wei 2020).

Inequalities between cities are increasing in the form of growth versus shrinkage (Nijman and Wei 2020). Shrinking cities are vulnerable places (Raco and Street 2012). This is a severe problem as urban shrinkage involves socio-demographic decline, retreating anthropogenic presence, the degradation of fixed social capital and the neglect of built and infrastructure heritage (Vendemmia and Këlçuku 2020).

Consequently, academic publications have studied urban shrinkage drivers and impacts (Oswalt 2005; Martínez-Fernández et al. 2012) in a research agenda on growth (Großmann et al. 2013). Shrinking cities in Europe have been analysed (Döringer et al. 2020; Haase et al. 2016), concluding that the forms of degrowth vary greatly. The intensity of urban shrinkage differs from country to country (Eva et al. 2021), as evidenced in studies conducted at the national level in France (Chouraqui 2021), Germany (Nelle et al. 2017), Italy (Vendemmia and Kërçuku 2020), the Netherlands (Hoekveld 2012), Portugal (Alves et al. 2016), etc.

This chapter focuses on urban shrinkage in intermediate cities in Spain. "Medium-sized European cities have been playing an increasingly significant role in the economic development of countries in recent decades, establishing themselves as genuinely specialized local production systems with great potential for stimulating the economy and generating added value" (Parrilla-González 2021, p. 1). Nonetheless, they are experiencing depopulation phenomena. In Spain, this is a novel process in the country's spatial system, in which depopulation has always been associated with rural areas and not cities, with the decline of the rural world being the *first wave* of Spanish depopulation (Collantes and Pinilla 2019). However, a second wave has commenced in the present century, now affecting medium-sized cities. Indeed, in Spain, with the 2008 crisis, medium-sized cities went from boom to recession (Bellet-Sanfeliu and Andrés-López 2021), moving from implosive growth to a significant decline as a result of an intense financial, economic and, finally, social crisis. At the same time, the recession increased the spatial concentration of population and economic activity. The most populated cities have gained more weight at the population level and are where recovery has occurred sooner and stronger (Royuela et al. 2017). In contrast, medium-sized cities have witnessed their development slow down and/or reach a complete standstill.

In studies on medium-sized cities on a national scale, there is a growing interest in urban shrinkage (Vendemmia and Kërçuku 2020). In Spain, however, this process has not generated great interest among academics, although some works have identified the phenomenon (González-Leonardo et al. 2022). This research aims to help remedy this lack of studies.

Furthermore, the literature has identified correlations between urban shrinkage and a number of processes. Of these, the most noteworthy is the evolution of employment (Bernt et al. 2012; Hartt 2018). Low activity and employment rates, high unemployment rates and the dwindling young population's difficulties in finding work characterise the social devitalisation of shrinking cities.

The main aim of this research is to study shrinking medium-sized cities in Spain in the twenty-first century, drawing on two key elements: depopulation and the evolution of employment. Our study explores possible relationships between the two processes. The study's initial hypothesis is that Spanish medium-sized cities experiencing degrowth, in line with the suggestions of the literature on urban shrinkage (Bernt et al. 2012; Hartt 2018), present a clear correlation between population loss and the evolution of employment.

The interest of this research lies in its being one of the few available on the subject in urban studies and, to the best of our knowledge, the only one that links depopulation and employment in medium-sized Spanish cities in the twenty-first century.

After this introduction, the structure of the chapter continues with a theoretical study of urban shrinkage. This is followed by a description of the object of study, data sampling and the analytical method of analysis. Finally, we present the research results, the discussion and conclusions.

15.2 Urban Shrinkage

Urban shrinkage can be described as a crisis triggered and characterised by job and population losses (Martínez-Fernández et al. 2012). The concept of urban shrinkage describes a complex phenomenon in which demographic decline and economic downturn are the results of the structural characteristics of the city (Fol and Cunningham-Sabot 2010). There is a shrinkage of cities and their dissipation at the regional level (Brenner 2014).

Many researchers have recognised the multidimensionality of urban shrinkage (Pallagst et al 2009; Martínez-Fernández et al. 2016). Haase et al. (2014) proposed a heuristic model of urban shrinkage (Fig. 15.1). Hoekveld (2012) and Hartt (2018) highlight that the processes contributing to urban shrinkage are highly complex. Population decline impacts on almost all areas of urban life: business and employment, housing, social and technical infrastructure, municipal finances, social cohesion, segregation, etc. Urban shrinkage thus results in a mismatch between supply and demand for built structures, urban space and infrastructure (Bernt et al., 2012).

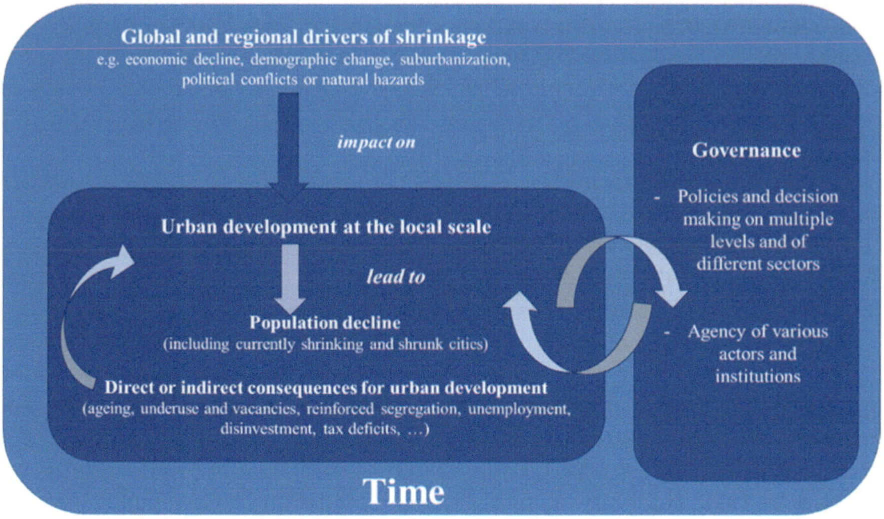

Fig. 15.1 Heuristic model of urban shrinkage (Haase et al. 2014; own preparation)

In the case of medium-sized cities, the crisis is typified, in the particular by impoverishment of the population and degradation of the built environment (Chouraqui 2021).

Planners have responded to population decline by initiating economic development strategies; a conventional approach that has failed in many places. Indeed, as Fol (2012) suggests, the proposed remedies have, on occasion, been worse than the disease, resulting in socio-spatial disparities. Additionally, in many cases, recovery and adaptation strategies have generated short-term impacts, with their long-term sustainability and ability to pave the way for a new course of events being called into question (Radzimski 2018).

The experience of cities affected by urban shrinkage shows there is no simple way to forestall the negative consequences of this process. At the same time, governance is an essential challenge for local authorities (Stryjakiewicz and Jaroszewska 2016). Going beyond growth strategies, Herrmann et al. (2016) propose sustainability policies for shrinking cities. Hospers (2014) calls for policy responses to urban shrinkage that shift from growth thinking to civic engagement.

Academia has responded to the urban shrinkage crisis by a reconceptualising decline as shrinkage, beginning to explore creative and innovative ways for cities to shrink successfully. This new approach to re-thinking decline provides a non-economic vision of the response to depopulation (Hollander et al. 2009). For example, Karp et al. (2022) suggest investing in green infrastructure for recreation and supporting ageing populations with retirement services as a response to urban shrinkage.

For the specific case of medium-sized cities, Chouraqui (2021) concludes that public policies must take account of their heterogeneity to adapt urban planning strategies to various contexts. Medium-sized cities constitute both a window of opportunity and a challenge for spatial planning (Güneralp et al. 2020). In this context, the COVID-19 crisis may represent an opportunity for shrinking medium-sized cities by relocating workers to less crowded and more economical locations (Besana 2021).

15.3 Object of Study, Data Sampling and Method of Analysis

Our study focuses on shrinking medium-sized cities in Spain in the twenty-first century. To define a medium-sized city, we used the criteria developed in previous works by the authors (Martínez-Navarro et al. 2020). The first and most decisive is the range of population size. To be considered medium-sized, their population in 2002 had to be between 50,000 and 300,000. This yielded a total of 137 cities, to which we have added a further four. These are Zaragoza and Córdoba, both with a population above the range, and Soria and Teruel, with populations below. These latter two were chosen as they are capitals of provinces, the Spanish sub-regional administrative division, which has no other medium-sized city in the provincial territory.

In the evolution of population, a period of prosperity can be seen, during which 131 of the 141 cities analysed had positive growth between 2001 and 2010, while 51 lost population in the recessionary period from 2011 to 2020. The criterion for defining shrinking medium-sized cities was to select only those that lost population in the overall period analysed for the twenty-first century (2001–2020). In other words, their 2020 population was lower than in 2001. A total of 17 cities met this requirement, of which 11 are provincial capitals, with six being in the Castilla-León region. All of them, except Santa Cruz de Tenerife, are in peninsular Spain and are mostly located in the north-western quadrant (Fig. 15.2). Of this sample, 10 of the intermediate cities have lost population across all the years of the twenty-first century: Cádiz, Ferrol, Getxo, Granada, León, Ourense, Salamanca, Santander, Torrelavega and Valladolid. The other seven cities grew between 2001 and 2010, but subsequent depopulation in the decade from 2011 to 2020 meant they fell within the established criteria.

Our methodology draws on an analysis of quantitative demographic and socioeconomic data. The population has been selected from the beginning of the century (2001–2020). For the labour market data, the unemployed and those affiliated to the Social Security (SS) up to 2020 have been selected. These data can be linked to the target employment in each city. The starting date has been established according to official statistical availability: from 2006 for the unemployed and from 2003 for those affiliated to the SS. This minor date mismatch is considered acceptable, given that the dynamics will be analysed using an average growth rate. In fact, it is not uncommon in econometric analysis to include variables from different reference periods; for example, Kim et al. (2000) and Noonan et al. (2021).

Fig. 15.2 Object of study: Spanish shrinking medium-sized cities 2001–2020 (INE 2022; own preparation)

For the analysis, an average annual growth rate (AAGR) was calculated for the three diachronic variables. This was used to study the possible link between the demographic decline and the employment variables explained. A percentage index has been developed that relates the indicator under analysis (P) in its final value for the period studied (t_a) and in its initial value (t_o). These data have been weighted by the number of years analysed ($t_a - t_o$), and with the average value of the final (t_a) and initial (t_o) values in the denominator with the following formula:

$$\text{AAGR}(t_o, t_a) = \frac{(P(t_a) - P(t_o))/(t_a - t_o)}{(P(t_a) + P(t_o))/2} * 100$$

To calculate the AAGR, we included sub-periods related to the Spanish economic context and the complete analysis for 2001–2020. Thus, the AAGR was calculated for three intervals. These periods are marked by the clear turning point the real estate bubble bursting and the subsequent economic and social crisis, which reached its peak in 2010. This enabled us to divide the 2001–2020 period into two ten-year sub-periods: 2001–2010 and 2011–2020. For this purpose, we used the official data as of 1 January provided by the continuous population register (INE 2022), the mean monthly municipal unemployment statistics (SEPE 2022) and the total number of workers affiliated to the social security system on the last day of January (SS 2022). Finally, Pearson's correlation coefficient was used to measure the possible links between the evolution of population and employment in Spanish shrinking medium-sized cities in the twenty-first century. Although this indicator is not valid to demonstrate cause and effect, it does determine whether there exists a relationship or association between various variables of interest (Schober et al. 2018).

15.4 Spanish Shrinking Medium-Sized Cities in the Twenty-First Century. Depopulation and Employment

The selection criterion used means that all the medium-sized cities in the sample lose population over the study period, but not at the same rate (Table 15.1), with the AAGR (2001–2020) of the population ranging from −0.04% in the case of Linares to −1.07% in that of Ferrol. The mean for the dataset is 0.37%.

The AAGR (2006–2020) of the unemployed presents positive values (Table 15.1), as is logical, since it is reasonable to expect that a greater population loss would

Table 15.1 AAGR (annual %) for population and unemployed (INE 2022; SEPE 2022; own preparation)

City	AAGR (2001–2020) population	AAGR (2006–2020) unemployed	AAGR (2001–2010) population	AAGR (2006–2010) unemployed	AAGR (2011–2020) population	AAGR (2011–2020) unemployed
Alcoy/ Alcoi	−0.08	0.82	0.21	4.21	−0.32	−0.68
Avilés	−0.38	0.61	0.09	2.14	−0.80	−0.10
Cádiz	−0.94	0.32	−1.02	1.25	−0.87	−0.23
Ferrol	−1.07	−0.08	−0.97	1.23	−1.19	−0.75
Getxo	−0.40	0.43	−0.50	1.77	−0.33	−0.39
Granada	−0.21	1.01	−0.19	3.10	−0.30	0.11
León	−0.54	0.35	−0.28	2.04	−0.75	−0.47
Linares	−0.04	0.72	0.65	2.24	−0.70	−0.07
Ourense	−0.17	0.18	−0.04	1.58	−0.25	−0.59
Palencia	−0.18	0.40	0.18	2.38	−0.47	−0.50
Salamanca	−0.48	0.52	−0.29	2.68	−0.64	−0.46
Santander	−0.12	1.08	0.43	3.97	−0.67	−0.17
Segovia	−0.35	0.89	−0.22	3.43	−0.41	−0.30
Sta. C. de Tenerife	−0.20	1.10	0.35	4.74	−0.66	−0.69
Torrelavega	−0.42	0.98	0.00	3.75	−0.82	−0.39
Valladolid	−0.32	0.46	−0.10	2.82	−0.51	−0.59
Zamora	−0.39	0.50	0.06	2.61	−0.80	−0.53
Average	*−0.37*	*0.61*	*−0.10*	*2.70*	*−0.62*	*−0.40*

correspond to a more significant number of unemployed. There is only one exception, Ferrol, which has a negative value, albeit a very low one, 0.08%. For this series, the positive mean of the AAGR is higher, 0.61%, than the negative mean of the population, with 7 cities above that value. Santa Cruz de Tenerife has the highest AAGR with a 1.10% average annual increase in unemployed inhabitants. The Pearson correlation between the AAGR of population and unemployed for the twenty-first century analysed is 0.61.

Let's analyse the AAGR of the population in the first sub-period that of the first decade of the twenty-first century (Table 15.1). We find the 10 medium-sized cities undergoing population decline throughout the twenty-first century. The number of inhabitants in 6 of the others grew between 2001 and 2010, while the population stagnated in the last one, Torrelavega. The arithmetic mean is −0.10% AAGR (2001–2010) of the population.

As regards the unemployed, in this first sub-period, which, due to the previously mentioned data availability, corresponds to 2006–2010, there are positive and high AAGRs (Table 15.1), with an average of 2.70%. In addition, up to 7 cities are above the average, including Santa Cruz de Tenerife, with an AAGR of 4.74%. The effects of the Great Recession meant that the difference between 2006, still a year of economic expansion, and 2010, at the height of the crisis, led to a significant increase in the number of unemployed. In this sub-period, the correlation coefficient between depopulation and unemployment is identical to that of the century (0.61).

In the second sub-period, from 2011–2020 (Table 15.1), all the Spanish shrinking medium-sized cities show a negative AAGR of population, ranging from −0.25% for Ourense and −1.19% for Ferrol once more, the intermediate city with the highest depopulation rate in Spain. The arithmetic mean is −0.62%, albeit with ten cities showing degrowth rates above this. The population decline was more severe in the second decade of the present century than in the first. It is a phenomenon that has escalated recently.

However, the evolution of the number of unemployed exhibits an opposite trend in this second sub-period of analysis. Here, the average AAGR is negative, −0.40%. In other words, in the second decade of the twenty-first century, that of the crisis and subsequent incipient recovery thwarted by the COVID-19 pandemic of 2020, the number of unemployed falls in all the shrinking medium-sized cities, except in Granada (Table 15.1). A comparison of the AAGRs of the population and unemployed between 2011 and 2020 yields a non-significant correlation coefficient (0.10). The relationship or association between depopulation and unemployment disappears in this second sub-period.

Table 15.2 shows the AAGRs for the population and those for objective employment, that is, the individuals affiliated to the SS. The correlation of the AAGRs for the two decades of the century, 2001–2020 in the population and 2003–2020 in the SS affiliates, the difference in dates being due to the data availability, lacks significance (0.08). Only five cities—Alcoy, Avilés, Ferrol, Linares and Torrelavega—have a negative AAGR of SS affiliates. The other cities combine population decline with increased employment, which contradicts our initial hypothesis that the relationship would be the opposite, lower population and lower employment. The mean AAGR (2003–2020) for SS affiliates is 0.34%.

The first decade of the twenty-first century in Spain was characterised by economic growth, and all the shrinking medium-sized cities have a positive AAGR (2003–2010) of persons affiliated to the SS (Table 15.2), except Alcoy and Avilés. The average for the sample is positive, 0.97%. In this sub-period, there is no significant correlation between population evolution and SS affiliation in the cities. Pearson's coefficient yields a negative value of −0.32.

The second sub-period of analysis (2011–2020) reveals six cities with an unfavourable evolution in the total number of SS affiliates (Table 15.2): Avilés, Ferrol, Linares, Salamanca, Torrelavega and Zamora. The other cities have positive values. The average AAGR of SS affiliates is also positive, but much more modest (0.22%) than between 2003 and 2010. There is now a higher correlation between the evolution of the population and that of individuals affiliated to the SS (0.60).

We analysed the Pearson correlation coefficient for the two datasets of socio-economic variables concerning the evolution of the population and in the periods

Table 15.2 AAGR (annual %) for population and persons affiliated to the social security system (SS) (INE 2022; SS 2022; own preparation)

City	AAGR (2001–2020) population	AAGR (2003–2020) registered with SS	AAGR (2001–2010) population	AAGR (2003–2010) registered with SS	AAGR (2011–2020) population	AAGR (2011–2020) registered with SS
Alcoy/Alcoi	−0.08	−0.04	0.21	−1.41	−0.32	1.12
Avilés	−0.38	−0.54	0.09	−0.57	−0.80	−0.28
Cádiz	−0.94	1.07	−1.02	2.01	−0.87	0.93
Ferrol	−1.07	−0.58	−0.97	1.16	−1.19	−1.41
Getxo	−0.40	0.23	−0.50	0.44	−0.33	0.43
Granada	−0.21	1.19	−0.19	1.59	−0.30	1.17
León	−0.54	0.32	−0.28	0.93	−0.75	0.06
Linares	−0.04	−0.74	0.65	0.65	−0.70	−1.22
Ourense	−0.17	0.47	−0.04	1.24	−0.25	0.28
Palencia	−0.18	0.75	0.18	1.78	−0.47	0.23
Salamanca	−0.48	0.42	−0.29	1.52	−0.64	−0.03
Santander	−0.12	0.90	0.43	0.57	−0.67	1.74
Segovia	−0.35	0.91	−0.22	1.35	−0.41	0.89
Sta. C. de Tenerife	−0.20	0.50	0.35	1.39	−0.66	0.10
Torrelavega	−0.42	−0.04	0.00	1.26	−0.82	−0.61
Valladolid	−0.32	0.67	−0.10	0.91	−0.51	0.77
Zamora	−0.39	0.30	0.06	1.74	−0.80	−0.47
Average	*−0.37*	*0.34*	*−0.10*	*0.97*	*−0.62*	*0.22*

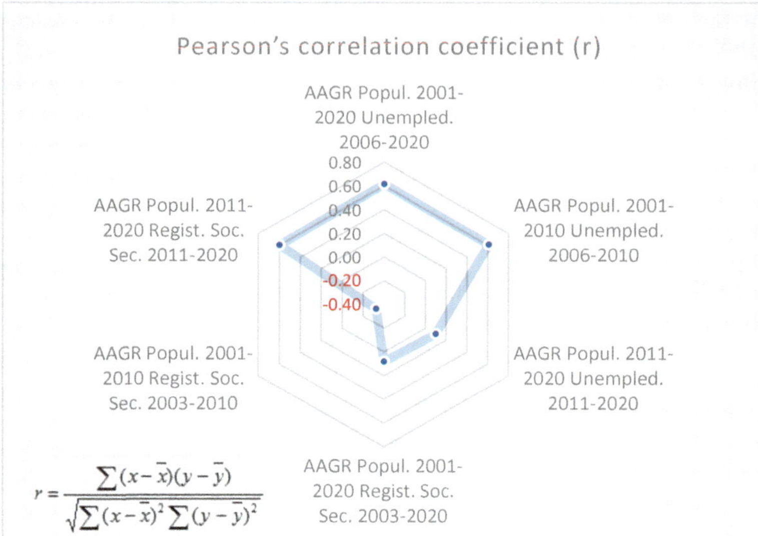

Fig. 15.3 Pearson's correlation coefficient between the AAGRs of population, unemployed and SS affiliates (INE 2022; SEPE 2022; SS 2022; own preparation)

studied (Fig. 15.3), which, to be significant, must be close to 1 or −1 (Schober et al. 2018). Kennedy (2008, p. 19) proposes that to be considered high; a correlation coefficient between two independent variables must be "0.8 or 0.9 in absolute value". None of the correlations obtained reaches these values. In the social sciences, however, values above 0.5 can be considered a large, statistically significant, correlational effect (Onwuegbuzie and Daniel 2002). With the latter criterion, a correlational effect can be seen between depopulation and unemployment in the twenty-first century and its first decade and between depopulation and SS affiliation in the second decade. The remaining correlations are non-significant in all cases.

We developed a representation in two scatter plots to better understand the possible groupings in the behaviour of the variables analysed. Both are drawn for the evolution across the complete first two decades of the twenty-first century, with the x-axis representing the AAGR (2001–2020) of the population. On the y-axis, the first graph l (Fig. 15.4) plots the AAGR (2006–2020) of the unemployed and the second (Fig. 15.5) the AAGR (2003–2020) of individuals affiliated to the SS.

Regarding the relationship between the AAGRs for population and unemployment (Fig. 15.4), the cases vary greatly between cities with no clear general correlation ($r^2 = 0.37$). However, the association is considerable in many examples, such as Alcoy, Avilés and Salamanca. On the other hand, in some cities, such as Ferrol, and especially Ourense, León and Cádiz, urban shrinkage has reached such a level of maturity that the growth in the unemployed has fallen or is even negative in the case of Ferrol. The jobless population is leaving these cities for other areas of employment; a factor that, in turn, also explains their depopulation. In Spain, an increase in internal

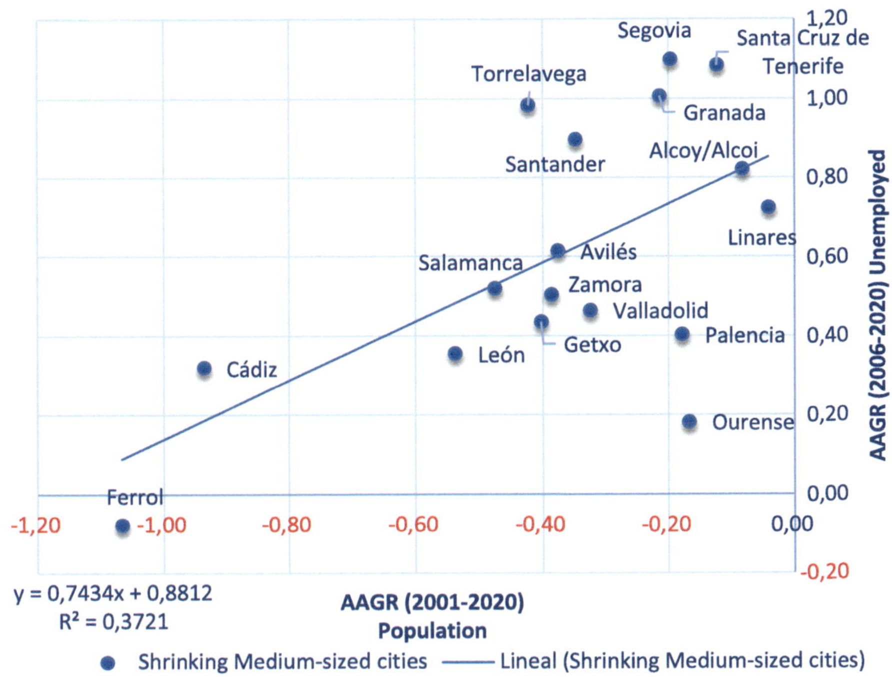

Fig. 15.4 Comparison of AAGRs of population growth, 2001–2020, and unemployed, 2006–2020 (INE 2022; SEPE 2022; own preparation)

migration of highly qualified young people from medium and small cities to large cities, mainly to the urban area of Madrid, has recently been detected (González-Leonardo et al. 2022). In other cities, urban shrinkage is in its infancy, with the associated phenomenon of an increase in the number of unemployed, for example, in Alcoy or Linares. However, it is difficult to generalise the conclusions due to the dispersion present in the graph.

A complex discussion also arises from the scatter plot between the AAGR of the population, 2001–2020 and SS affiliates, 2003–2020 (Fig. 15.5). There is no significant correlation between the two variables ($r^2 = 0.006$). Only five cities—Alcoy, Avilés, Ferrol, Linares and Torrelavega—follow the logic of lower population and fewer workers. For the rest, the point cloud suggests no clear conclusions.

15.5 Discussion and Conclusions

We have studied Spanish shrinking medium-sized cities in the twenty-first century using two key elements: depopulation and employment, the latter through the number of unemployed and objective employment, exploring the relationships between the two. The initial hypothesis of the research was that these urban centres would present

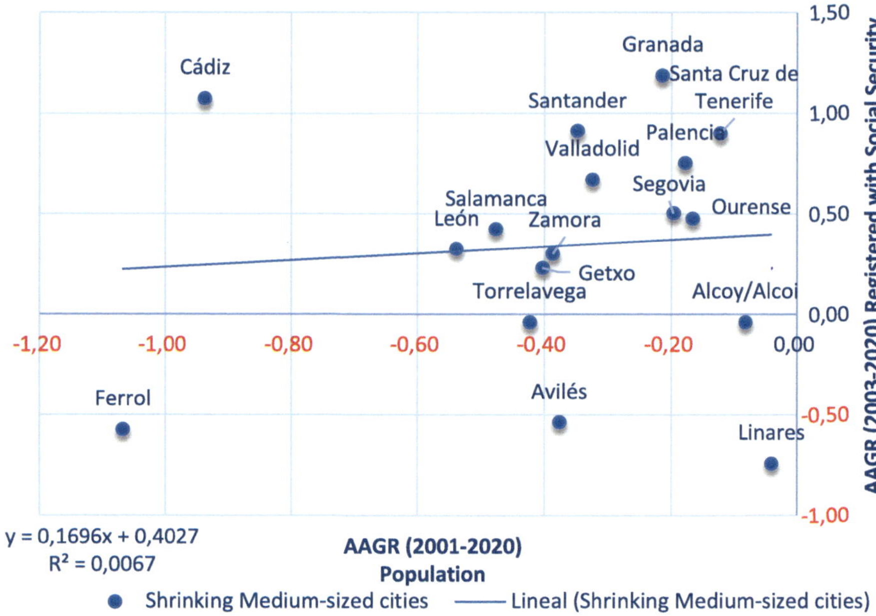

Fig. 15.5 Comparison of AAGRs of population growth, 2001–2010, and individuals affiliated to the SS, 2003–2020 (INE 2022; SS 2022; own preparation)

a correlation between population loss and falling employment. However, we conclude that this hypothesis is false for Spanish shrinking medium-sized cities as a whole, although it may explain the reality of some specific cases. Such cases are the cities where population declines, the number of unemployed increases and the number of SS affiliates falls, namely, Alcoy, Avilés, Linares and Torrelavega.

The Pearson correlation coefficients revealed no relationship or association between the AAGRs of population, unemployment and SS affiliation at a highly significant level. However, there is a correlational effect between population and unemployment.

Urban shrinkage is a complex and multi-causal phenomenon with many possibilities. A number of medium-sized Spanish cities have been losing population continuously across the twenty-first century. The depopulation-employment relationship is not sufficiently explicit to understand the complete process. Part of the population loss is likely related to centrifugal movements resulting from suburbanisation or demographic phenomena, such as ageing. More explanatory variables need to be explored.

Economic activity, studied in this research through the optics of the unemployed and individuals affiliated to the SS, is but one aspect of a process, urban shrinkage, with multiple facets and complex reality. Our research results reflect this circumstance. The evidence shows that the economic aspect has a greater impact in some cities than others but is neither exclusive nor decisive. Therefore, future research should address the second wave of depopulation in Spain by including new demographic indicators, such as ageing, birth rates and migratory movements, relating them to the process of suburbanisation and to policy and governance questions.

References

Alves D, Barreira AP, Guimarães MH, Panagopoulos T (2016) Historical trajectories of currently shrinking Portuguese cities: a typology of urban shrinkage. Cities 52:20–29. https://doi.org/10.1016/j.cities.2015.11.008

Bernt M, Cocks M, Couch C, Großmann K, Haase A, Rink D (2012) Shrink smart. The governance of shrinkage within a European context. Research brief no. 2: policy response, governance and future directions. Retrieved June 19, 2022, from https://www.ufz.de/export/data/400/39031_ResearchBrief2_.pdf

Besana F (2021) Proposals of European citizens for reviving the future of shrinking areas. Quaestiones Geographicae 40(4):15–28. https://doi.org/10.2478/quageo-2021-0039

Brenner N (ed) (2014) *Implosions/explosions. Towards a study of planetary urbanization.* Jovis Verlag

Bradbury KL, Downs A, Small KA (1982) Urban decline and the future of American cities. Brookings Institution.

Bellet-Sanfeliu C, Andrés-López G (2021) Urbanización, crecimiento y expectativas del planeamiento urbanístico en las áreas urbanas intermedias españolas (1981–2018). Investigaciones Geográficas, 76:31–52. https://doi.org/10.14198/INGEO.18054

Chouraqui J (2021) Medium-sized cities in decline in France: between urban shrinkage and city centre devitalisation. Raumforschung Und Raumordnung. Spatial Research and Planning 79(1):3–20. https://doi.org/10.14512/rur.26

Collantes F, Pinilla V (2019) ¿Lugares que no importan? La despoblación de la España rural desde 1900 hasta el presente. Prensas de la Universidad de Zaragoza.

Döringer S, Uchiyama Y, Penker M, Kohsaka R (2020) A meta-analysis of shrinking cities in Europe and Japan. Towards an integrative research agenda. European Planning Studies 28(9):1693–1712. https://doi.org/10.1080/09654313.2019.1604635

Eva M, Cehan A, Lazăr A (2021) Patterns of urban shrinkage: a systematic analysis of Romanian cities (1992–2020). Sustainability 13(13):7514. https://doi.org/10.3390/su13137514

Fol S (2012) Urban shrinkage and socio-spatial disparities: are the remedies worse than the disease? Built Environment 38(2):259–275. https://doi.org/10.2148/benv.38.2.259

Fol S, Cunningham-Sabot E (2010) Urban decline and shrinking cities: a critical assessment of approaches to urban shrinkage. Annales de Géographie 674(4):359–383. https://doi.org/10.3917/ag.674.0359

González-Leonardo M, López-Gay A, Esteve A (2022) Interregional migration of human capital in Spain. Reg Stud Reg Sci 9(1):324–342. https://doi.org/10.1080/21681376.2022.2060131

Großmann K, Bontje M, Haase A, Mykhnenko V (2013) Shrinking cities: notes for the further research agenda. Cities 35:221–225. https://doi.org/10.1016/j.cities.2013.07.007

Güneralp B, Reba M, Hales BU, Wentz EA, Seto KC (2020) Trends in urban land expansion, density, and land transitions from 1970 to 2010: A global synthesis. Environ Res Lett 15(4):044015. https://doi.org/10.1088/1748-9326/ab6669

Haase A, Athanasopoulou A, Rink D (2016) Urban shrinkage as an emerging concern for European policymaking. Eur Urban Reg Stud 23(1):103–107. https://doi.org/10.1177/0969776413481371

Haase A, Rink D, Großmann K, Bernt M, Mykhnenko V (2014) Conceptualizing urban shrinkage. Environ & Plan A: Econ Space 46:1519–1534. https://doi.org/10.1068/a46269

Hartt MD (2018) How cities shrink: complex pathways to population decline. Cities 75:38–49. https://doi.org/10.1016/j.cities.2016.12.005

Herrmann DL, Shuster WD, Mayer AL, Garmestani AS (2016) Sustainability for shrinking cities. Sustainability 8(9):911. https://doi.org/10.3390/su8090911

Hoekveld JJ (2012) Time-space relations and the differences between shrinking regions. Build Environ 38(2):179–195. https://doi.org/10.2148/benv.38.2.179

Hollander JB, Pallagst K, Schwarz T, Popper FJ (2009) Planning shrinking cities. Prog Plan 72(4):223–232

Hospers GJ (2014) Policy responses to urban shrinkage: from growth thinking to civic engagement. Eur Plan Stud 22(7):1507–1523. https://doi.org/10.1080/09654313.2013.793655

Instituto Nacional de Estadística (INE) (2022) Estadísticas de padrón continuo. Retrieved June 21, from https://www.ine.es/dyngs/INEbase/es/operacion.htm?c=Estadistica_C&cid=1254736177012&menu=ultiDatos&idp=1254734710990

Karp DN, Bagchi-Sen S, Rogerson P (2022) Not all shrinking places are similar: the variegated nature of population decline in the United States. Appl Geogr 138:102581. https://doi.org/10.1016/j.apgeog.2021.102581

Kennedy P (2008) A guide to econometrics, 6th edn. Blackwell Publishing

Kim Y, Barkley DL, Henry MS (2000) Industry characteristics linked to establishmentconcentrations in nonmetropolitan areas. J Reg Sci 40(2):234–259. https://doi.org/10.1111/0022-4146.00173

Martínez-Fernández C, Audirac I, Fol S, Cunningham-Sabot E (2012) Shrinking cities: urban challenges of globalization. Int J Urban Reg Res 36(2):213–225. https://doi.org/10.1111/j.1468-2427.2011.01092.x

Martínez-Fernández C, Weyman T, Fol S, Audirac I, Cunningham-Sabot E, Wiechmann T, Yahagi H (2016) Shrinking cities in Australia, Japan, Europe and the USA: from a global process to local policy responses. Prog Plan 105:1–48. https://doi.org/10.1016/j.progress.2014.10.001

Martínez-Navarro JM, García-González JA, Escudero-Gómez LA (2020) Las ciudades medias de España y sus coronas en el siglo XXI (2002–2017): dinámica demográfica y desarrollo inmobiliario. Urbe. Revista Brasileira de Gestão Urbana 12:e20190202. https://doi.org/10.1590/2175-3369.012.e20190202

Nelle A, Grobmann K, Haase D, Kabish S, Rink D, Wolff M (2017) Urban shrinkage in Germany: an entangled web of conditions, debates and policies. Cities 69:116–123. https://doi.org/10.1016/j.cities.2017.02.006

Nijman J, Wei YD (2020) Urban inequalities in the 211st century economy. Appl Geogr 117:102188. https://doi.org/10.1016/j.apgeog.2020.102188

Noonan L, O'Leary E, Doran J (2021) The impact of institutional proximity, cognitive proximity and agglomeration economies on firm-level productivity. Journal of Economic Studies 48(2):257–274. https://doi.org/10.1108/JES-07-2019-0345

Onwuegbuzie AJ, Daniel LG (2002) Uses and misuses of the correlation coefficient. Res Sch 9:73–90

Oswalt P (ed) (2005) Shrinking cities. Hatje Cantz

Pallagst K, Aber J, Audirac I, Cunningham-Sabot E, Fol S, Martinez-Fernandez C, ... Wiechmann T (2009) The future of shrinking cities: problems, patterns and strategies of urban transformation in a global context. UC Berkeley. Retrieved June 19, 2022, from https://escholarship.org/content/qt7zz6s7bm/qt7zz6s7bm.pdf?t=krnmp5

Parrilla-González JA (2021) Does the tourism development of a destination determine its socioeconomic development? An analysis through structural equation modeling in medium-sized cities of Andalusia, Spain. Land 10(4):378. https://doi.org/10.3390/land10040378

Raco M, Street E (2012) Resilience planning, economic change and the politics of post-recession development in London and Hong Kong. Urban Studies 49(5):1065–1087. https://doi.org/10.1177/0042098011415716

Radzimski A (2018) Involving small landlords as a regeneration strategy under shrinkage: evidence from two East German cases. Eur Plan Stud 26(3):526–545. https://doi.org/10.1080/09654313.2017.1391178

Royuela V, Castells-Quintana D, Melguizo C (2017) Las ciudades españolas tras la Gran Recesión: una mirada a la población, el empleo y los precios de la vivienda. Papeles de Economía Española 153:23–33

Schober P, Boer C, Schwarte LA (2018) Correlation coefficients: appropriate use and interpretation. Anesth Analg 126(5):1763–1768. https://doi.org/10.1213/ANE.0000000000002864

Seguridad Social (SS) (2022) Afiliación y alta de trabajadores. Retrieved June 21, from https://www.seg-social.es/wps/portal/wss/internet/EstadisticasPresupuestosEstudios/Estadisticas/EST8

Servicio Público de Empleo Estatal (SEPE) (2022) Paro registrado y contratos por municipios. Retrieved June 21, from https://www.sepe.es/HomeSepe/que-es-el-sepe/estadisticas/datos-estadisticos/municipios.html

Stryjakiewicz T, Jaroszewska E (2016) The process of shrinkage as a challenge to urban governance. Quaestiones Geographicae 35(2):27–37. https://doi.org/10.1515/quageo-2016-0013

Vendemmia B, Kërçuku A (2020) Internal migration in Italy: mapping shrinking in medium-size cities. In CIST2020—population, time, territories. Paris. Retrieved June 19, 2022, from https://cist2020.sciencesconf.org/319254/document

Chapter 16
The Peripheries of Spanish Developmentalism. An Enduring Model of Urban Fragmentation

Jesús M. González-Pérez, Juan M. Parreño-Castellano, and Dolores Sánchez-Aguilera

Abstract One of the urban spaces that best identify Spanish developmentalist cities lies on their outlying residential fringes. Built between the 1960s and the 1980s, they are dense, mono-functional areas, largely developed at the initiative of the public authorities during the Fordist stage. Areas initially planned to accommodate marginalized domestic migrants, 50 years later, they stand out for their serious shortcomings, with a resident population made up of vulnerable social classes. Within the above framework, the aim of this article is to analyse processes of urban fragmentation, examining the sociodemographic changes experienced by three neighbourhoods all representative of this peripheral growth, located in Barcelona, Palma (Mallorca) and Las Palmas de Gran Canaria. An analysis is made of biodemographic and socioeconomic indicators in order to characterize the said neighbourhoods according to their level of vulnerability and the changes that have taken place. Broadly speaking, from the analysis, two characteristic typologies were identified: neighbourhoods where substitutionary migratory processes have occurred and others still inhabited by vulnerable groups of local origin, partly due to difficulties in the upward social mobility of their inhabitants. Despite the time that has passed, as a whole, they continue to be areas for the social reproduction of the most vulnerable social classes in cities that are even more fragmented in today's current post-pandemic scenario.

J. M. González-Pérez (✉)
Department of Geography, University of the Balearic Islands, Cra. de Valldemosa km. 7.5, 07122 Palma, Majorca, Spain
e-mail: jesus.gonzalez@uib.es

J. M. Parreño-Castellano
Department of Geography, University of Las Palmas de Gran Canaria, Pérez del Toro, 1, 35003 Las Palmas de Gran Canaria, Spain
e-mail: juan.parreno@ulpgc.es

D. Sánchez-Aguilera
Department of Geography, University of Barcelona, c/ Montalegre 6, 08001 Barcelona, Spain
e-mail: dsanchez_aguilera@ub.edu

E. Navarro-Jurado et al. (eds.), *Urban Dynamics in the Post-pandemic Period*, The Urban Book Series,
https://doi.org/10.1007/978-3-031-36017-6_16

Keywords Urban fragmentation · Housing estates · Vulnerability · Immigration

16.1 Introduction. The Urban Peripheries of Spanish Developmentalism. The Working Class City

The urban peripheries of the post-Fordist cities are the outcome of the use of urban planning models of diffuse urbanization, composed of discontinuous low-density urban fabrics made up of segregated or fragmented urban parts. In general terms, these areas are mono-functional and are specialized in different uses, and they are mainly the result of privately-financed development projects (Bauer and Roux 1976; Hall 2014; Indovina 1990; Monclús 1999).

The peripheries of Spanish developmentalist cities were nonetheless based on a completely different set of principles and goals. Their function was mainly residential and they were often developed as a result of public sector initiatives, whether directly or indirectly. In any case, their location and function are based on the logic of urban segregation. In the case of the private developers of shopping centres, private hospitals or residential estates deliberately promote segregated developments to give them a more gentrified feel. In the case of public developers, during the Franco regime, the Ministry of Housing, the Obra Sindical del Hogar (a state body in charge of building public housing) or the Ministry of Labour, among others, created residential areas for the poor migrant working classes through high-density development projects which were segregated or delimitated due to the limitation of costs and the development of a policy oriented primarily to the business of the construction companies and to the satisfaction of the interests of some landowners (García Cuesta 2000; Parrilla Recuero 2020).

Consequently, one of the urban spaces that best identifies Spanish developmentalist cities are their residential peripheries, mainly represented by housing estates. From an architectural point of view, the Instituto Nacional de la Vivienda (the National Housing Institute) advocated an urban model influenced by the Modern movement and by garden cities, extolling the construction of apartment buildings with a double-loaded corridor inspired by the Hygienist movement, fragmented open buildings and areas where wheeled traffic was restricted, with the recommendation of Radburn's superblock (De Terán 1999). From a socio-urban perspective, the aim was to create working class housing at the lowest possible cost, locating it on the peripheries. These areas were hard to integrate into the urban fabric and they stood out for their substandard services and poor-quality buildings (Capel 2002). They were segregated places designed to accommodate industrial workers or low-skilled service-sector ones working in the field of tourism, especially in the case of the archipelagos (González-Pérez and Piñeira-Mantiñán 2020; González-Pérez 2022) and, on some occasions, intentionally segregated from the rest of the city when they were conceived to act as containers for the ethnic and economic-led socially excluded population (Rio Ruiz 2019).

Half a century later, evident improvements have been made to most of these neighbourhoods, particularly in terms of their urban planning and infrastructure, but they are still places with serious shortcomings (Beltrán Abadía 2002). Although they are no longer located on the peripheries and they have gained in centrality with the growth of the city, generally speaking, they continue to be marginal spaces fragmented from the rest of the city in social, perceptual and symbolic terms.

Within this framework, the aim of this chapter is, first, to study the urban fragmentation of three neighbourhoods or medium-sized housing estates (Ciutat Meridiana in Barcelona, Son Gotleu in Palma and Las Rehoyas in Las Palmas de Gran Canarias) representative of the aforementioned peripheral growth and domestic migratory dynamics of the 1960s and 70s. These neighborhoods shared, from their origin, a high rate of immigrant population, low income levels and high indicators of poverty in the context of their respective cities. Second, it analyses the evolution of this fragmentation through the spatial, demographic and socioeconomic changes that the three neighbourhoods have undergone with the passing of the years.

The study uses secondary sources of urban planning, demographic and socioeconomic data. The issue of the urban fragmentation of these spaces is tackled from a diachronic perspective, analysing locations, the morphological and housing characteristics of the built spaces, the population's structure by age, sex and nationality and the mean income per person.

16.2 Urban Fragmentation as an Exponent of Social Inequality

Urban fragmentation is a polysemic concept (Szupiany 2018), used as an analytical category in recent decades in combination with different terminology (quartered city, layered city, rebundled city, fractal city, etc.) (Marcuse 1989; Soja 2008). Although for Kozak (2018), it is a term that is not supported by a particularly broad, well-developed theoretical framework, it is also true that, in recent years, a new pattern of segregation with urban fragmentation has emerged that accounts for the structure of today's cities and which seems to give the term greater sense. This new pattern is based on the increasing proliferation of separate urban areas with limited permeability, on a small scale, and on organization around nodes, on a large scale (Burgess 2018).

Since its origins, fragmentation has been associated with social segregation and, by extension, with poverty and vulnerability and hence also with cultural and ethnic inequalities (Veiga 2004). These inequalities tend to be reflected in urban spaces through the existence of residential segregation, based on income, nationality, culture, ethnicity, race, life style or religion.

Similarly, because parallels can be drawn among urban fragmentation, dual cities, and the social division of space (Duhau 2013; Pérez-Peñuelas 2021), the terms urban segregation and fragmentation are often used as synonyms, which is not terminologically correct, prompting theoretical responses from different standpoints (Kozak

2018). Academic literature has highlighted the importance of land markets and income in explaining patterns of residential segregation. However, the consequences of urban fragmentation have not been systematically explored in order to gain a better insight into the structure and dynamics of residential segregation (Burgess 2018).

Urban segregation is sometimes linked to the existence of physical or symbolic intra-urban limits or to the presence of a structure based on scattered, segregated urban areas that hinder the population's mobility. When this occurs, residential segregation, a term alluding to the distribution or spatial patterns of certain groups,turns into urban fragmentation. The first examples of this model of fragmentation can be tied in with the segregated development of specific areas for vulnerable communities of people; for instance, segregated areas of poor-quality housing, created as a result of housing policies, or informal settlements conspicuous not only for their morphological characteristics but also for poverty, marginality and insecurity. These processes, whose maximum exponents can be found in Latin-American cities (Sabatini 2006; Usach and Freddo 2016), have also featured in the urban configuration of Spanish cities (Parreño-Castellano and Moreno-Medina 2006).

Although other areas with a high concentration of vulnerable people might not be physically fragmented, they can still be symbolically fragmented, with their conversion into ghettoes or the stigmatization of their inhabitants. This is the case of ethnic neighbourhoods or areas that stand out for their poverty, high level of insecurity or sharp drop in the value of housing, even though access to them is not a problem. Both fit in with the dynamics of the three areas studied in this paper.

16.3 Fragmentation and the Built Space: The Disconnected City

The chosen case studies are three neighbourhoods that exemplify fragmentation processes in Spanish urban contexts. They were conceived, right from the outset, to be set apart from the rest of the city, both in terms of their urban characteristics and those of their inhabitants. With the passing of the years, certain factors that differentiate them have become less acute and so, by extension, their fragmentation has lessened, although other aspects still persist, despite the social and even morphological dynamism of some of the neighbourhoods. In this section, we will analyse the fragmentation of these areas from the perspective of their location, morphology and the quality of their housing.

16.3.1 Location

The first factor in the fragmentation of these neighbourhoods is their peripheral or even segregated location in the city where they were created (Fig. 16.1). Ciutat Meridiana stands on the fringes of the municipality of Barcelona in a relatively isolated area, close to Collserola Nature Reserve—Barcelona's green lung—flanked by fast roads that partly cut off the neighbourhood from the urban sprawl of Barcelona's built-up areas. Las Rehoyas was also planned to occupy a peripheral, segregated part of Las Palmas de Gran Canarias on several parcels of land that belonged to a big agricultural estate in Barranco de Mata, on marginal areas of a steep slope.

Son Gotleu was also at a distance from the city's most dynamic areas when it was created, albeit not to such an outlying extent. A peripheral urban area on the outer limits, at some distance from the city's initial growth area, today it is hemmed in by Palma's ring road. As with the other cases, its peripheral location was mainly attributable to a search for cheap land, but also to a desire to keep the working class population separate and far from the rest of the city (González-Pérez 2017).

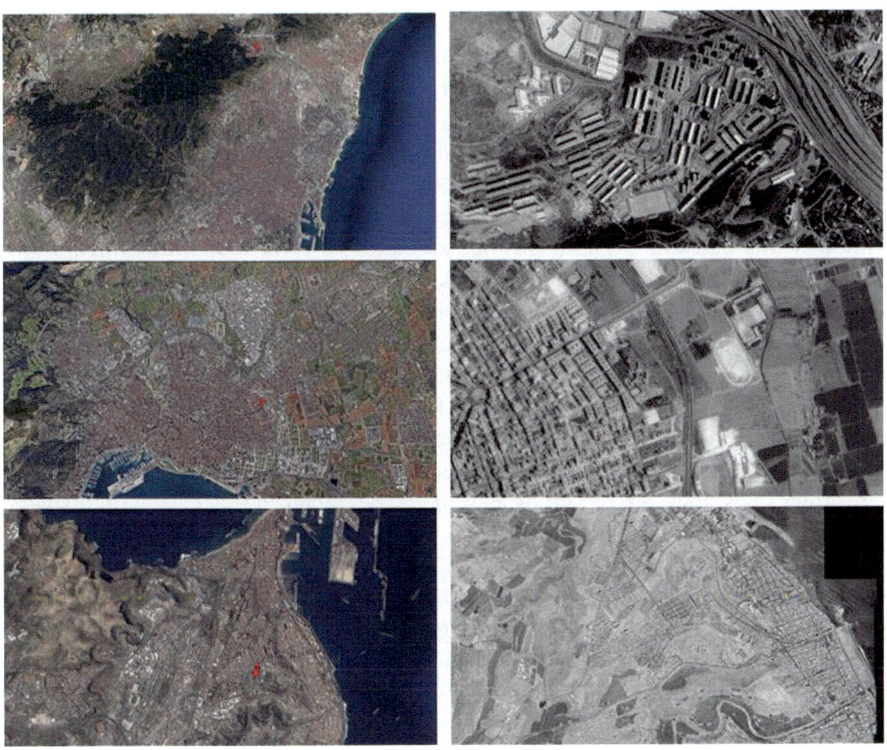

Fig. 16.1 Location of Ciutat Meridiana (top), Son Gotleu (centre) and Las Rehoyas-Arapiles (bottom). Aerial photo taken in 1983 and orthophotos taken in 1984 and 1966, respectively. *Source* Own using Google Earth, Geoportal AMB, IDEIB and IDECanarias

Some of the first factors in the fragmentation of these neighbourhoods include a separate non-central position, physical constraints due to the location of transport infrastructure (mainly roads) or the presence of steep slopes, leading to hindered access to the rest of the city. This fragmentation, as a result of their locations, has been partly offset as the city has spread, coming to surround some of these neighbourhoods, and by projects to overcome certain physical obstacles, for instance by bringing the underground to the neighbourhood, by building escalators and a funicular in Ciutat Meridiana or by creating lifts to lead to street level in Las Rehoyas.

16.3.2 Urban Morphology and the Quality of Housing

Generally speaking, these are neighbourhoods with a similar morphology, with a predominance (and sometimes this is the only type) of high-rise housing. Son Gotleu is a good example (Fig. 16.2, top). Although a small part of the neighbourhood was developed under the guidelines of the Palma General Alignment and Reform Plan (1943), explaining why the urban fabric of Palma's initial growth area was respected in a small south-western area, Son Gotleu has a predominance of social housing promoted by the Spanish Housing Ministry with an urban and built morphology based on the concept of a housing estate (González-Pérez and Piñeira-Mantiñán 2020).

The oldest social housing (Ses Coromines) dates back to 1960. The type of architecture is reminiscent of the housing built by the Vivienda Sindical del Hogar, with building blocks of five-storey housing and open spaces between the buildings. Although this housing is based on the principles of early modern architecture, it stands out for the poor construction work and small size of the dwellings (55 m^2 according to cadastral records). Another big area of social housing, created between 1964 and 1970, appears separated by the street Indalecio Prieto. Housing of this area was built to be rented out under the Spanish Limited Income Housing Act of 15 July 1954 and Spanish Subsidized Housing Act of 13 November 1957. The area comprises nineteen blocks of apartments along a street known as Calle Tomás, which cuts through them from north to south. Although the buildings are bigger than the former ones, they tend to be of a similar type, albeit in this case with a shared dividing wall (González-Pérez 2022).

Las Rehoyas in Las Palmas de Gran Canaria (Fig. 16.2, bottom) also takes the form of a housing estate, contrasting with traditional settlements in Risco and the growth areas of its immediate surroundings. The neighbourhood is made up of two different areas, developed by the Patronato Benéfico de la Construcción Francisco Franco between 1959 and 1965: Las Rehoyas Bajas or El Polvorín and Las Rehoyas Altas-Arapiles, united by the barracks of the BRIPAC paratroopers brigade. The first, which adjoins the central neighbourhoods of Arenales and Risco de San Antonio, is made up of 23 four or five-storey long rectangular buildings with a double-loaded corridor and 686 very small dwellings (between 38 and 42 m^2). The buildings were constructed on a steep slope, organically positioned alongside a zigzagging road,

Fig. 16.2 Morphology of Ciutat Meridiana, Barcelona (top), Son Gotleu, Palma (centre) and Las Rehoyas-Arapiles, Las Palmas de Gran Canaria (bottom). *Source*: Own based on Google Earth, http://socialarqui.blogspot.com/2014/12/viviendas-de-la-etapa-franquista-en-las.html

with an evident lack of open space and community services. The building quality was very poor and so problems soon arose in their storage and sanitation facilities and in the waterproofing of roofs, with the emergence of aluminosis among other problems.

The Rehoyas Altas-Arapiles area began to be developed shortly afterwards, leading to 90 buildings and 2,558 dwellings. The urban development model included buildings similar to those of El Polvorín and tripod or star-shaped buildings, once again on hillside terraces where there was a steep slope in order to occupy as much land as possible. The dwellings were between 52 and 59 m^2. Unlike Las Rehoyas Bajas, different forms of access to housing were possible; for instance, through their purchase or rental, or by paying rent deductible from their purchase price. In the development of the area, the remaining sloping areas were transformed into most of its open space, with communal gardens between two parallel blocks when possible.

Lastly, Ciutat Meridiana is a similar example (Fig. 16.2, top). It is the outcome of the privatization of public housing through its sale from the late 1950s in Spain. The neighbourhood, built between 1963 and the early 1970s, stands on plots of land initially earmarked for the construction of a cemetery although the land was found to be too wet. Architecturally, it is based on building blocks of open housing, with higher buildings positioned so as to take full advantage of the steep sloping land. The outcome is a densely built area, originally lacking in services. From the very start, the housing conditions were not good, with chronic situations such as a lack of insulation in buildings or serious functional problems like an absence of lifts.

Over the years, in all cases, the quality of the housing has improved in terms of its services and facilities. For instance, in Rehoyas, there is now a big urban park with sports facilities where the paratroopers' barracks used to be. The same is also true of the housing conditions, since in all cases refurbishment projects have been carried out with greater or lesser success. However, the state of the buildings is generally poor and ongoing maintenance is needed. To offer a better insight, in the case of Rehoyas-Arapiles, 55% of the housing was estimated to be in a bad, poor or ruinous state (Spanish Ministry for Transport, Mobility and Urban Affairs 2021). Costly replacement projects have even been carried, as in Rehoyas Bajas (Gago Vaquero 2010; Parreño-Castellano and Suárez-Naranjo 2020), and new projects are underway, like the one that will affect all the buildings in Rehoyas Altas-Arapiles.

Despite all this, their morphological homogeneity and poor-quality housing still persist today as factors in the fragmentation and social stigmatization of these neighbourhoods.

16.4 Fragmentation Due to the Demographic Features That Set Them Part: The Ethnified City

In this section, we will analyse the fragmentation of these areas due to the urban factors that set them apart in terms of the origins, nationality and biodemographic characteristics of their populations, examining the issue from a temporal perspective. These distinctive demographic characteristics have boosted their vulnerability.

16.4.1 Migratory and Ethnic Factors That Set Them Apart

As we indicated earlier, the origin of these neighbourhoods can be traced back to the increase in the population due to strong natural growth and to the rural exodus both within and between provinces from 1955 to well into the 1960s in Spain. From their initial creation, the three neighbourhoods were used to accommodate low-skilled, low-income migrant workers, employed in the industrial, port and service sectors. Las Rehoyas mainly absorbed families from Las Palmas and from rural areas in the province, while Son Gotleu and Ciutat Meridiana mainly attracted migrants from other Spanish provinces. These people's origins and low skills set them apart socially, determining ensuing fragmentation processes.

This situation lasted until at least the late 1990s, when a change started to occur in the neighbourhoods' biodemographic makeup. With the growing international migratory flows between the twentieth and twenty-first centuries, particularly during the first decade of the twenty-first century, largely motivated by the attraction of employment brought about by the booming real estate sector, a process of diversification and ethnic specialization began, particularly in Barcelona and Palma.

The economic growth of the first decade of the twenty-first century was founded on a high influx of foreigners, with these immigrants making up 20.5% of the population of Palma and 17.2% in Barcelona in 2011 (Fig. 16.3). In this scenario, the neighbourhoods of Son Gotleu and Ciutat Meridiana had even higher percentages—40% and 35.7%, respectively—confirming the swift process of social substitution that had occurred in just over a decade. In the case of Las Rehoyas, the replacement of the neighbourhood's original groups was in the early stages that year and so the area had a low percentage of foreign residents, below that of the average figure for the city of Las Palmas de Gran Canarias.

The effects of the economic crisis explain the slowdown in new migratory flows and the increase in Spanish citizenship or the return home of foreigners who had arrived some years earlier. As a result, in Son Gotleu and Ciutat Meridiana, a drop in the relative share of the foreign population could be noted in 2021, although the neighbourhoods still stood out for their strong ethnic mix. In contrast, in this second decade of the twenty-first century, there has been a slow but sustained increase in foreigners in Las Rehoyas in contrast with the whole city of Las Palmas de Gran Canarias. In this case, the new trend is related to changes in the origin of migratory flows, gentrifying dynamics in areas ethnified during the previous decade and the replacement of part of the population with other nationalities in Las Palmas de Gran Canarias in recent years (Parreño-Castellano et al. 2021).

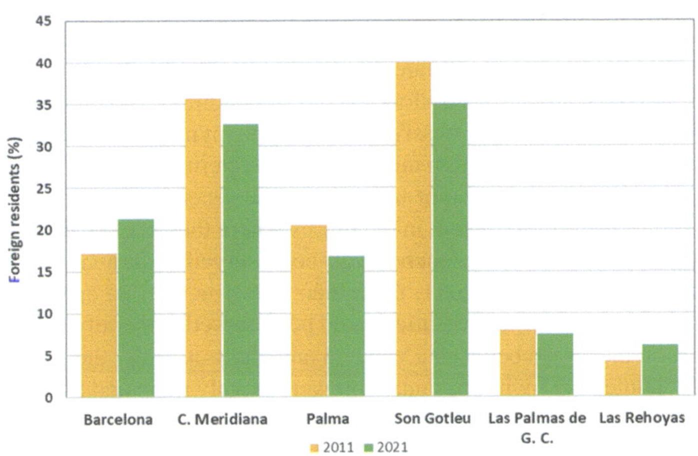

Fig. 16.3 Percentage of foreign residents in 2011 and 2021. *Source* Own based on the Spanish Statistics Office, Continuous Population Census 2011 and 2021

Son Gotleu and Ciutat Meridiana's strong ethnification is a recent factor in their urban fragmentation and it can be better understood if we bear in mind the nationality of their foreign residents. In the first case, in 2011, there was a predominance of Africans (from Nigeria, Morocco and Senegal), together with a high presence of Ecuadorians, Pakistanis and Chinese people. In 2021, these groups were much the same, with the emergence of an Asian community who grew proportionally. By nationality, the makeup of Son Gotleu's foreign residents differs from that of the city of Palma as a whole, where Europeans predominate (Italians, Bulgarians, Germans) in addition to Latin-Americans (from Ecuador and Colombia). Son Gotleu stands out as having a foreign resident population with lower incomes, more job instability and the biggest language difficulties.

Ciutat Meridiana is a similar case. In 2011, it had a predominance of Ecuadorians, followed by a mix of different nationalities (from Pakistan, Nigeria and the Dominican Republic). No groups of European origin could be noted although Italians did make up a specific share of the population of Barcelona. In 2021, the most prevalent group of foreigners in Ciutat Meridiana was from Honduras—a recent flow that has trebled in size in Barcelona in recent years—together with communities from Pakistan, Morocco and other Latin-American countries (the Dominican Republic, Ecuador), whose share has dropped. The concentration of recent flows of immigrants from developing countries is further evidence of the urban vulnerability and fragmentation associated with ethnification.

16.4.2 Factors that Set Them Apart in Terms of Age and Sex

Due to the residential impact of developmentalism and the baby boom associated with it, the structure of the population of these neighbourhoods by age and sex is characterized by the presence of progressive pyramids, buoyed up by the young population's high demographic share. The demographic evolution of these neighbourhoods is characterized by a slow ageing and feminization process, revealed through the consolidation of regressive demographic pyramids in the late 1990s.

At that point, the arrival of high numbers of foreign immigrants from developing countries marked the starting point in Spain of the second demographic transition (Delgado and Martínez 2019) and, in the places where they settled, the rejuvenation of their populations. This phenomenon can be observed in Son Gotleu and Ciutat Meridiana (see Fig. 16.4). In 2011, from their age structure, an ageing population lower than that of their respective cities could be observed, generally speaking with a higher share of men aged between 25 and 44 and children under the age of ten. This could be accounted for by immigrants who were largely male and by an increase in the fertility rate, mainly due to the reproductive model of a large part of the immigrant population of Asian and African origin.

In 2021, the halt in immigrant arrivals, the increase in immigrants returning home and a drop in the fertility rate with the economic crisis brought the population pyramids of these two neighbourhoods more in line with their respective cities,

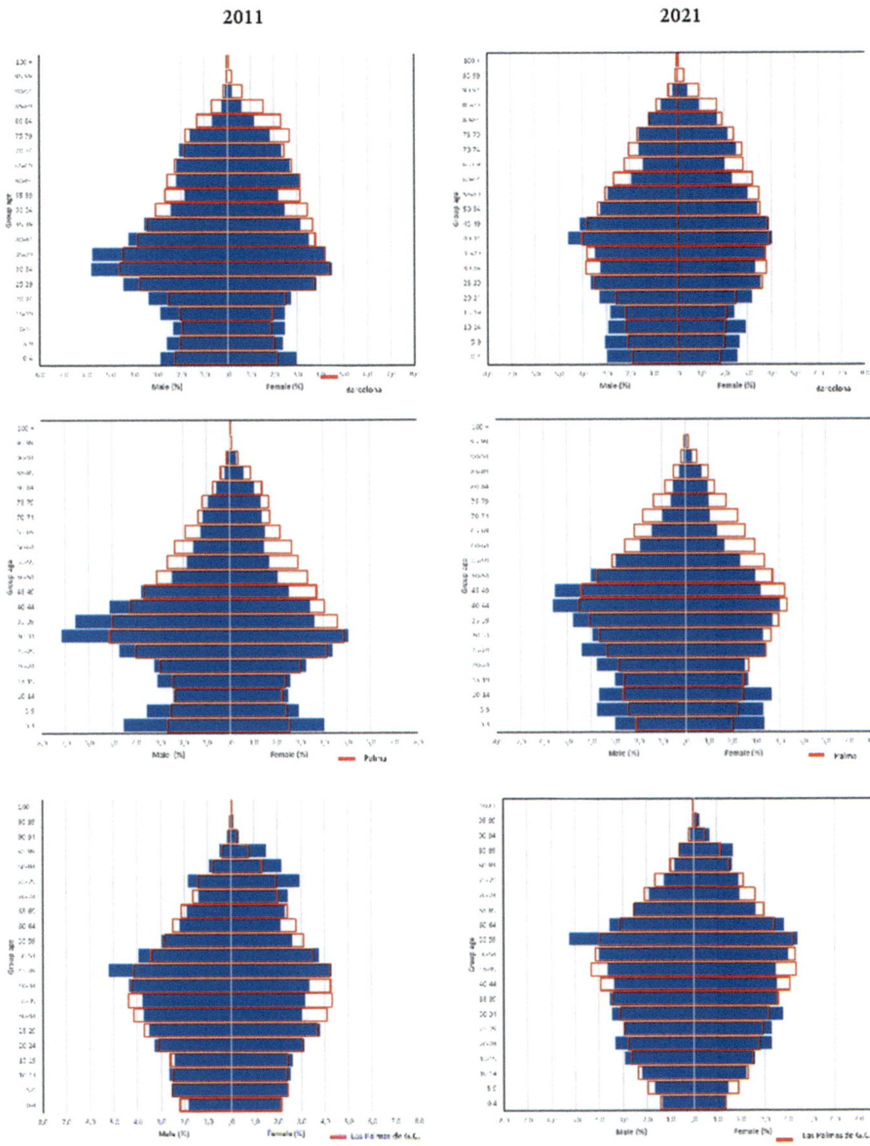

Fig. 16.4 Structures by age and sex in 2011 (left) and 2021 (right). *Source* Own based on the Spanish Statistics Office; Continuous Population Census 2011 and 2021. The respective municipalities are shown in red

with newly regressive models, above all in Ciutat Meridiana. Nonetheless, in both neighbourhoods, the under 15 age group continues to have a greater relative presence.

In the case of Las Rehoyas, its lower level of immigrants led to a more aged pyramid in relation to that of the whole municipality in 2011, with a high relative share of women aged 70 or over. Ten years later, although the structure by age is even more regressive, with a lower relative presence of children, a slight drop in the population's ageing process can be noted due to the intensification of migratory flows and, above all, to some older people's move away to other areas of the city with the replacement processes that occurred in El Polvorín.

As a result, in addition to the differences in the built space that set these neighbourhoods apart, a clear process of biodemographic fragmentation can also be observed, brought about, firstly, by the very high presence of Spanish migrants in the three analysed cases during the first five-year periods and of foreigners from developing countries latterly, with consequent integration problems due to their origins or recent arrival in Son Gotleu and Ciutat Meridiana. Secondly, it is also the outcome of the greater vulnerability represented by the very high presence of children and young sectors of the population during the early years of these neighbourhoods and the higher feminization and growing ageing population that have ensued, particularly in the case of Las Rehoyas.

16.5 Fragmentation and the Socioeconomic Reality: The Segregated City

From a socioeconomic viewpoint, these neighbourhoods were created to house disadvantaged populations. Decades later, they continue to be impoverished neighbourhoods and characterized by their higher socioeconomic vulnerability. The data for the mean net income per person confirms this situation (see Table 16.1): the three neighbourhoods are among the poorest in their respective cities. In absolute terms, in 2019, their census sections had incomes ranging between 6,153 and 9,624 euros, meaning, in some cases, that the income level did not even reach half the mean figure for the municipality. Son Gotleu stood out in this sense, with incomes below 40% of the mean figure for Palma in four of its five census sections.

The low-income levels and their evolution between 2015 and 2019 can be accounted for by the reverse relationship with the percentage of African and Asian foreigners. In Son Gotleu, for instance, the census section with the highest mean income is the only one in which there is a greater share of Latin-American immigrants, while in the rest there is a predominance of Nigerians, Moroccans and Pakistanis. In Ciutat Meridiana, on the other hand, the lower increase in income over the last two years is associated with the increase in the share of Pakistani and Moroccan people and the lower share of Latin-American immigrants.

In Las Rehoyas, the income values are related to the incapacity of the labour market to find employment for a low-skilled population traditionally employed in

Table 16.1 Mean income per person by census section

	Mean income per person (€)		Income by census section/mean city (%)		Mean income per person indexed evolution (2015 = 100)	
AREA	2015	2019	2015	2019	2015	2019
Palma	16,000	18,968	100.0	100.0	100.0	118.6
Son Gotleu						
0704004019	5,413	7,178	33.8	37.8	100.0	132.6
0704004022	4,791	6,153	29.9	32.4	100.0	128.4
0704004023	5,262	6,580	32.9	34.7	100.0	125.0
0704004024	4,510	6,200	28.2	32.7	100.0	137.5
0704004026	8,056	9,624	50.4	50.7	100.0	119.5
Barcelona	14,946	16.894	100.0	100.0	100.0	113.0
Ciutat Meridiana						
0801908110	6,688	7,474	44.7	44.2	100.0	111.8
0801908111	7,795	8,468	52.2	50.1	100.0	108.6
0801908112	7,520	8,612	50.3	51.0	100.0	114.5
0801908113	7,907	8,547	52.9	50.6	100,0	108.1
0801908114	8,359	9,020	55.9	53.4	100.0	107.9
0801908115	7,897	8,893	52.8	52.6	100.0	112.6
Las Palmas de G.C	10,536	12.429	100.0	100.0	100.0	118.0
Las Rehoyas						
3501604007	5,519	7,000	52.4	56.3	100.0	126.8
3501604008	5,655	7,246	53.7	58.3	100.0	128.1
3501604016	5,855	7,272	55.6	58.5	100.0	124.2
3501604017	6,070	7,879	57.6	63.4	100.0	129.8
3501604018	7,432	–	70.5	–	100.0	–
3501604019	6,954	8,276	66.0	66.6	100.0	119.0
3501604022	6,286	7,589	59.7	61.1	100.0	120.7
3501604023	6,861	8,345	65.1	67.1	100.0	121.6
3501604024	5,956	7,791	56.5	62.7	100.0	130.8

Source Based on the Spanish Statistics Office, Atlas of household income

the construction sector (GEURSA 2013) with a low level of education and to the high proportion of pensioners, widows and women with no income. The employment rate in 2012 was just 22.5%.

Concentrations of low-income residents and the presence of impoverished groups are a clear factor in fragmentation through differentiation, and the insecurity that comes with this further accentuates this fragmentation.

16.6 Conclusions

The urban peripheries built during the years of Spanish developmentalism under the 2nd and 3rd Spanish National Housing Programmes (1955–1976) represented a breakaway from the growth models of Spanish cities back them, both morphologically and in terms of their urban landscapes. The "ensanche" as a form of urban growth became a thing of the past; concepts associated with functional cities and Scandinavian satellite cities were incorporated (green areas, the hierarchization of roads, open isolated blocks, decongestion, extreme zoning, etc.) and strategically planned segregated housing blocks were built, leaving undeveloped spaces between them. This policy did indeed help to mitigate the need for housing brought about by the exodus of the rural population of the time, but it was an authentic failure as it led to the segregation and exclusion of large groups of the most vulnerable sectors of the population. Six or seven decades after the construction of these peripheral areas, their fragmentation persists to a greater or lesser extent, both in terms of their built space and in their social differences, contributing to continuing negative perceptions of them by other residents of the city.

Barcelona, Palma and Las Palmas de Gran Canarias are good case studies of this kind of urban development. Through this comparative study of the three housing estates, a more in-depth analysis has been made of current debate on urban fragmentation. Irrespective of the dynamics that have affected the three neighbourhoods under study; they still retain three of the original characteristics that impact on their continued fragmentation: impoverishment, exclusion and a peripheral location.

From the perspective of the fragmentation of the built space, the three case studies stand out for their segregated peripheral location and for a morphology and housing characteristics that clearly set them apart. Some of the physical factors that contribute to their fragmentation have been partly mitigated over the years through refurbishment projects or the replacement of buildings or else the neighbourhoods have been equipped with services and facilities, but this has still not overcome the legacies of the mass development of dense estates of poorly built, low-quality housing.

Improvements to the built space have not managed to prevent these neighbourhoods from being marginal areas, rejected by most of the population of their cities, and demographic and income-related factors continue to set them apart, partly due to a lack of efficient plans in the field of social action. The demographic fragmentation of the said neighbourhoods can be traced back to their origins, and they are spaces of social reproduction that first served to accommodate the working class population from mainland rural areas (Ciutat Meridiana and Son Gotleu) or from the island of Gran Canarias' own rural areas (Las Rehoyas). In the late 1990s, there were changes in their biodemographic makeup in the first two cases, with the widespread replacement of part of the population, when Latin-Americans (Ciutat Meridiana) and Africans (Son Gotleu) became the most numerous groups of foreigners. Despite the change in the nationality and origins of the resident population, they continued to be fragmented spaces.

They also continue to stand out from the rest of the city for other biodemographic reasons. Originally, they were places with a younger demographic structure than the rest of the city. Over the years, they have become neighbourhoods with regressive dynamics, with a very high ageing or feminized population in the case of Rehoyas. The characteristics of their residents have changed, but the fragmentation still exists.

From a socioeconomic point of view, Son Gotleu, Ciutat Meridiana and Las Rehoyas have always been vulnerable, impoverished areas. Indeed, today, they are the neighbourhoods with the lowest incomes in their respective cities. In the census section with the lowest mean income in Son Gotleu, this figure is 37.8% of the mean value for Palma; in the case of Ciutat Meridina, it is 44.25% of the figure for Barcelona; and in the case of Las Rehoyas, it is 56.3% of the figure for Las Palmas de Gran Canaria. Poverty and a lack of policies clearly designed to tackle the problem have aggravated possibilities of social mobility for their residents while also hindering their integration.

In short, during the developmentalist period, the approach that was taken to the housing problem of the said cities' most vulnerable groups resulted in a form of urban development and a social fabric that, in the absence of any efficient subsequent solution, have played a key role in preventing the replacement of these groups of people with others of a higher social standing. In other words, the persisting fragmented role of these neighbourhoods in today's post-pandemic city can be traced back to their very origins.

Acknowledgements Grant "Housing and international mobility in cities of the Balearic Islands: the emergence of new forms of urban inequality" (RTI2018-093296-B-C22) funded by MCIN/AEI/10.13039/501100011033 and by "ERDF A way of making Europe"

Grant "Housing and international mobility in cities of Canary Islands: the emergence of new forms of urban inequality" (RTI2018-093296-B-C21) funded by MCIN/AEI/10.13039/501100011033 y "ERDF A way of making Europe". Grant "Cities in transition. Urban fragmentation and new socio-spatial patterns of in-equality in the post-pandemic context. The case of the urban area of Palma (Mallorca)" (PID2021-122410OB-C31), funded by MCIN/AEI/10.13039/501100011033 and by "ERDF A way of making Europe".

References

Bauer G, Roux JM (1976) La rurbanisation ou la ville éparpillée. SEUIL

Beltrán Abadía RB. (2002) De aquellos barros, estos lodos: la política de vivienda en la España franquista y postfranquista. Acciones e investigaciones sociales 16:25–67. https://doi.org/10.26754/ojs_ais/ais.200216233

Burgess R (2018) Segregación y fragmentación urbana: algunos comentarios. Revista Iberoamericana de Urbanismo 14:79–86

Capel H (2002) La morfología de las ciudades. Sociedad, cultura y paisaje urbano. Ediciones del Serbal

De Terán F (1999) Historia del urbanismo en España III. Siglos XIX y XX. Cátedra

Delgado JM, Martínez LC (2019) Composición y cambio de los comportamientos sociodemográficos en España en los inicios de una Segunda Transición Demográfica. Cuadernos Geográficos 58(1):253–276. https://doi.org/10.30827/cuadgeo.v58i1.6750

Duhau E (2013) La división social del espacio metropolitano. Nueva Sociedad 243:79–91

Gago Vaquero JL (2010) El Polvorín. Reposición en historia de un barrio. Ediciones del Umbral

García Cuesta JL (2000) De la urgencia social al negocio inmobiliario: promoción de viviendas y desarrollo urbano en Valladolid (1960–1992). Ayuntamiento de Valladolid

GEURSA (2013) Diagnóstico social del barrio de Las Rehoyas-Arapiles 2012 y opinión sobre su reposición. Ayuntamiento de Las Palmas de G.C

González-Pérez JM (2017) Les ciutats de les Balears. Processos d´urbanització i urbanisme. Lleonard Muntaner

González-Pérez JM (2022) Evictions, foreclosures, and global housing speculation in Palma, Spain. Land 11(2):293. https://doi.org/10.3390/land11020293

González-Pérez JM, Piñeira-Mantiñán MJ (2020) La ciudad desigual en Palma (Mallorca): geografía del confinamiento durante la pandemia de la COVID-19. BAGE 87. https://doi.org/10.21138/bage.2998

Hall P (2014) Cities of tomorrow: an intellectual history of urban planning and design since 1880. John Wiley & Sons

Indovina, F. (ed.) (1990). La città diffusa, Daest-IUAV.

Kozak, D. (2018). Revisitando el debate sobre la fragmentación urbana: una década y media después de "Splintering Urbanism". Revista Iberoamericana de Urbanismo 14:15–22. https://doi.org/10.18682/cdc.vi134.5016

Marcuse P (1989) Dual city: a muddy metaphor for a quartered city. Int J Urban Reg Res 13(4):697–708. https://doi.org/10.1111/j.1468-2427.1989.tb00142.x

Ministerio de Transportes, Movilidad y Agenda Urbana (2021) Catálogo de Barrios Vulnerables de España 2011. Retrieved February 12, 2022, from https://www.mitma.gob.es/arquitectura-vivienda-y-suelo/urbanismo-y-politica-de-suelo/observatorio-de-la-vulnerabilidad-urbana/analisis-urbanistico-de-barrios-vulnerables/catalogos_bv

Monclús FJ (1999) Ciudad dispersa y ciudad compacta. Perspectivas urbanísticas sobre las ciudades mediterráneas. D'Humanitats 7:95–110

Parreño-Castellano JM, Domínguez-Mujica J, Moreno-Medina C (2021) Etnificación, desahucio por impago de alquileres y desplazamiento residencial en una ciudad turística insular (Las Palmas de Gran Canaria - España). Estudios Geográficos 82(291):e081. https://doi.org/10.3989/estgeogr.202192.092

Parreño-Castellano JM, Moreno-Medina C (2006) La creación de asentamientos públicos segregados en los setenta. El caso de Jinámar (Islas Canarias, España). Eria 70:175–190. https://doi.org/10.17811/er.0.2006.175-190

Parreño-Castellano JM, Suárez-Naranjo M (2020) El difícil camino para la reposición de los polígonos de vivienda de promoción pública en Las Palmas de Gran Canaria. In: Parreño JM, Moreno C (eds) Geografías Urbanas de Gran Canaria y Fuerteventura. Mercurio, pp 59–73

Parrilla Recuero A (2020) Urbanismo y vivienda en el Madrid de posguerra. La Falange, entre el derecho a la ciudad y la segregación espacial. URBS. Revista de Estudios Urbanos y Ciencias Sociales 10(1):117–126

Pérez-Peñuelas BO (2021) Problemática en la delimitación de fragmentación urbana por compartir procesos con otros impactos urbanos. Revista Ciudades, Estados y Política 8(2). https://doi.org/10.15446/cep.v8n2.91937

Rio Ruiz MÁ (2019) El tratamiento de la cuestión residencial gitana en el tardofranquismo: desigualdad, segregación y conflicto étnico. Sociología Histórica 10:115–152. https://doi.org/10.6018/sh.451221

Sabatini F (2006) La segregación social del espacio en las ciudades de América Latina. BID

Soja E (2008) Postmetrópolis. Estudios críticos sobre las ciudades y las regiones. Traficantes de sueños

Szupiany EB (2018) La ciudad fragmentada: una lectura de sus diversas expresiones para la caracterización del modelo latinoamericano. Revista de Estudios Sociales Contemporáneos 19:101–116

Usach N, Freddo BV (2016) Dispersión y fragmentación socioespacial en el crecimiento reciente de una ciudad petrolera de la Patagonia argentina. Papeles de población 22(90):265–301. https://doi.org/10.22185/24487147.2016.90.041

Veiga D (2004) Desigualdades sociales y fragmentación urbana: obstáculos para una ciudad democrática. In Torres Ribeiro AC (Comp.) El rostro urbano de América Latina. CLACSO, pp 193–209

Chapter 17
Housing, a Problem Perpetuated Over Time in Spain. New Initiatives to Promote Access to Affordable Housing in Madrid

María José Piñeira, Ramón López Rodríguez, and Francisco R. Durán

Abstract Ensuring access to affordable housing with quality standards is becoming one of the main urban policies on a global scale. In Spain, housing is a chronic problem that has been accentuated as a result of the real estate, health and environmental crisis that we have been suffering for the last twelve years. For four decades, housing policies were based on neoliberal private development, reducing investment in the development of public housing to historic lows. The fact that there are eleven million people in a situation of housing emergency has made it urgent to design new strategies that favour access to quality housing. Throughout this chapter we will approach the recommendations that are being issued from Europe in terms of housing policies, and then contextualise the problem in Spain and analyse the initiatives that are being carried out in its capital to mitigate residential vulnerability based on semi-structured interviews conducted with policy makers, municipal housing companies and non-profit organisations.

Keywords Housing emergency · Housing policies · Madrid · Spain

17.1 Introduction

Housing has become an increasingly pressing economic, social and environmental challenge (OECD 2021). The causes include rising housing prices and rents, which have undermined affordability and favoured social exclusion processes.

M. J. Piñeira (✉) · R. L. Rodríguez · F. R. Durán
Department of Geography, University of Santiago de Compostela, Santiago, Spain
e-mail: mariajose.pineira@usc.es

R. L. Rodríguez
e-mail: ramonlopez.rodriguez@usc.es

F. R. Durán
e-mail: francisco.duran@usc.es

E. Navarro-Jurado et al. (eds.), *Urban Dynamics in the Post-pandemic Period*,
The Urban Book Series,
https://doi.org/10.1007/978-3-031-36017-6_17

According to Eurostat (2021) in Europe 70% of the population reside in homes that they own, with only 30% opting to rent. The problem is that since 2010 housing prices have increased by 26% and rents by 14%. A circumstance that highlights the widening gap between the income level of the population and housing prices. The socioeconomic impacts of the 2008 economic crisis have been accentuated by the COVID-19 health crisis, generating large losses of employment and income, which have been concentrated in the most vulnerable groups, for whom the increase in housing costs represents a disproportionate burden. Consequently, a series of problems are accentuated such as the late payment of mortgages, rents or bills; over-occupation of housing; material deprivation of housing such as keeping the house warm, lack of sanitation or deterioration of buildings such as a leaking roof. A report from the European Parliament in 2021 echoed the large number of Europeans living in unhealthy, substandard, energy-poor or overcrowded spaces; or worse, facing the risk of eviction or homelessness. It highlighted that for 80 million people, housing costs are an excessive burden. Faced with this situation, it drafted a report on the need to facilitate access to decent, affordable and energy-efficient housing. It also highlighted the urgency of eradicating homelessness—the levels of which have increased in the last decade—and developing new housing and housing market policies. In this sense, they recommend member states to invest in social, public and affordable housing and to guarantee secure tenure, as well as to promote a more inclusive market by halting evictions, eliminating old mechanisms related to speculative movements (investment funds) or greater control over abusive rent increases favoured by gentrification and/or touristification processes (European Parliament 2021).

Throughout this chapter we will focus on the analysis of the housing problem in Spain. A country where the housing affordability rate reached 46% in 2006 (when it is advisable for it not to exceed 35%), where despite the fact that housing prices fell after the housing bubble and there is still a large stock of unsold new housing, there are thousands of families who cannot access affordable, quality housing. We will analyse the factors that have caused the housing problem to become chronic in this country, and then study the case of Madrid and the policies that its municipal government is implementing to favour access to housing, both for the most vulnerable population and for the middle-income population whose income is not enough to buy an apartment or pay rent, so that their quality of life standards are affected.

17.2 Housing: A Chronic Problem in Spain

In Spain, there are eleven million people in a situation of housing emergency. According to the National Institute of Statistics (INE), the impacts of the 2008 crisis have led to 2.5 million people affected by evictions. The triggering factor for these evictions are non-payment of mortgages or non-payment of rents. A circumstance that has worsened with the COVID-19 health pandemic, because although the number of housing evictions carried out by the courts fell by 46% in 2020, according to the General Council of the Judiciary (2020), 18% more proceedings were initiated for non-payment of mortgage loans. Currently, there are still 100,761 people affected by the Temporary Employment Regulation Expedient COVID-19 (ERTE), many of whom have been out of work for more than two years and have no expectations of returning to the labour market soon, which places them in a situation of vulnerability and greater insecurity in terms of paying for housing. In this sense, we must bear in mind that Spain is still a country of homeowners (with 75.1% of the population preferring ownership compared to 24.9% who opt for renting), and that housing is associated with feelings of shelter, security and protection (Pasca 2014).

The problem is not that there are not enough houses, as more have been built than ever since the early 1990s (2,050,000 housing units), but that the supply does not meet the needs of demand. Large segments of the population are unable to access them—either because of their reduced purchasing power, the tightening of credit or the decrease in the formation of new households—or they do so under poor conditions or at the cost of great sacrifices (Pérez 2020). There is not enough housing supply at reasonable prices and in line with the purchasing power of the middle class. Nor is public housing sufficient for groups that, due to their economic situation, are excluded from the housing market. Whilst in 1985 the promotion of subsidised housing (51%) exceeded free housing, from 1988 onwards it began to decline reaching 7.2% in 2002 and just over 1% at present (Housing Europe 2021a). This is a tiny percentage when compared to that recorded in the Netherlands (30%) or France (15%). In addition, housing prices have not stopped rising since 2014. In the period 2014–2019 the price of m2 at the national level increased by 12.2%. An increase that was not accompanied by a rise in salaries, which implied a greater effort in access to housing. Likewise, while it is true that during the pandemic there was a price freeze, after the pandemic, there is again an increase in housing prices in coastal cities such as Malaga, Seville and Valencia (with increases of 7.6%, 3.9% and 3.4% respectively), large cities such as Madrid and Barcelona (with 2.9% and 1.3% respectively) (Everett 2021), as well as in the periphery of cities, in the face of the demand for larger homes. In addition, the real estate market is being affected by an increase in construction costs of almost four points above inflation (Appraisal Company 2022).

Thus, if it is advisable that the price of housing should not be more than five times the gross annual salary, that 35% of its total value should be saved—which would be between 50,000 and 100,000 euros for a standard home—and that spending on housing should not account for more than 30–35% of household spending, the result is that in Spain we have a pocket of insolvent or semi-solvent housing demand (Pérez 2020) that should be given some alternative. According to Idealista in our country it takes an average of 11 years' salary to pay off a house, whilst in Finland it takes 7 years, in Portugal 12 and in France 13 years (Idealista 2021). In the Balearic Islands this period increases to 16.5 years, followed by the Community of Madrid and Catalonia, 8.6 and 8.5 years respectively. La Rioja and Murcia are the regions with the lowest purchase price and sale effort, with a period of around 4.5–4.7 years (Fig. 17.1). In this context, only those who have sufficient financial capacity and meet the credit quality standards required by banks to apply for mortgages can buy a home.

Fig. 17.1 Years of full salary needed to pay for a house in Spain as of January 2021 (*Source* Statista [2022])

But it is not only the price of homes for sale that is on the rise. Rents have also followed the same trend, which further aggravates the situation. It is not only the availability of financial resources that influences the choice of renting. According to the Economic and Social Council, it is also directly related to social aspects. Young people, households headed by foreigners and single-parent families are the ones who most resort to this form of tenure (Economic and Social Council 2021), which implies a great economic effort for them. According to the Fotocasa and Infojobs report of 2020, the average price of housing was €10.65/m2 whilst the gross salary was €25,173 (€2,098 monthly). The percentage of gross salary required to pay the rent in relation to income was 41% (when in 2015 it was 28%). The situation was more serious in Catalonia (53%), Madrid (50%) and the Basque Country (49%). However, unlike the rise in sale prices during the pandemic, rents registered a decrease in seven of the seventeen autonomous communities, which meant a reduction in the economic effort needed to pay. The largest decline occurred in the Canary Islands, which went from 43% in 2019 to 36% in 2020, followed by the Balearic Islands, Madrid, Catalonia, the Basque Country, Andalusia and the Valencian Community (Infojobs and Fotocasa 2020). The reason for this sharp decline lies in the oversupply of housing in the market during the health crisis as a result of mobility restrictions that affected tourist rentals—according to the INE in February 2021, the number of tourist homes in Spain was around 294,000, or 8.3% less than in August 2020 (INE 2021)—due to the return of university students to their families homes during the lock-down, and the increase in working from home.

Even so, we are faced with very high percentages of effort, which not only reduces the opportunities of young people and forces many to return to their parents' home, but also leads to overcrowding in housing, with several families having to live under the same roof, or having to rent rooms or even balconies to sleep on.

In this context, new housing policies are urgently needed. It is time to think and act differently. In this sense, the #Housing2030 report focuses on four topics to implement effective affordable housing policies in the UNECE region (Fig. 17.2): housing governance and regulation; access to finance and funds; access to and availability of land for housing construction; and climate-neutral housing construction and renovation (United Nations 2022).

In the case of Spain, housing policies focus their attention on the promotion of public housing, the supply of affordable rental housing, the renovation of the built stock and the introduction of empty housing on the market (Pareja and Sanchez 2012). In this sense, the 2009–2012 Housing Plan already included the need to facilitate that subsidised housing can be obtained both through new development and through the renovation of the existing stock, allowing the qualification as subsidised housing of that which is unoccupied and has a legal regime of free origin (BOE 2008). It also

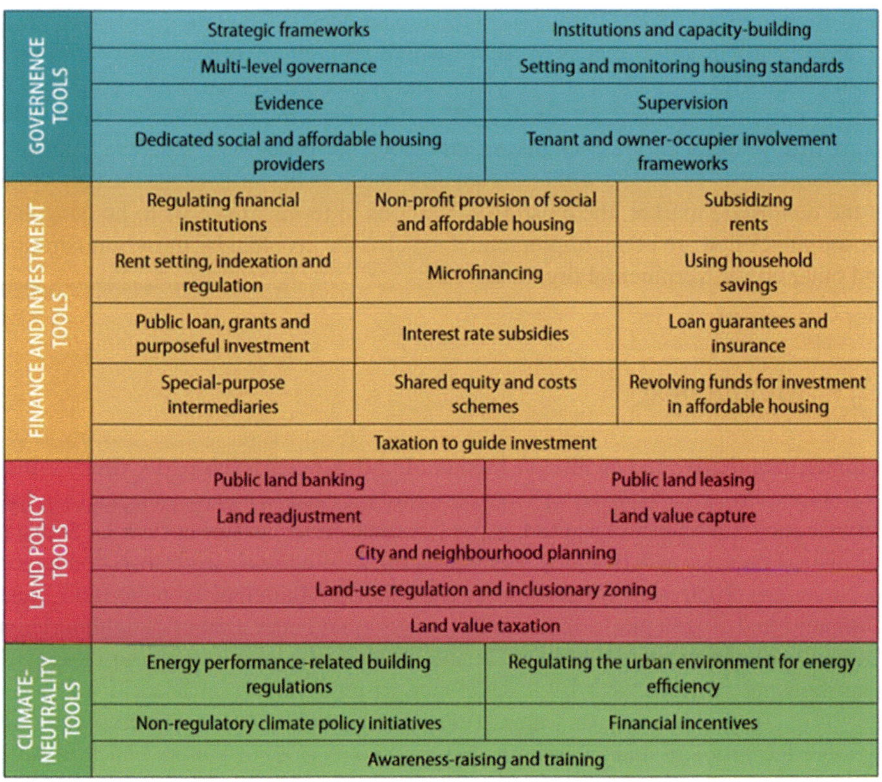

Fig. 17.2 Policy tools for affordable housing (*Source* United Nations [2022])

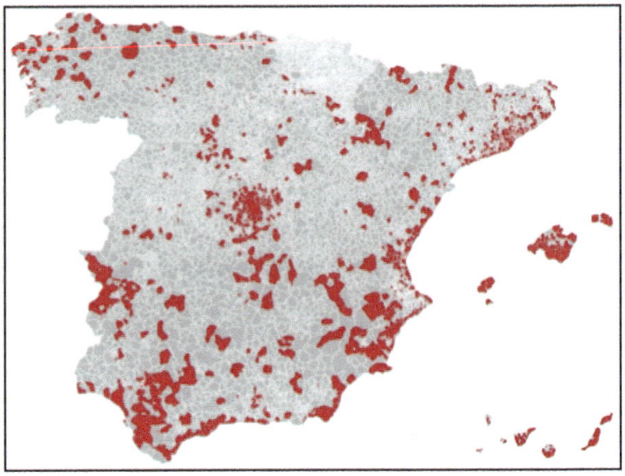

Fig. 17.3 Map of municipalities under stress due to rents (*Source* Dolz and Galarza [2021])

states that the financial effort made by families to access housing should not exceed one-third of their income, and that both in new developments and in the renovation of subsidised housing, at least 40% should be allocated to rent. On the other hand, the draft Housing Law of 2022 arose from the urgent need to regulate the price of rents in stressed urban areas, that is, in those where rents have increased by 5% above the regional CPI over the last five years, and where more than 30% of income is used to pay for housing (La Moncloa 2022). Amongst the provinces with the most municipalities affected by rises in rented housing are Barcelona, Valencia, Baleares and Madrid (Fig. 17.3). We will focus our case study on the latter to see the behaviour of the real estate market, the existing limitations in terms of access to housing and the initiatives that are being implemented to solve the problem by the administration and other non-governmental organisations.

17.3 Methodology

A mixed quantitative and qualitative methodology has been developed. Official statistical sources from the Ministry of Transport, Mobility and Urban Agenda, and Madrid City Council have been consulted, as well as data from platforms such as Statista, Idealista, Fotocasa and Infojobs. Semi-structured interviews were also conducted in November 2020 with officials from the Madrid Municipal Housing and Land Company, and citizen platforms such as the Regional Federation of Neighbourhood Associations of Madrid (FRAMV). The interviews were organised in four blocks: (i) what housing policies have been developed in recent years in Madrid, (ii) what are the main problems in this area; (iii) what initiatives are being implemented to solve the problem and where are they located, (iv) what factors can limit/stop them. This information was completed in the same month as the seminar "Measuring and analysing metropolitan inequalities. Topics, methods, hypotheses and results", in which experts from the Autonomous University of Barcelona and the Polytechnic School of Madrid took part, addressing the problems of housing as a factor of segregation, price increases and market tensions.

17.4 Madrid, One of the Most Expensive Cities in Spain

According to data from February 2022 from Idealista, Madrid is the fourth most expensive Spanish city for housing sales, after Ibiza (€5,163/m2), San Sebastian (€4,958/m2) and Barcelona (€3,929/m2) (Idealista 2022). Currently, the selling price of a home in Madrid is €3,766/m2, a price that exceeds that recorded in 2007 (€3,517/m2) and represents an increase of 40.6% over that of 2015 (€2,678/m2). The most expensive districts are those located around the Centre (Salamanca, Chamartín, Arganzuela, Moncloa and Retiro) (Fig. 17.4). However, districts in the south of the capital such as Puente Vallecas, San Blas, Villaverde or Carabanchel registered values of around €3,000–3600 in 2008 (when the effects of the crisis were not yet so evident), despite being urban sectors that are home to a low-income population.

In terms of rents, Madrid leads the ranking of rental prices in Spain (€780) and is amongst the European cities where rents grew the most in 2021 (around 14.22%), along with Berlin, Hamburg, Reykjavik, Vienna or Milan. If in 2020 the average rental price was around €1,104 in the third quarter of 2021 it reached €1,261. Amongst the factors that explain this increase are the increase in the stock of holiday rental

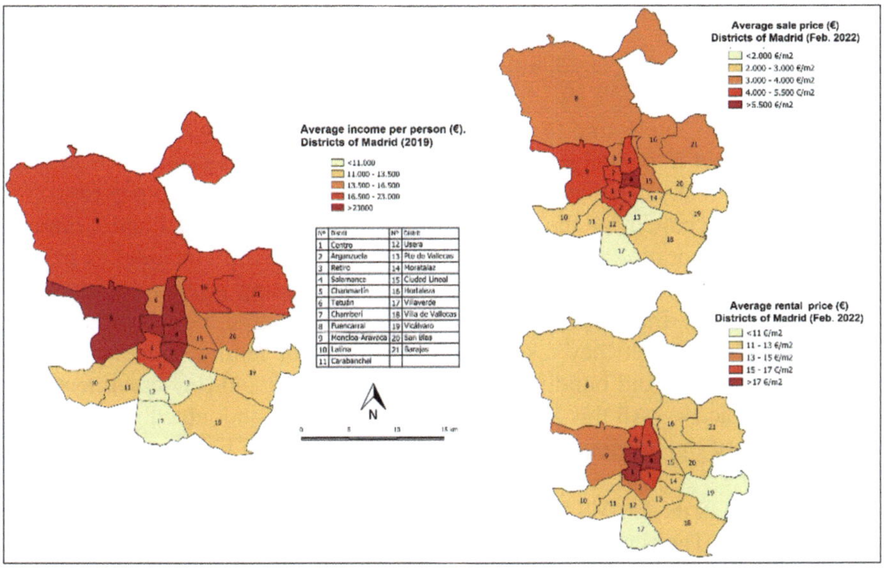

Fig. 17.4 Sale–rent prices of housing in the districts of Madrid (*Source* Madrid City Council and Plataforma Idealista)

housing after the relaxation of the restrictions imposed by the pandemic, and the municipal sanctions imposed on non-legalised tourist housing, which considerably reduced the supply. Although rental prices fell slightly with the pandemic, of the 2,404 census tracts that make up the urban area of Madrid, 1,465 (61%) spend more than 30% of their income on housing and are considered stressed areas (Fig. 17.5). The district with the highest prices for collective housing is Chamberí, although it is not amongst the stressed areas given that the high incomes of its inhabitants mean that it is not such an effort to make these payments. The same can be seen in the districts of Salamanca, Retiro or Moncloa-Aravaca, although these are closer to 30%, with rents exceeding €2,000. This price increases in relation to the rent of single-family homes, for which one can pay between €3,000 and 4,500 in the district of Moncloa-Aravaca, but in this case the degree of effort is higher despite the high rents. In this regard, it is worth bearing in mind that after the confinement caused by COVID-19, there has been a greater demand for modern housing, with larger spaces and outdoor areas in the last quarter of 2020. The purchase of single-family homes reached 22.5% of the total number of transactions, and in the third quarter of 2021 it reached 21% (Registradores de España 2021).

The southeast area of the city, specifically the districts of Vicálvaro and Villa de Vallecas where, despite the fact that collective housing rentals prices range between €700 and 1,000, the effort that families have to make to pay their rent exceeds 40% of household income.

17.5 New Housing Policies: Access to Affordable Housing

The regional and municipal administrations of Madrid have expressed their opposition to the new Housing Law 2022, stating that they will not apply it. In view of this situation, it is interesting to analyse what actions are being taken by the local government to tackle the problem of access to affordable housing and the effort made by families—in many cases vulnerable sectors of the population—to pay for it.

A publication edited by the Madrid Municipal Housing and Land Company (EMVS) commemorating its 40 years of history (EMVS 2021) includes an excerpt from a 1981 report in which it was stated that "the housing problem in Madrid is serious, sometimes dramatic. Result of declining speculation, prices and rents are so high, in general, that they take large percentages of family budgets. Housing shortages, unrestored old buildings, cracking new blocks and tiny apartments that are difficult to access. All this is well known". Words that undoubtedly show us the chronification of the housing problem in the capital, as can be extrapolated to the current period.

In all this time, the EMVS has built more than 27,000 homes and currently owns 6,300 properties. The current objective, according to those interviewed, is to increase the municipal housing stock, either through the construction of new housing or through the incorporation of vacant housing in the real estate market. In both cases they will be offered as public housing for rent, as opposed to what was done in the past

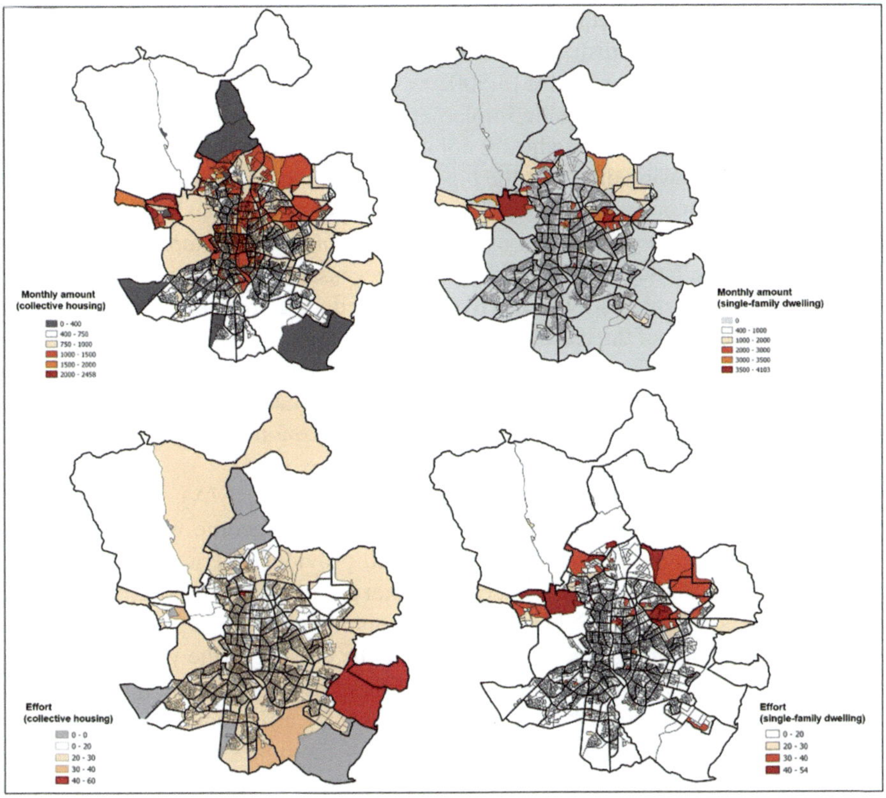

Fig. 17.5 Monthly rent and level of effort for collective and single-family housing (*Source* Ministry of Transportation, mobility and urban agenda [2020])

when they were intended for property acquisition. It was in 2020 when the largest acquisition of housing by the local administration took place, with the purchase of 108 homes from SAREB and 19 from Social Security for a total of €18.2 million (EMVS 2020). In order to access them, an awarding regulation was developed in which priority is given to: having an income lower or equal to 3.5 times the Public Indicator of Multiple Effect Income (IPREM), residing or working in Madrid and not being the owner or tenant of free or publicly managed housing. Priority is given to groups of people in a situation of residential vulnerability; people under 35 years of age and over 65; single-parent or large families; people with disabilities; victims of gender violence, trafficking and / or sexual exploitation. This offer is complemented by other social programs aimed at people in situations of extreme residential vulnerability such as: women in vulnerable situations (immediate temporary housing for female victims of gender violence); solidarity housing (people and families immersed in a process of social intervention); or housing first (for homeless people or people in a situation of very serious exclusion). According to data from EMVS itself, between 2015 and 2020 a total of 3,735 rental housing units have been awarded, 65% of which have been accessed through the allocation regulations. More than half of the total housing has been awarded in the districts of Vallecas and Villaverde where the average net rent per person does not exceed €13,500.

Other initiatives are also being implemented aimed at sectors of the population with a certain level of income, who are not in a situation of vulnerability, but who are unable to access housing due to its the high cost. The Integral Municipal Rental Plan includes the Reviva program, whose objective is to put vacant housing on the market. The owners assign the property in usufruct to the municipal company for a period of 4 years, and the latter is responsible for the necessary renovations (advancing up to €45,000 without interest, which will then be deducted from the rent) and for the integral management of the lease (ensuring the payment of the rent to the owner even if it is not occupied). The tenant is guaranteed not to spend more than 30% of their income to pay the rent. Associated with the Reviva plan is the Housing Voucher, which will vary in amount, depending on the socioeconomic characteristics of the tenants (income and household size).

Other initiatives are the Rental Intermediation Service (SIA), which makes it easier for tenants to find apartments that adapt to their income without requiring them to pay more than 40% of their income and provides a series of advisory and insurance services for tenants to reduce the market price by 10%. In addition, there are the following programs: CONFIANZA, for people under 35 years of age. Through this program, the municipal company advances the deposit to the young tenant and gives him 24 months for its repayment; the REHABILITA program, through which aid is obtained for the conservation of facades, energy efficiency and improvement of accessibility; or the ADAPTA program, which facilitates the refurbishment of housing for people with reduced mobility, sensory or intellectual disabilities, etc.).

According to EMVS data, a total of 22,629 people were registered as housing applicants in November 2019. Two years later, that number had increased to 29,859, noting an increase in family units composed of only one member (from 41.14% in 2019 to 46.83% in 2021). As for the age structure, no significant changes are seen, although the 25–45 age group accounts for 60% of applicants. Finally, if we look at income level, there is a slight increase in those in the 1.5 to 2.5 IPREM brackets. By location, the district of Puente de Vallecas stands out with more than 20% of the applications, followed far behind by Carabanchel 9%, Usera 8.81% and Latina 8%, which represents more than 46% of the applications.

However, these programs were not enough to cushion the vulnerability resulting from the COVID-19 pandemic. The confinement highlighted the need for social and economic measures at different levels of government to ensure the maintenance of housing (owned and rented). The approval of Royal Decree 11/2020, of March 31 (BOE 2020) put on the table the possibility of reducing rents by 50% during the state of alarm; and to defer payment during the same time, to be paid back in installments later. In March, the EMVS had 5,626 rental contracts and took the following decisions: to suspend evictions until June 2020; to reimburse all tenants for the month of April and to reduce rental payments by 50% from May to December, to help people/families in a situation of vulnerability. These measures cost €1,377,764.81.

In any case, in Madrid it is not only the Municipal Housing Company that is trying to provide access to housing. Innovative solutions are emerging that try to find alternatives to the traditional real estate market models. One example is Entrepatios, an initiative that promotes self-sufficiency and near-zero energy consumption (passive house), and participatory processes for the management and design of buildings. It has arisen as a right-to-use housing cooperative that aims to build alternatives to the speculative real estate market with social and environmental criteria. Once the construction work is completed, the cooperative is not dissolved, so the housing does not become the property of the inhabitant, but remains the property of the cooperative, which decides on the entry and exit of tenants. In this case, the tenant can stay indefinitely, with a fee or "soft rent" of around 650 euros for a 70 m2 apartment and 20 m2 of common spaces, plus an entrance fee (40,000 euros) which is refundable in the event of leaving the cooperative. On the economic front, one of its red lines is that the price per m2 of rent should not exceed €10. The cooperative is currently made up of 66 cooperative members, and works in coordination with social economy companies: Lógica'eco -manager of collective initiatives- and sAtt Arquitectura -an architectural studio-. Its objective is to move from being a community of owners to a community of neighbours through the implementation of three developments: two already have land for construction in Las Carolinas (Usera District), in Villa de Vallecas District (Fig. 17.6), and a third is looking for a place to build.

Fig. 17.6 Housing cooperative in the Villa de Vallecas District (*Source* Authors)

17.6 Conclusion

In Spanish cities, it is increasingly difficult to access affordable and quality housing. The increase of prices above income has meant that housing has become a factor of segregation and social exclusion. In this context, housing has become a key aspect for the urban agendas of the main Spanish cities. There is an unprecedented effort by public administrations to modify housing policies both in terms of access and market, in order to make them more efficient, inclusive and sustainable. They are aware that affordability is especially challenging for low-income households, as well as for a middle class that spends more than 35% of its income on housing payments.

Various initiatives have been launched on a global scale to achieve this, such as the provision of housing through housing associations with limited benefits in Austria, tax reforms in the Netherlands or macroprudential measures promoted in Sweden and Canada. In the Spanish case, the approval of the Housing Law of 2022 has been a turning point by limiting the price of rent in stressed areas, applying fiscal rewards to rent and price reductions and recommending that new property developments provide 30% for subsidised housing and social rentals. On the other hand, local administrations have begun to promote initiatives that involve new logics of co-development, co-management and co-design, both between regional and local administrations and between public–private agents (cession and use of municipal land to promote co-housing initiatives, co-living, development of promotions with ethical construction companies, intermediation between owners–tenants, rehabilitation of housing for social rental, etc.).

In the case of Madrid, unlike other Spanish cities such as Barcelona—recognised by Housing Europe as a pioneer in housing policy—(Housing Europe 2021b), the actions promoted by the local administration are still limited. Although there is concern about the housing issue and innovative solutions have been launched, the truth is that we have observed certain weaknesses: (i) the promotion of housing for people in housing emergency situations is insufficient; (ii) there is a lack of awareness amongst the population of initiatives promoted by the administration; (iii) there is a mismatch between the supply of publicly promoted housing and the real demand for it, since the latter rejects the allocation that has been granted, despite the fact that the housing has an attractive design and is energy efficient. Amongst the factors that cause this are, on the one hand, that the price of rent is still too high and is an excessive burden; the high construction densities that are being built; and, on the other hand, the fact that the price of buying a property is still high and is an excessive burden.

Thus, in order to promote efficient housing policies, we consider that (i) it is necessary to know the social reality of cities and their neighbourhoods and promote housing that is truly affordable for them (according to the average income per household registered in each district), of quality and energy efficient, since this will result in a greater capacity to deal with situations of energy poverty. (ii) it is vital to make the population aware of the alternatives available to them. In this sense, it is not enough to publicise them in the media. It is essential to bring the administration closer to the neighbourhoods and their inhabitants to explain initiatives that, due to their lack of knowledge, may generate insecurity and/or rejection. This improvement in information is necessary to put an end to the stigmatisation of social or public housing by the public, historically associated with lower classes, without resources, vulnerable or for certain social groups. (iii) initiatives to improve access to housing (such as co-housing) should be an alternative for all social strata. A high initial contribution and rents which, although below market rents, are still high for a large part of society, is not a solution that favours affordability and social integration. In this regard, the usufruct cession of municipal land reduces social costs, the work with developers and the establishment of soft rents facilitates the integration of the most vulnerable groups in these developments. (iv) the administration should accompany the entire process of access to housing, and not limit itself to the construction work. It should mediate with construction companies and financial services cooperatives with ethical and solidarity principles in order to obtain better credit conditions for the population with a limited income.

However, despite these weaknesses, the fact is that we are in a new period in which there is a concern to change the way of making the city and attending to the needs of the population. Let's hope that in the medium term the administrations will be able to correct these imbalances that are now being observed and bet on affordable, quality, environmentally sustainable housing as well as on an inclusive market.

Acknowledgements This research is part of the project "The housing problem in fragmented metropolises in Spain. Perpetuation over time, new housing markets and solutions from the public administration" (PID2019-108120RB-C31) funded by the Ministry of Science, Innovation and Universities.

References

Appraisal Company (2022) El mapa del tiempo urbanístico. https://tools.st-tasacion.es/productos/Mapa-del-Tiempo-Urban%C3%ADstico/Mapa-del-Tiempo-Urban%C3%ADstico

BOE (2008) Real Decreto 2066/2008, de 12 de diciembre, por el que se regula el Plan Estatal de Vivienda y Rehabilitación 2009–2012. (309), 51909–51937 https://www.boe.es/buscar/doc.php?id=BOE-A-2008-20751

BOE (2020) Real Decreto-ley 11/2020, de 31 de marzo, por el que se adoptan medidas urgentes complementarias en el ámbito social y económico para hacer frente al COVID-19. https://www.boe.es/buscar/act.php?id=BOE-A-2020-4208

Dolz C, and Galarza G (2021, 7 de octubre) El mapa de las zonas tensionadas a las que va a afectar la nueva Ley de Vivienda. El Mundo. www.elmundo.es/economia/2021/10/07/615e31f8e4d4d852148b457d.html

Economic and Social Council (2021) Economía, trabajo y sociedad. Memoria sobre la situación socioeconómica y laboral en España 2020. Madrid: Consejo Económico y Social

European Parliament (2021) European Parliament resolution of 21 January 2021 on access to decent and affordable housing for all. https://www.europarl.europa.eu/doceo/document/TA-9-2021-0020_EN.html

Eurostat (2021) Housing in Europe. 2021 Interactive edition. https://ec.europa.eu/eurostat/cache/digpub/housing/index.html?lang=en

Everett Allen K (2021) Global residential cities index—Q3 2021. Knight Frank. www.knightfrank.com/research/report-library/global-residential-cities-index-q3-2021-8670.aspx
General Council of the Judiciary (2020) Effect of the crisis on judicial bodies. https://www.poderjudicial.es/cgpj/es/Temas/Estadistica-Judicial/Estudios-e-Informes/Efecto-de-la-Crisis-en-los organos-judiciales/
Housing Europe (2021a) The state of housing in Europe 2021. Bruselas: Housing Europe. www.stateofhousing.eu/The_State_of_Housing_in_the_EU_2021.pdf
Housing Europe (2021b) Effective policies for affordable housing in the UNECE region. www.housing2030.org/reports/
Idealista (2021) Spaniards are among the Europeans who make the greatest effort to buy a house: they spend 11 years of their salary on it. https://www.idealista.com/news/inmobiliario/vivienda/2021/11/12/793297-los-espanoles-entre-los-europeos-que-mas-esfuerzo-hacen-para-comprar-casa-destinan
Idealista (2022) Evolución del precio de la vivienda en venta en España. www.idealista.com/sala-de-prensa/informes-precio-vivienda/
INE (2021) Viendas turísticas. Estadística Experimental. www.ine.es/experimental/viv_turistica/experimental_viv_turistica.htm
Infojobs and Fotocasa (2020) Relación de salarios y vivienda en alquiler en 2020. https://nosotros.infojobs.net/prensa/notas-prensa/los-espanoles-cobran-mas-y-destinan-el-mismo-dinero-al-pago-del-alquiler-en-2020-segun-infojobs-y-fotocasa
La Moncloa (2022) Proyecto de Ley por el derecho a la vivienda. https://www.lamoncloa.gob.es/consejodeministros/referencias/Paginas/2022/refc20220201.aspx
Madrid Municipal Housing and Land Company (EMVS) (2020) Memoria de Gestión 2020. Madrid: www.emvs.es/Transparencia/Calidad/Documents/MemoriasGestion/Memoria%20de%20Gesti%C3%B3n%20EMVS%202020.pdf
Madrid Municipal Housing and Land Company (EMVS) (2021) 40 años Creando hogares, transformando Madrid. EMVS, Madrid
OECD (2021) Brick by brick : building better housing policies. https://www.oecd-ilibrary.org/economics/brick-by-brick_b453b043-en
Pareja M, Sánchez MT (2012) La política de vivienda en España: lecciones aprendidas y retos de futuro. Revista Galega De Economía 21:2
Pasca García L (2014) La concepción de la vivienda y sus objetivos. Universidad Complutense de Madrid, Madrid
Pérez Quintana V (2020) La vivienda: problema y propuestas. Papeles de de relaciones ecosociales y cambio global(148), 41–49
Registradores de España (18 de 03 de 2021) Estadística registral inmobiliaria. www.registradores.org/documents/33383/0/ERI_1T_2021.PDF
Statista (2022) Número de años de sueldo íntegro necesarios para pagar una vivienda en España a enero de 2021, por comunidad autónoma. https://es.statista.com/estadisticas/544760/indice-de-esfuerzo-para-comprar-una-vivienda-en-espana-por-cc-aa/
United Nations (2021) #Housing2030: Effective policies for affordable housing in the UNECE region. United Nations Economic Commission for Europe. https://unece.org/sites/default/files/2021-10/Housing2030%20study_E_web.pdf

Chapter 18
Residential Expectations in a Neoliberal Perspective: A Sociological View of Social Classes and the Right to Housing

Félix Rojo-Mendoza, Claudia Mercado-Cerroni, and Voltaire Alvarado-Peterson

Abstract The neoliberal system not only implies the establishment of a model based on the role the market plays in society at the expense of the state, but also a paradigm that impacts people's livelihoods. Due to this, the social class structure currently includes a significant number of middle classes and a far smaller segment of workers' groups. Regarding livelihoods, the neoliberal system conditions the ways of conceiving and reproducing consumption of various material and symbolic goods, including the residential area for living. What happens when you want to live in certain areas of the city but the economic capital available is not enough to fulfill that desire? In other words, what mechanisms to different social classes use to adapt their residential desires to their objective position in the social system (class position)? This chapter describes the expectations and mechanisms of residential adaptation of people from different social classes in Temuco and Padre Las Casas, one of Chile's largest conurbations with the highest population growth in recent decades. For this, following the assumptions of grounded theory, 60 semi-structured interviews of people from different social classes in these cities were conducted and analyzed. The residential expectations analyzed express different needs according to social class.

This work was supported by the National Research and Development Agency (ANID) and the Regular Fondecyt Project No. 1201255, "Spatial tastes in the production of urban spaces in neoliberal Chile: the cases of Temuco-Padre Las Casas and Iquique-Alto Hospicio."

F. Rojo-Mendoza (✉)
Departamento de Sociología, Ciencia Política y Administración Pública, Universidad Católica de Temuco, Temuco, Chile
e-mail: frojo@uct.cl

C. Mercado-Cerroni
Programa de Doctorado en Ciencias Sociales, Universidad de La Frontera, Temuco, Chile
e-mail: c.mercado04@ufromail.cl

V. Alvarado-Peterson
Departamento de Geografía, Universidad de Concepción, Concepción, Chile
e-mail: voalvarado@udec.cl

E. Navarro-Jurado et al. (eds.), *Urban Dynamics in the Post-pandemic Period*, The Urban Book Series,
https://doi.org/10.1007/978-3-031-36017-6_18

Keywords Socio-spatial structure · Habitus · Class identity · Middle city · Residential taste

18.1 Introduction

Contemporary capitalism has blurred relations between social classes, capital and habitus, which have an impact on the intensification of divisions in consumption, including those related to space. In this sense, it is currently more frequent for people to consume out of the economic sense of wanting to be part of the mass (Butler 2005) and not necessarily in a way that is consistent with the capitals that are part of their personal history.

The neoliberal model's ability to permeate the deepest parts of social life is among the causes of growing internal inconsistencies within social classes in the current form of capitalism. The neoliberal project installed in different societies is not only related to the economic aspect, but has important consequences regarding the values, practices and discourses that people have regarding the world and themselves (Akerlof and Shiller 2016).

This means that the expectations of participation in the economic system are very high, which in turn implies that all people, regardless of the class they belong to, can long for access to a particular service or good despite not having the economic capital to obtain it. This is something that Bourdieu (1998) himself noted when he observed that, in addition to the expansion of consumption in different social groups, discomfort arises among people who cannot get what they really want.

The same is true in residential terms. People of different social classes profile residential expectations and try to make the identity principles they define themselves by agree with the inhabited spaces or those they wish to inhabit. However, to the extent that this adjustment is not materialized, different types of residential strategies are activated to achieve the goal of living where they want to live. The literature on these issues has paid much more attention to the strategies used by the middle class to adjust its residential identities (Savage et al. 2005; Watt 2009; Matthews 2015) than to the lower classes, which in the context of the neoliberal model's massification also dream of fulfilling their residential expectations.

This chapter describes the residential expectations of people from different social classes in Temuco and Padre Las Casas, one of Chile's largest conurbations with the highest demographic growth in recent decades. For this, following the assumptions of grounded theory, 60 semi-structured interviews of people from different social classes in these cities were conducted and analyzed. Among the sample features, age diversity, forms of home possession (own, rented or people living with relatives), and specifical residential sectors, stand out. The dimensions consulted in these interviews were related to three aspects of residential trajectories: memories and experiences in the places inhabited in childhood, tastes and strategies implemented for choosing the current residence, and potential future residential movements. It is worth mentioning that the results that here presented are part of a broader project aimed at exploring people's discourses from different social classes regarding housing and the possibility of obtaining property of their own.

18.2 Social and Physical Space Interaction in City's Production

Social stratification models consolidate the idea that societies are distinguished by social classes. In turn, these classes are formed around the shared position people have in the labor market and, therefore, the material and symbolic goods that they can access depend on this location in the social structure. In this sense, social classes differ in that people control variable volumes of resources and powers associated with economic, cultural and social capital from their place in the labor market. The first capital is represented by wealth and income, the second by cultural assets and the institutional credentials from the education system, while the last capital refers to contacts that allow access to social networks. As a result, the social space is only the area in which these capitals relate in everyday life, a fact that, according to their availability, determines the distance or proximity between people (Bourdieu 1984).

However, social space is not just inscribed in the subjectivity of mental structures due to the embodiment of objectified structures, but also in the objectivity of spatial structures (Bourdieu 2018). That is, social space tends to coincide with geographic space; the closer they are, the more likely social groups, agents and institutions will share traits. This means that some territories might be advantageous to one group and, on the other, certain subjective aspects may be decisive to residing in certain physical spaces (Levy 2000). It prompts people to move within the physical space in search of residential areas that are consistent with the position they have achieved in social space, thus generating a capital of movement in cities (Kaufmann et al. 2004).

In no way this implies that socially distinct people (with differentiated capitals) cannot occasionally meet in physical spaces, but rather that the tendency is toward a consistency between inhabited space and group social belonging (Bourdieu 1998). Thus, given unequal social relations, access, control and disposition of physical space also depend on existing social power differences. In this sense, the physical space in social dispute represents a type of symbolic capital in a field whose space and materiality are continually at stake (Centner 2008).

However, the correspondence between social and physical space described above is less clear today. This is primarily because people currently tend to be unaware of their location in the social system in terms of their occupations and therefore seek to position themselves in a social sphere that differs from the habitus and capitals they administer. This does not mean that occupation has lost strength as a predictor variable of life opportunities, but rather that the class structure is not necessarily related to class identity (Bourdieu 2005). Savage et al. (2005) noted this situation in United Kingdom when they observed that employees, including those highly committed to their work, did not relate much with other workers in their same condition and, therefore, do not build an identity related to their work. This new scenario leads people to move within the social and physical space toward areas that are consistent with their class identity, regardless of whether they have the concrete possibility of making that movement. In this context, the differences created between what is desired in subjective terms and what can be achieved in objective terms, what Bourdieu (1998) called hysteresis, determine people's processes of identity reflexivity, something correlated with patterns of residential aspiration and choice (Savage 2010).

18.2.1 Residential Hysteresis: Reconfiguring Class Identity in Cities

In spatial terms, the distance between the dream place to live and the material resources available to fulfill it, something we call residential hysteresis in this work, creates social malaise among those unable to move within the city as they would like. This means that the fewest material possibility of choosing one's place to live, the greater discomfort of people who subjectively feel that their physical space no longer represents them in terms of their social position.

Under these terms, people with more economic capital have a greater likelihood of reducing the residential hysteresis produced by occupying a physical space that disagrees with the social space achieved. They do this by using the instances that the private managers of urban neoliberalism offer (Brenner et al. 2010). For their part, the lower classes without economic capital are destined to compete to obtain a place to live through the actions of the institutional structure of a subsidiary state and its social housing policies (Vergara 1984; Hojman 1993; Richards 1993). This does not mean that these classes do not have desires to project their lives in certain spaces in the city, but rather that they have scant chances of materializing it due to the lack of economic resources.

To the extent that the upper classes are guaranteed the economic capital to fulfill their dreams and the lower classes have none, this kind of mismatch between what is desired and what is achieved is most clearly seen in the middle classes. This internally diverse group (Erikson and Goldthorpe 1993; Savage et al. 2013; Espinoza et al. 2013) shows consumption patterns aimed to replace the lack of resources in other capitals (Bridge 2001). From the residential perspective, these movements, especially the ones concerning accommodated people, reveal that they aim finding places that are consistent with their class identities and, therefore, where they can develop social relationships with others considered equals. This ultimately structures a so-called elective identity belonging (Savage et al. 2005) guided by principles of adjustment of residential hysteresis, regardless of whether people have prior connections to the area they have chosen to live in. Elective belonging can operate under principles of symbolic and practical selectivity of inhabited places, something related to incorporating only those aspects that offer consistency between areas and habitus in the new identity configuration (Watt 2009; Matthews 2015; Méndez 2018; Stillerman 2017).

This logic related to the residential choice of the middle classes is very different from the scant possibilities available to the lower classes to reduce their residential hysteresis, which is only related to sustaining their residential identity discourses in terms of nostalgia for place inhabited (Savage et al. 2005). In this context, the lower classes, and even those middle classes with diminished spending capacity due to economic crises, cannot resolve the residential hysteresis and therefore merely limit themselves to structuring their identities around the memories evoked from the space they inhabit. Similarly, the salaried employees in middle-class groups cannot resolve the positional discomfort of being forced to go into debt for housing that is inadequate in terms of both quality and location within the city (Bourdieu 2005). Their only choice is to reconfigure their residential identities by resorting to nostalgic adjustment that evokes the space currently inhabited.

However, different people can build future residential fantasies regardless of the class they belong to. By severing the occupational structure from class identity, residential projections can alleviate the residential hysteresis produced in the present, even though it may never actually be achieved. In particular, the basis of these projections may be related to achieving access to home ownership.

18.2.2 My Home, my Identity: Residential Expectation in Neoliberal Code

A series of international studies have shown the persistent aspiration for housing as a superior social and material good beyond people's possibilities of access and materialization (Mckee et al. 2017; Benson and Jackson 2017; Jansen 2014) or the economic crises that make their goals unattainable (Aramburu 2016; Rowan and Doling 2017). Thus, under the principle of reducing residential hysteresis, taste for housing property is established as a cornerstone for evaluating the residential expectations of different social classes beyond the needs of conquest of certain consolidated spaces in the city. In this sense, displacement toward certain spaces in the city would be negotiable if it ensured full property ownership, making the place-roots relationship a functional one considering the objective of ownership. This lays the foundation for any future residential movement that relates the longing for a space with the possibility of materializing the material appropriation of a home, positively correlating both dimensions to the extent that the latter is achieved (Rojo-Mendoza 2021).

Given the above, the spatial preferences illustrated by the fantasies of future residence can make anyone dream of living in places that are currently inaccessible, especially in terms of property ownership. This can be seen in the work by Groot et al. (2013), who relate the particular features that the course people's lives take, the logic that the housing market adopts and the formation and materialization of preferences regarding homeownership. Their results indicate that people tend to adapt and adjust their tenure preferences over housing to the perceived possibility of being able to actually move to a dwelling in the condition of owner or tenant, in what is called adaptive preference formation in psychology (Groot et al. 2013). In a similar vein, Benson and Jackson (2017) argue that, regardless of the restructuring that the price of housing has undergone in recent years in London, expectations regarding future acquisition of a home in the middle class, a group whose identity in terms of status and distinction is associated with real estate ownership, have not been abandoned. They show how new economic conditions cause personal and place identities to mutate, placing emphasis on the fact that changing economic and social processes also generate new strategies and positions in the class reproduction process (Benson and Jackson 2017).

This relativity in the adaptation of aspirations on housing is so great that in the case of young professionals in Scotland, this hysteresis is seen as a significant problem that not only has an impact on the relationship between housing and social class, but also on the entire political order due to the divisions produced between the fields of economic and cultural power (Crawford and McKee 2018).

In Chile, neoliberal welfare also translates residential space as a spatial taste, driving the desire for property ownership in a broad range of classes and regardless of economic crises and the real possibility of buying a home (Vergara 1984; Richards 1993; Hidalgo et al. 2017). This manifest interest in home ownership has a long tradition, especially since the 1930s, when housing policy was systematized and institutionalized, with housing recognized as a citizen's right exercised in cooperation with the state (Hidalgo 2010; Imilan 2016). The Popular Housing Fund (*Caja de la Habitación Popular*) that operated in Chile between 1939 and 1952 is an example of this (Fig. 18.1). Currently, and considering the neoliberal incentives that lower-class people are given, the state transforms these lower-class groups into sources of demand for housing (Hidalgo et al. 2017; Sabatini et al. 2012). This makes the Chilean lower classes clearly express the idea of home ownership (Rojo-Mendoza 2021), in stark contrast with what Hamnett and Butler (2013) observed in the British working classes, where it is less evident.

The combination of the class dimension and a trajectory of subsidies dating back from the present to the 1920s has made Chilean subsidy policy a fundamental agent in the production of taste for property in cities. Currently, subsidies as a contribution to the savings and mortgage capacity of those applying for them have reached unprecedented breadth and coverage compared to previous historical periods. For example, the addition of the so-called emerging sectors places the foundation of the middle class on access to subsidies for the acquisition of housing whose value surpasses USD 85,000, always with a share of individual savings (Alvarado 2019). Thus, housing subsidies could channel a reduction of residential hysteresis and therefore become a neoliberal welfare device. This situation has installed subsidies as a mirage of the liberties that state neoliberalism as a political-territorial ideology has conceived to guarantee freedom of choice and the attainment of a housing subsidy. The sums involved in these housing subsidies averaged USD 2.5 billion a year between 2010 and 2020 (Table 18.1).

Fig. 18.1 Homes built by the Popular Housing Fund (1939–1952). "Miguel Dávila" neighborhood in Santiago and "Quinta" neighborhood in La Serena. Photographs from 1950 (*Source* Popular Housing Fund Report. Imprenta Universo, Republic of Chile, 1950)

Table 18.1 State investment in subsidies for the purchase, rental and self-construction of housing 1990–June 2022[1]

Region	Homes 1990–2022 purchase, rental and self-construction	Public expenditure 1990–2022, USD
Arica y Parinacota	36,725	634,332,043
Tarapacá	102,678	1,175,368,718
Antofagasta	99,637	1,020,580,042
Atacama	65,443	797,183,164
Coquimbo	203,175	1,682,717,705
Valparaíso	448,320	3,874,602,057
Metropolitan Region of Santiago	1,405,911	10,607,341,785
O'Higgins	274,620	2,516,374,483
Maule	472,570	3,756,697,711
Ñuble	35,049	483,888,433
Biobío	664,405	5,814,585,305
Araucanía	380,544	3,305,292,701
Los Ríos	78,239	765,167,204
Los Lagos	255,286	2,398,791,556
Aysén	42,136	538,530,027
Magallanes	48,113	616,901,406
Total	4,612,851	39,988,354,340

Source Urban Observatory, Ministry of Housing and Urban Planning, Chile, (2022)

18.3 Residential Ideal and Hysteresis by Social Class: A Case Study

The sample of 60 people interviewed was determined according to intentioned criteria associated with three search variables in the cities of Temuco and Padre Las Casas: the spatial profile by class, occupation in the labor market and the type of tenure in the inhabited dwelling. The idea was to have people who represented diverse realities regarding these variables (Table 18.2).

After analyzing the discourses contained in 60 interviews conducted in the Temuco-Padre Las Casas conurbation (Chile), a first unit of analysis that stands out is the one associated with the residential ideals that people from different social classes express, in addition to the mechanisms and actions that they use to materialize it. In some cases, this results in both the residential hysteresis produced by

[1] Information available at the Urban Observatory of the Ministry of Housing and Urban Planning https://www.observatoriourbano.cl/estadisticas-habitacionales/#subsidios (Consulted 18 August 2022).

Table 18.2 General characteristics associated with the sample interviewees

Spatial profiles of social class	Number of urban areas	Number of interviews conducted	Characteristics of the areas	Profile of the sample interviewed
Very high	15	6	Significant presence of high-income horizontal developments and a concentrated offering of educational and varied services	People between the ages of 35 and 79. Service professionals, retirees. Proprietors
High	12	10	High-income vertical neighborhoods that concentrate a combination of public and private services infrastructure	People between the ages of 27 and 75. Professionals, people with higher technical skills and retirees. Proprietors and tenants
Medium	7	18	Forms of habitation associated with vertical housing developments and old peri-central areas of the city	People between the ages of 29 and 76. Service professionals, people with higher technical skills and retirees. Proprietors and tenants
Low	39	14	Combination of social housing and housing subsidized with mechanisms such as DS19[2]	People between the ages of 33 and 71. Occupational activities in trades (mechanics, carpenters), housewives. Proprietors and tenants
Very low	24	12	Significant concentration of self-built neighborhoods and new homeless camps. Scant public and private infrastructure	People between the ages of 29 and 78. Occupational activities in unskilled activities, independent workers, housewives. Proprietors and people living with relatives

Source Authors' compilation based on information processed from the Regular Fondecyt project 1201255

this scenario of residential choice as well as the activation of different strategies to address it.

First, higher social classes with greater available economic capital have more freedom of residential choice. This generates class distinctions in terms of the different groups' capacities to materialize their residential expectations in the city. In social classes with larger volumes of economic resources, the elements highlighted in the decision regarding where to buy and live involve the functional features of

[2] This is a Chilean state housing policy aimed at subsidizing homes for middle class families with low economic capital.

the spaces offered. In the middle classes, this freedom of choice is more limited and restricted, but certain minimum principles of eligibility for space and housing are nevertheless established. Thus, regarding the question as to why they live in the current place, the higher and middle social classes' replies were as follows:

> You can't choose the place, but you can choose the house. They gave me three choices. At the north exit, another one here in Padre, but higher up, and this one here. We chose here, because it was smaller, the neighborhood was smaller, and, I don't know, I found it nicer; we liked it. (75-year-old man, retired primary school teacher, middle-class homeowner).
>
> In those days I was a real football fan and I considered being near the stadium. My wife wanted to buy over in Prieto Street. So, we had a few conversations and came to a happy conclusion. It was quiet, near the stadium and it brought me memories because it was the place where we would come to play when we were kids. (74-year-old man, technical-administrative professional, upper-class homeowner).

On the other hand, and considering the same question, in the lower classes without economic capital, the principle of ownership of a home provided with state assistance is something that allows reducing the residential hysteresis caused by living in a space they never dreamed of. Thus, and regardless of whether the state facilitated the distribution of land—or home ownership, the fact is that these classes reduce their residential displeasure by replacing their expectations of a good habitation with one in which the fact of obtaining a home of their own neutralizes any identity discomfort associated with the place where they live. This is expressed in replies like:

> Because this land had just become available. A group of people was organized, and we applied. My husband was accepted, and we stayed here. (80-year-old woman, housewife, lower-class homeowner).
>
> My parents' dream was for me to go to a house of my own. And the possibility came up and we were able to apply for this house, so we left there to come here. (56-year-old man, carpenter, lower-class homeowner).
>
> [...] but because of the circumstances of life, I'm happy here. I wanted a more central place, but that was because of transportation, for my daughters, a more central place where everything was more central for them. But I like this place; I like it a lot. (56-year-old woman, saleswoman, lower-class homeowner).

In the case of the upper classes that rent a home, the desire for home ownership is secondary to the construction of identity associated with the place they occupy in the city. The discourses of these classes regarding the space they inhabit are more built on the foundation of a utilitarian principle of the city, in which the services provided are most important (especially educational ones), and not so much on the possibility of controlling that area through ownership of a particular property. From this vantage, the coexistence between economic and cultural capital within this social class allows it to reduce the malaise that may arise from an eventual inconsistency between the social position attained and the physical space inhabited. And when we speak of cultural capital, we refer to the projected educational assets that families want to give their children. This is expressed in phrases like:

> We chose this neighborhood because of Facu [Facundo], my son, who goes to a public school about a 10-minute walk from here, and the previous house that we rented was also the same distance from the school. (38-year-old man, designer, upper-class renter)

Thus, the supplementary action of cultural capital over economic capital in residential terms is related to the following fact: the residential hysteresis produced in people belonging to the upper classes is not always resolved with access to home ownership, but also with the chance to rent an adequate space. That is, one that is consistent with the social sector conquered in terms of identity. For this, cultural capital operates as a crystallizing vehicle that legitimizes the elective criteria of the place of residence beyond ownership, which simultaneously adjusts the residential identities of these social groups. Thus, social classes with a high cultural capital but without sufficient economic capital to fulfill their residential ideal can reduce or adjust their residential expectations associated with becoming homeowners if the rented property and its surroundings meet their immediate needs. This does not mean that they have renounced the possibility of owning a place to live, but rather that they adjust and expect that option, focusing elective criteria on others that are entail relevant issues, such as children's education. Under these principles of residential choice, part of these classes even dreams of acquiring ownership of a home far from the city, thus highlighting the high value that the countryside has as a lifestyle in the stage after childrearing (Rojo-Mendoza 2022). In this way, and considering a projected residential scenario, some of the replies were as follows:

> I plan to buy a house in Pumalal (rural area) or near a lake, with the same conditions as the house. (30-year-old woman, speech therapist, upper-class renter).
>
> My partner and I plan to buy a space that allows us to build a house. I don't think it will be in Temuco, I think it will perhaps be further south, more in the countryside […] We will have our house when we aren't moving anymore. (38-year-old man, designer, upper-class renter).

In the lower classes that are also renters, the criteria for a space's eligibility are related to the economic resources available to them and their proximity to the workplace. Thus, the chances to reduce the residential hysteresis produced in this group do not involve adapting their expectations to the diversity of services that the city offers, but rather to the functionality associated with where they work. In this sense, and considering the residential future, some of the replies associated with this social class were as follows:

> Still, I try to make sure they're not too far, too peripheral, when it comes for me to move somewhere. I usually work downtown. (36-year-old woman, independent worker, lower-class renter).
>
> I chose it because it was in a place I already knew, I found it quiet and financiallywithin my budget. (33 year-old man, mechanic, lower-class renter).

Thus, the adjustment of residential expectations on the part of lower social classes that do not yet have access to home ownership involves the cost-proximity ratio of the inhabited space. In this context, it is worth asking whether residential hysteresis in these social groups would decline or increase if they were to achieve home ownership, but in a location that is far from the workplace. In other words, when one acquires ownership of a home far from the desired location, does the right to property reduce the residential hysteresis that is generated? This is a question that opens the debate for future exploration of the subject.

18.3.1 School and Residential Choice Based on Children

As mentioned earlier, in middle and upper classes, discourses sustaining decision making regarding where to live, are associated to a far more complex and elaborate narrative than in lower classes. Eventhough these discourses mention the importance of connectivity and dwelling space sizes, aspects such as recreational supply and children educational services are predominant. This last idea has been mentioned in other research works in Chile (Méndez and Gayo 2019). In this sense, the mechanisms used to reduce residential hysteresis in this group are aimed at gaining domain over a space that is far larger than the inhabited dwelling. Therefore, in the middle and upper classes the process of identity reflexivity associated with space attempts to remedy any type of inconsistency in the comprehensive dominion over space and it features, including the control of strategic services for these families, such as proximity to their children's school. As shown in the previous section, in this broader control of the space the condition of owner or tenant does not lead to significant differences in the discourses analyzed. Thus, in response to the question related to the aspects that were considered when it came to deciding where to live, some of the interviewees said:

> We were looking for a place with bedrooms, since I have daughters; so, we were looking for a little house with more than two bedrooms [...] And then we sold the plot and came for my daughters, because they were studying. (67-year-old woman, housewife, middle-class homeowner).
>
> The house is old, but the area is very nice. Lots of squares, a bunch of bike lanes, supermarkets, a selection of services, public primary and secondary schools, and good public and subsidized secondary schools. So, we thought about it and we said that we wanted our children to go to public schools and that's it. (46-year-old man, teacher, upper-class homeowner).
>
> I don't know if it's my ideal place, but it is for my daughter, who is bigger now, in terms of having a life more like the one I had when I was a girl. In the street or the neighborhood, so I'd say yes. And it is close, super close, to a lot of municipal schools, public schools. (50-year-old woman, journalist, upper-class renter).

Thus, finding and living in spaces near educational establishments imagined to be optimal for children accounts for a significant portion of the reduction in middle- and upper-class residential hysteresis. In this sense, there is a clear relationship between the projection of their children's cultural capital and the residential choice patterns that these groups follow, all of which lays the foundation of a significant part of class identities associated with space.

It should be noted that this dimension of residential choice related to children's education is interesting when analyzed in the context of the residential segregation in Chilean cities. The proliferation of subsidized and paid establishments, in addition to the widespread discourse regarding the differences in quality between public and private administrators (Canales et al. 2016), has meant that a significant portion of the supply of non-state education is concentrated in spaces where, in theory, there is greater demand for private educational alternatives. This, together with families' freedom to choose their children's schools and the sustained decline in public school enrollment (Paredes and Pinto 2009), prompted an essential part of the upper- and middle-class's residential mobility to revolve around a discourse associated with securing the cultural capital of their descendants. Under this model of residential mobility, part of the socio-spatial transformations are directly related to the diversification of the educational model in Chile and its spatial expression in cities, which implies an important correlation between residential expectations and school choice (Méndez and Gayo 2019).

18.4 Conclusion

The residential expectations analyzed express different needs according to social class. The lower classes seek to achieve home ownership, which is why discourses clearly express the efforts made over time to achieve that result. In addition, the expectation of home ownership is materialized in a hazardous way, where the combination of practices such as saving, applying for certain lands and state actions lead people to live in given spaces in both cities. Thus, any identity mismatch that may arise within this class is resolved in spatial terms with actions aimed at attaining home ownership and with it, relativizing any non-conformity that living in a place that was not entirely desired may create.

Renters who belong to these same lower classes have better alternatives to adjust certain dimensions of their expectations, especially those related to living close to the workplace. This functional option of the inhabited space allows regulating these people's expectations now, pausing future residential desires. However, and with the background of home ownership as a central mechanism in the identity adjustment of these classes, making the transition from renter to owner would appear to ultimately snatch away the few options that these groups have to inhabit functionally desired areas. The question to be asked in these cases is whether the subsequent attainment of home ownership by these current tenants will positively or negatively adjust their residential expectations. As this chapter has shown, conformism with ownership would appear to be hard to overcome in the Chilean neoliberal model, which can dangerously fuel social policies or subsidies for poor quality housing in unwanted places.

Regarding the middle and upper classes, their residential expectations are met as the functions that the spaces offer are appropriate to their needs. Home ownership is not an imperative for people in these classes who have not yet attained it and instead the important thing is the satisfaction produced by functional control and domain over inhabited spaces. According to these results, these classes' access to the urban system does not necessarily involve home ownership, but also enjoyment of the diversity of alternatives offered. The potential residential hysteresis in this group is therefore resolved through mechanisms that ensure access to services considered indispensable. One of these is the supply of educational establishments for their children, an important matter when it comes to deciding where to live. Thus, and considering the form taken by residential expectations in these middle and upper classes, they would appear to be the ones that, at least at the discursive level, adjust their identities based on the benefits offered by the city, which helps to sustain the social position attained. The lower classes do not enjoy the benefits that the city offers and only reinforce their identity in residential terms through home ownership, which often does not in the slightest way fulfill the dreams they always had.

References

Akerlof G, Shiller RJ (2015) Phishing for phools: the economics of manipulation and deception. Princeton University Press, New Jersey

Alvarado V (2019) El bienestar en el Estado neoliberal: escenarios de la propiedad en el Gran Santiago. CUHSO 29(2):13–35. https://doi.org/10.7770/0719-2789.2019.cuhso.04.a02

Aramburu M (2016) Rental as a taste of freedom: the decline of home ownership amongst working-class youth in spain during times of crisis. Int J Urban RegNal Res 39(6):1172–1190. https://doi.org/10.1111/1468-2427.12218

Benson M, Jackson E (2017) Making the middle classes on shifting ground? residential status, performativity and middle-class subjectivities in contemporary London. Br J Sociol 68(2):215–233. https://doi.org/10.1111/1468-4446.12256

Bourdieu P (1984) Distinction. Cambridge University Press, Cambridge, A Social Critique of the Judgement of Taste

Bourdieu P (1998) Practical reason. Stanford University Press, Stanford, On the Theory of Action

Bourdieu P (2005) The social structures of the economy. Polity, Cambridge

Bourdieu P (2018) Social space and the genesis of appropriated physical space. Int J Urban Reg Res 42(1):106–114. https://doi.org/10.1111/1468-2427.12534

Brenner N, Peck J, Theodore N (2010) After neoliberalization? globalizations 7(3):327–345. https://doi.org/10.1080/14747731003669669

Bridge G (2001) Bourdieu, rational action and the time-space strategy of gentrification. Trans Inst Br Geogr 26(2):205–216. https://doi.org/10.1111/1475-5661.00015

Butler T (2005) Gentrificaiton and globalization: the emergence of a middle range theory. Sciences Po, Programme Villes et Territories, Paris

Canales M, Bellei C, Orellana V (2016) ¿Por qué elegir una escuela privada subvencionada? Sectores medios emergentes y elección de escuela en un sistema de mercado. Estudios Pedagógicos 42(3):89–109

Caja de la Habitación Popular (1950) Memoria año 1950. Universo, Santiago

Centner R (2008) Places of privileged consumption practices: spatial capital, the dot-com habitus, and San Francisco's Internet Boom. City Community 7(3):193–223. https://doi.org/10.1111/j.1540-6040.2008.00258.x

Crawford J, McKee K (2018) Hysteresis: understanding the housing aspirations Gap. Sociology 52(1):182–197. https://doi.org/10.1177/0038038516661263

De Groot C, Manting D, Mulder CH (2013) Longitudinal analysis of the formation and realisation of preferences to move into homeownership in the Netherlands. J Housing Built Environ 28:469–488. https://doi.org/10.1007/s10901-012-9320-7

Erikson R, Goldthorpe JH (1993) The constant flux: a study of class mobility in industrial societies. Oxford University Press, Oxford

Espinoza V, Barozet E, Méndez ML (2013) Estratificación y Movilidad Social Bajo un Modelo Neoliberal Maduro: El Caso de Chile. Revista Laboratorio 25:169–192

Hamnett C, Butler T (2013) Re-classifying London:a growing middle class and increasing inequality. City 17(2):197–208. https://doi.org/10.1080/13604813.2013.765719

Hidalgo R, Alvarado V, Santana D (2017) La espacialidad neoliberal de la producción de vivienda social en las áreas metropolitanas de Valparaíso y Santiago (1990–2014) ¿Hacia una construcción

ideológica de un rostro humano? Cadernos Metrópole 19(39):513–535. https://doi.org/10.1590/2236-9996.2017-3907
Hidalgo R (2010) El papel de las leyes de fomento de la edificación obrera y la caja de habitación en la política de vivienda social en Chile 1931–1952. INVI 15(39):92–120. https://doi.org/10.5354/0718-8358.2000.62105
Hojman D (1993) Chile: The political economy of development and democracy in the 1990's. The University of Pittsburgh Press, Pittsburgh
Imilan W (2016) *Políticas y luchas por la vivienda en Chile: el camino neoliberal*. Working paper series Contested Cities, Available in http://repositorio.uchile.cl/bitstream/handle/2250/141198/Politicas-y-luchas-por-la-vivienda-en-chile-el-camino-neoliberal.pdf?sequence=1
Jansen SJ (2014) Why is housing always satisfactory? a study into the impact of cognitive restructuring and future perspectives on housing appreciation. Soc Indic Res 116:353–371. https://doi.org/10.1007/s11205-013-0303-1
Kaufmann V, Bergman M, Joye D (2004) Motility: mobility as capital. Int J Urban RegNal 28(4):745–756. https://doi.org/10.1111/j.0309-1317.2004.00549.x
Levy J (2000) Les Nouveaux Espaces de la Mobilité. In: Les Territoires de la Mobilité, Presses Universitaires de France, pp 155–170
Matthews P (2015) Neighbourhood belonging, social class and social media-providing ladders to the cloud. Hous Stud 30(1):22–39. https://doi.org/10.1080/02673037.2014.953448
Ministerio de Vivienda y Urbanismo (2022) Observatorio Urbano, Estadísticas Habitacionales. Available in: https://www.observatoriourbano.cl/estadisticas-habitacionales/#subsidios
Mckee K, Moore T, Soaita A, Crawford J (2017) Generation rent' and the fallacy of choice. Int J Urban Reg Res 41(2):318–333. https://doi.org/10.1111/1468-2427.12445
Méndez ML, Gayo M (2019) Upper middle class social reproduction. Wealth, Schooling, and Residential Choice in Chile. Palgrave Pivot Series, New York
Méndez ML (2018) Neighborhoods as arenas of conflict in the neoliberal city: practices of boundary making between "us" and "them." City Community 17(3):737–753. https://doi.org/10.1111/cico.12326
Paredes R, Pinto JI (2009) ¿El fin de la educación pública en Chile? Estudios De Economía 36(1):47–66
Richards B (1993) Poverty and housing in Chile: the development of a neo-liberal welfare state. Habitat Int 19(4):515–527. https://doi.org/10.1016/0197-3975(95)00043-F
Rojo-Mendoza F (2022) Spatial preferences and counterurbanization in Temuco, Chile: between the pleasure of the natural and residential anonymity. Geografiska Annaler: Series B, Human Geography. https://doi.org/10.1080/04353684.2022.2082312
Rojo-Mendoza F (2021) Los gustos espaciales y las expectativas residenciales en Chile: el principio del derecho a la vivienda. Bitácora Urbano Territ 31(2):87–99. https://doi.org/10.15446/bitacora.v31n2.89491
Rowan A, Doling J (2017) The end of mass homeownership? changes in labor markets and housing tenure opportunities across Europe. J Housing Built Environ 32:649–672. https://doi.org/10.1007/s10901-017-9551-8
Sabatini F, Rasse A, Mora P, Brain I (2012) ¿Es posible la integración residencial en las ciudades chilenas? Disposición de los grupos medios y altos a la integración con grupos de extracción popular. Revista EURE 38(115):17–34. https://doi.org/10.4067/S0250-71612012000300008
Savage M (2010) The politics of elective belonging. Hous Theory Soc 27(2):115–161. https://doi.org/10.1080/14036090903434975
Savage M, Bagnall G, Longhurst B (2005) Globalization and belonging. SAGE, London
Savage M, Devine F, Cunningham N, Taylor M, Li Y, Hjellbrekke J, Le Roux B, Friedman S, and Miles A (2013) A new model of social class? Findings from the BBC's great british class survey experiment. Sociology 47(2):219–250. https://doi.org/10.1177/0038038516661263
Stillerman J (2017) Housing pathways, elective belonging, and family ties in middle class chileans' housing choices. Poetics 61:67–78. https://doi.org/10.1016/j.poetic.2017.01.005

Strauss A, Corbin J (2002) Bases de la Investigación Cualitativa: Técnicas y Procedimientos Para Desarrollar la Teoría Fundamentada. Universidad de Antioquia, Medellín

Vergara P (1984) Auge y caída del neoliberalismo en Chile: un estudio sobre la evolución ideológica del régimen militar. FLACSO, Santiago

Watt P (2009) Living in an oasis: middle-class disaffiliation and selective belonging in an english suburb. Environ Plan A 41:2874–2892. https://doi.org/10.1068/a41120

Chapter 19
(Re)Thinking Gentrification Processes. The Place of Religion

Víctor Albert-Blanco

Abstract Gentrification has been defined by urban sociologists and geographers as a process of appropriation of neighborhoods by middle and upper classes, provoking the displacement of poorer inhabitants. This sociodemographic change also implies a reconfiguration of urban landscapes due to the opening of new shops, services, and activities according to the tastes of the new neighborhood inhabitants. Gentrification, therefore, can be understood as a dispute on the access to housing and the uses of public places, producing a new spatial regime that determines what and who is acceptable in urban space. However, the role of religion in this process of disputing the public space remains almost untouched. Based on the spatial turn in the study of religion in social sciences, I examine the construction of these ambivalent discourses by different local actors (public authorities, local associations, religious actors…) and their implications for expressions of religious minorities in conflictual urban spaces. On the one hand, I show that the public presence of some religious minorities is rejected by dominant social groups who consider it as a source of urban degradation and as an incompatible element with the supposed secular uses of public space. This often implies new spatial regulations which redefine the place and form of religious manifestations. On the other hand, I illustrate how religious pluralism is strategically used by these same actors labeling the neighborhood as a "multicultural" and "cosmopolitan" site.

Keywords Gentrification · Religion · Urban space · Diversity · Neighborhood

V. Albert-Blanco (✉)
Université Paris 8, CRESPPA-GTM, Paris, France
e-mail: valbert.blanco@gmail.com

Universitat Autònomca de Barcelona, Barcelona, Spain

E. Navarro-Jurado et al. (eds.), *Urban Dynamics in the Post-pandemic Period*,
The Urban Book Series,
https://doi.org/10.1007/978-3-031-36017-6_19

19.1 Introduction

During the last decades, European cities have known an important increase of their diversity. Despite have been longtime characterized as the epicenter of secularization process, closely tied to modernization and urbanization, European cities have seen new forms of religion (Becci et al. 2013). These religious expressions are the consequence of immigration, the arrival of individuals and groups coming from other countries and geographical areas such as Africa, Latin America, or Asia. These groups display spatial strategies shaped by the national and local contexts and they open new worship places or shops with religious markers in the neighborhoods where they live or frequent. These areas can be affected by other urban process such as pauperization, segregation or, also, gentrification, the appropriation of a neighborhood by an upper-middle class who did not habit it previously. Which are the place and form of minority religious expressions in these gentrified urban enclaves? How religious expressions are perceived by local actors participating in urban transformations? Do religious manifestations participate in urban imaginaries on these areas?

Drawing on a long qualitative fieldwork carried out between 2016 and 2020 in two gentrified neighborhoods in Paris (Goutte d'Or) and Barcelona (Raval), in this chapter I try to answer to these questions, and I claim that religious expressions have an ambivalent role in these contexts and conflictive urban processes. Based on the spatial turn in the sociological study of religion, I examine the construction of these ambivalent discourses by different local actors, as well as their implications for expressions (places of worship, celebrations on public spaces, specialized shops, images, and symbols…) of religious minorities (especially Islam) in these urban spaces. First, I show that the public presence of these minorities can be rejected by the dominant social groups, who see them as one source of urban "degradation" and as an incompatible element with the supposed secular uses of public space. This entails a process of regulation and limitation of these expressions, pushing them to conform to what is expected from the ongoing urban transformation. Finally, I point out how, paradoxically and after these regulation mechanisms, religious pluralism and its inscription in urban geography are strategically used by certain actors in the dynamics of symbolic rebranding of the neighborhood coupled with gentrification, especially in its promotion as a "multicultural" and "cosmopolitan" site.

19.2 Gentrification and Religion: Toward a Proposal of Analysis

Although the social sciences have scrutinized the links between religions and urban space (Kong and Woods 2015; Saint-Blancat 2019), the effects of gentrification in this association remain almost untouched. At some point, gentrification could be understood as the zenith of modernization (Ley and Martin 1993), a particular phase of the urbanization process implying the removal of working classes who would be replaced by new modern, upper, and creative groups. However, the literature has shown that religion has not disappeared in this phase of the modernization: it is still (if not more) visible in urban spaces (Casanova 1994; Becci et al. 2013), also in the neighborhoods undergoing a gentrification process. Neighbors inhabit in these districts have different religious identities, both those who were already there and those who arrive with urban transformation. Moreover, gentrification sometimes takes place in neighborhoods with high rates of ethnic diversity, urban enclaves which became the place where immigrants coming from other countries have settled because of their affordable apartments and where, later or in parallel, upper-middle classes also install themselves.

Nevertheless, gentrification is not just a sequence of departures and arrivals of populations with different social identities (class, gender, national and/or ethnic origin, religion…). As scholars have shown, gentrification is accompanied by a set of "urban products", such as the shape and aesthetics of buildings, apartments or shops (Chabrol et al. 2016). These "products", whether they are the work of public or private actors, allow for the revaluation of the place and constitute a market of economic and symbolic goods. However, the value of these goods reflects the actors' position in the social space and its urban mirror. Thus, gentrification can be apprehended as a "field" (Bridge 2001; Ley 2003), understood as a terrain of relationships, social practices or material and historical traces inscribed in the urban space, which are the result of a "game" between different actors (Bourdieu 2002 [1984]); Ley 2003). Here, I claim to take religion into consideration in the analysis of these "urban products" and in the process of devaluation or revaluation of the territories in which they are located. In the neighborhoods characterized by religious diversity, there are places of worship, shops, and restaurants with religious markers (halal, kasher, Hindu…), rites and commemorations that sometimes occupy streets and squares, or even clothing and body expressions. These manifestations are not alien to urban changes but are part of them, especially in a context of gentrification.

Some actors, particularly those endowed with greater resources (economic, cultural, symbolic, or political), possess the ability to enforce their perspectives and categorizations of urban products (Bridge 2001), including religious expressions. The urban transformations derived from gentrification operate a reconfiguration of religious and secular spaces, materializing into particular urban products. This is how, for instance, the closure or demolition of certain places of worship is ordered by public administration, while others, on the contrary, are appreciated and valued, even being the object of promotion (Albert-Blanco 2019).

In order to analyze how minority religious expressions are perceived and regulated in gentrified neighborhoods, I suggest to apply a temporal and spatial dissection. I propose to distinguish four phases, each of them correlated to the temporality of gentrification advanced by scholars and literature. These stages do not be seen as static and delimitated, but as a theoretical tool allowing us to understand the interrelation between religious diversity and gentrification. At the empirical level, they can overlap one to another, depending on the particularities of each context. In this model I suggest the urban area becomes, at the beginning, a "super-religious diversity" enclave, correlated to its function as an immigrant centrality and working-class neighborhood. This sociodemographic transformation allows the stigmatization of the sector. Immigrant and religious minorities visibility, especially Islam, can be placed at the core of discourses designating the area as a "ghetto" and a dangerous site. In a third moment, public powers begin an urban transformation with the official aim to resolve social problems and improve the quality of live, a set of interventions that often facilitate (and even promote) the neighborhood gentrification. This transformation entails new public space regulations, some of them affecting religious practice and its aesthetics and visibility. Finally, as it was advanced in the introduction, religious diversity can be placed as a main element in the discourses celebrating the neighborhood "renaissance", and especially its diversity and cosmopolitanism. This does not remain at the discursive level but includes the possibility to "taste" the religious diversity through exhibitions, urban tours, restaurants and, even, to take part in some celebrations.

19.3 Religious Super-Diversity Enclaves

Some neighborhoods in the city-centers of different European cities became the place where immigrants coming from other countries and geographical areas settled. These districts were characterized by a dense urban plot and architecture, and they offered affordable (and degraded) apartments. Their transformation as immigrant centralities represented a continuity from their function as working-class areas. However, the immigration is not only a sociodemographic variable in the neighborhood composition but translates into material and symbolic markers inscribed in urban space. In these neighborhoods, immigrants and their descendants are visible through specialized shops or communitarian associations. Religion is another dimension that becomes visible through the presence of new worship places and celebrations in public space. Areas such as Tower Hamlets in London, some sectors of East Berlin, Javastraat in Amsterdam, Lavapiés in Madrid, or Martin Moniz in Lisbon, have known this process of diversification.

Despite being in different national contexts (France and Spain), Goutte d'Or and Raval, where I conducted my fieldwork, are two neighborhoods with very similar history and sociodemographic structures. Like the neighborhoods mentioned above, they are paradigmatic cases to study the imbrication of ethnic and religious pluralism and urban transformation. Both territories have historically been located on the symbolic "margins" of their cities, being working-class and industrial enclaves. Since the second half of the twentieth century, the two neighborhoods also became urban enclaves in which percentages of immigrants were concentrated, exceeding the city average. Currently, around 35% of the population in Goutte d'Or comprises immigrants (compared to 20% in all Paris), while in Raval, the immigrant population amounts to 50% (compared to 20% in all Barcelona). This is not a homogeneous population, and the ethnic and national profile differs in each context. In Goutte d'Or, immigrants came mainly from the former French colonies, such as Algeria and Sub-Saharan African countries. In Raval, Moroccans were the most important community in the neighborhood during the 80s and 90s, but some years later they conceded this place to Pakistanis, Bengalis, and Filipinos.

This diversity gives rise to significant religious plurality, which Becci et al. have identified as enclaves of 'religious super-diversity' (2017). For instance, in Raval there are now 7 Islamic worship places, 7 evangelical churches, and 1 Sikh gurdwara temple. In Goutte d'Or, there are currently 2 Islamic worship places, but also 1 Jewish oratory and 1 Nazarene church. Catholicism, the historical and hegemonic religion in both contexts, is still visible and active in these enclaves, not only with important and prominent parish churches but also with private schools and other material and symbolic markers, such as the name of some streets referring to Catholic saints or public shrines. Moreover, some immigrant groups, such as the Filipinos in Barcelona or Africans in Paris, are also attached to Catholicism. They use the neighborhood parishes, but they maintain some rites and celebrations.

19.4 Stigmatizing Neighborhood and Religious Minorities

The neighborhood becomes the object of a certain media, social and political stigmatization because of its populations and the uses of its public space. We can identify then a pattern of urban "pauperization" or "degradation". These phenomena are not necessarily a consequence of the presence of immigrants, but their cause. In this context, a decline discourse of these neighborhoods is constructed, qualifying them as chaotic and dangerous places. These representations are not new, but they can integrate the "foreign" presence and its religious dimension. Racist and Islamophobic imaginaries, relationships of otherness, find a physical and spatial representation in these territories. The construction of the image of the "ghetto" neighborhood is carried out by staging various "problematic" phenomena and certain Islamic manifestations can occupy a central place in this process. If these controverses mobilize a particular rhetoric, they share at the same time a vocabulary that serves to describe the set of "problematic" phenomena in the sector. For instance, "insalubrity" is used both to qualify housing stock or "immigrant" businesses and places of worship, while "incivility" allows to designate both a street prayer and street vending. The religious (especially Islam) is constructed as one of the elements that nourishes the "bad" image of the neighborhood, reinforcing the idea of a "closed" enclave.

This second phase, where impoverishment and stigma go hand in hand, is key to understanding the subsequent gentrification of the sector. The description of the decline of the "ghetto" neighborhoods has allowed (and still allows in some places) to mark the old popular territories of the city center. In turn, this has facilitated, or even served as a pretext, for the intervention of public and private authorities to transform them. In neo-Marxist approaches to gentrification, the concept of rent gap explains the link between the degradation of a sector and its subsequent revaluation. According to the geographer Neil Smith (1996), it is at the height of impoverishment that a neighborhood becomes more "gentrifiable", because of it is at this point that the gap between the price of current and potential land becomes greater and, therefore, more profitable for investors.

In Goutte d'Or and Raval, this stigmatization includes some religious expressions. In Paris, a significant public controversy refers to the visibility of Islam in the neighborhood. Since the 1990s, Muslims used to pray in the street because of the lack of space to do so in two local mosques. Local and external actors criticized the street prayers. They referred to a religious practice that would breach the secular understanding of public space, especially in the French context of *laïcité*. However, these critics also pointed out that this religious expression was a trouble for the neighborhood, a source of degradation, an obstacle to its "normalization" and to become "like any other Parisian district". In Raval, Barcelona, there was not a huge controversy about Muslim worship practices such as the street prayers as it was in Paris. However, the presence and visibility of Islam, and specially the significant number of mosques and halal shops, have garnered the attention of media and local actors. For instance, in 2004, some neighbors demonstrated against the opening of a new worship place in the sector. They collected signatures and put banners on the balconies, but the Islamic community succeeded to open its new place. Some years later, this kind of local controversies continue, and in 2017 another neighbors group tried to prevent the activity of another Islamic center by attending public meetings and presenting an administrative appeal. In these controversies, the rhetoric mobilized by some neighbors, media and politicians mix a fear of Islam and a claim for a "normal" neighborhood where religious expressions (and other phenomena) would be limited and controlled. A clear illustration of this rhetoric was displayed by the conservative People's Party (PP) during the campaign for the 2015 local elections, when they distributed posters and brochures stating that "Raval would not become an Islamic ghetto".

19.5 Transforming Urban Space, Regulating Religion

These discourses on urban decline claim for the intervention of public powers to resolve social dysfunctionalities. Thus, the administration initiates a huge urban transformation by demolishing old buildings, constructing new social (and private) apartments, opening new streets and public spaces, or establishing cultural centers. These interventions are often justified in a hygienist discourse that advocates the benefits of "social diversity" (Charmes and Bacqué 2016), understood it as the possibility that upper-middle classes come to live, visit, or use the neighborhood. This is presented as the remedy to the sector's social problems but achieving an "ideal" public space also requires other measures in the areas of security, education, culture… and religion.

The regulation of religion, especially the expressions considered as a "problem" or disruptive, is not isolated from the set of actions regarding the urban space transformation. Administrations implement different instruments to frame religious expressions. The governance of religious plurality constitutes a relatively autonomous domain of public policies at local level (Griera 2012; Martínez-Ariño 2021), but in its realization in a gentrifying neighborhood it is inserted and consistent with the set of interventions in other areas (Albert-Blanco 2022a). Moreover, the work of regulating religious, of classifying and promoting its objects, is not the exclusive task of the public powers. It concerns other actors involved in urban transformation, such as neighbor' associations, traders, or cultural and artistic entrepreneurs. It is linked to the "multidimensional work" of gentrification, (Bidou-Zachariasen and Poltorak 2008) articulated in the space and time of the neighborhood. These actors can claim for the closing of a particular worship place, or demand to put an end to some religious expressions they consider disruptive or incompatible with the accepted uses of public space. At the same time, paradoxically, they can include the religious pluralism in communitarian activities, such as the local festivals, or even as an object that can be tasted in cultural centers, restaurants, or boutiques.

This inclusion is not dissociated to the regulation itself and implies the transformation of religious expressions into "acceptable" objects or events, selecting those that can be seen as cultural and open to non-religious people (Martínez-Ariño and Griera 2020). This strategy is similar to that scholars of gentrification have observed regarding other expressions of ethnic minorities in these urban enclaves. As Summers has pointed out, "racialized expressions are more marketable in the emerging 'creative city' that emphasizes cultural consumption and creative, aesthetic practices" (Summers 2021, p. 118).

In the neighborhoods I conducted fieldwork, the regulation of religion concerns different expressions, and it is promoted by different actors. In Goutte d'Or, the French government forbids the street prayers in 2011. In concertation with local administration and religious communities, they were allowed to pray in a former fire station situated outside the neighborhood. At the same time, the city council promoted the construction of an Institute of Islamic Cultures (ICI). This center, opened in 2013, has two branches, one dedicated to culture and open to a non-religious public, and the other one is consecrated to praying and it is managed by an Islamic association. The official aim of this center was to offer a "dignified" space for praying, by showing at the same time a "good face of Islam" to non-Muslim people through exhibitions, cultural events, a restaurant, a patio, and a hammam. This space is an important element of the neighborhood urban transformation (Albert-Blanco 2022b), and it seems to be coherent to other interventions that contribute to the aestheticization of ethnic minorities and popular cultures (Chabrol et al. 2016). For instance, public authorities made an important effort in the field of retail with the aim to shrink the presence of street vending and ethnic shops. This policy does not only imply a limitation of these stores, but their transformation into acceptable "boutiques" selling design objects and clothes supposed to be inspired by neighborhood ethnic minorities (Chabrol et al. 2014).

In Raval, the efforts of public authorities in the field of religious governance are focused on the contention of the controversies mentioned above. The Barcelona City Council has a Religious Affairs Office (OAR) who accompanied religious groups in the process of opening new worship places. OAR officials, together with other public servants from other areas such as urbanism or security, ensure that worship places comply with the law and do not generate disturbance for neighbors. Moreover, in Raval, they support religious communities in the organization of some commemorations in public space. This is especially visible during the Ramadan, when local Islamic communities organize iftars, the ritual which breaks the fasting in the evening, in different places and squares. This activity increased in the last years and has become an important element in the neighborhood communitarian life. Far from being only a Muslim celebration, the Raval' iftars become an activity open to non-Muslim neighbors and visitors, where one can taste the traditional dishes from Arab and Muslim countries, but also hear oriental music and learn about the Ramadan meaning and Muslim traditions (Clot-Garrell et al., 2022). Similar to the Parisian case, this effort aims to ensure religious freedom and normalize the presence of minorities, especially Islam. However, these activities allow to symbolically rebrand the neighborhood through positive values such as "diversity", "multiculturality", and "cosmopolitanism".

19.6 Tasting Religious Flavors in a Trendy Neighborhood

The urban transformation promoted by public authorities entails the revalorization of the neighborhood and facilitates the installation of upper-middle classes. They can come to inhabit the sector, but they also can only frequent or visit it because of its new cultural centers, boutiques, or trendy coffees. These new inhabitants and visitors, the gentrifiers, value the "authenticity" of the neighborhood. As Zukin has noted, "downtown farmers' markets and ethnic food stores underline [its] image as oases of authenticity" (Zukin 2008), including the (limited and regulated) presence of working class and ethnic minorities. As scholars have shown, it operates as an inversion of collective imaginaries or, at least, a coexistence of contradictory representations. Where it was once expected to be a "degraded" and "ghetto" neighborhood, there has emerged a transformed and revitalized enclave, characterized by its "vibrant" and "cosmopolitan" nature. As Sylvie Tissot has pointed out, "these categories of apprehension of urban spaces are socially constructed and promoted in parallel to population displacements (…) and important economic transformations" (Tissot 2012, p. 7). These imaginaries are promoted by different actors, among them, first, the public authorities with discourses and statements celebrating the "diversity" in its multiple forms. These representations are also transmitted by other agents, such as the media (especially the cultural press) or even by private investors (such as local traders, real estate agencies). These imaginaries can also be spread by the inhabitants themselves: by gentrifiers and visitors who exhibit a taste for diversity that reflects, at the same time, their position in the social space (Tissot 2011), but also by "old" inhabitants who claim cultural and religious diversity to maintain a relative place within the urban space.

Religion is part of these imaginaries, becoming an important element in the representations of the neighborhood that describe it as a "multicultural" site. The taste for diversity and authenticity exhibited by some gentrifiers intersects here with the positive assessment of certain religious expressions if they are, as it was noted above, regulated, and delimited. It does not mean that these inhabitants and visitors are attracted to any "religious" aspect of Islam or other minority groups. Rather, it suggests the celebration of their (regulated) presence in the neighborhood that reinforces the urban imaginaries around the idea of "diversity". This taste is not isolated but is part of the actors' social and cultural dispositions and practices (Bourdieu 2021 [1979]). It is neither homogeneous nor uniform but learned and practiced according to a certain habitus acquired in the context of gentrification and in the residential, associative, and political trajectories of individuals and groups. However, this assessment of cultural and religious plurality does not mean the alterity and power relationships have disappeared. It coexists with them by developing apparently contradictory discourses and practices that, despite a formal and symbolic recognition of religious minorities, do not alter the subordinate position of Islam and Muslims in the social space.

In Goutte d'Or and Raval, local actors I met express this taste for diversity. This taste is reproduced and reinforced through different supports. Media, such as *Time Out* or other cultural magazines, use to publish stories and reportages about these neighborhoods, presenting them as "vibrant" and multicultural enclaves, including often the regulated religious expressions that one can find it. For instance, the cultural activities organized in the Parisian Institute of Islamic Cultures are promoted in the pages of these publications. The center is depicted as a 'hybrid' space where individuals can literally savor and explore the essence of Islam. These descriptions do not remain at the discursive level but are translated into the ordinary practices by some local actors and visitors. During my interviews, some of them referred to the Institute as a "positive" space, enhancing certain attributes such as "open", "nice", or "pleasant". These adjectives seem to certify the capacity of the center, and the public authorities who promote it, to transform religion into an acceptable set of objects and elements disposed to non-Muslim. Furthermore, engaging with the Institute and its cultural dimension is closely associated with participating in other activities in the neighborhood. Some neighbors I met integrated the attendance to the Institute in their social and cultural practices, articulating it in a set of different activities such as collaborate in a common garden, actively participate in the parents' school association, organize popular festivals or, even, frequent local and "authentic" coffees and markets.

In Raval, the taste for diversity is also reinforced by the general and cultural press, who presents the neighborhood as the "alternative" Barcelona city-center, or as the "doors to the Orient". Moreover, some activities such as the iftars mentioned above, are presented as an opportunity to reverse the neighborhood stigmatization, enhancing the effort made by local actors to improve conviviality and, especially, create "multicultural" experiences in the urban space. Again, these representations do not remain at the discursive level, but are incorporated by local actors and visitors who express their attachment to diversity. Some actors I interviewed expressed their "pleasure" to see, even participate, in these activities, but they also referred to the neighborhood "vibrancy" during Ramadan. Beyond these celebrations, they also value that some religious local communities participate in the neighborhood festivals, bringing "multiculturality" and "colors" into local traditions.

19.7 Conclusion

This chapter has examined the complex social process that determines the place and form of religious expressions (especially minority ones, such as Islam) in gentrified neighborhoods. Specifically, and drawing on the analysis of the empirical material collected in Goutte d'Or, Paris, and in Raval Barcelona, it proposes a theoretical approach to understand this process. In this sense, the coexistence of both phenomena (pluralization of the religious landscape and gentrification) gives rise to a process of regulation of religious expressions, which runs in parallel, and it is coherent, with the urban transformations. An ambivalent situation is thus produced, in which religious expressions can be constructed as "public problems" (near to other urban social phenomena) while, after a process of regulation and delimitation, they can take part of the representations and imaginaries that describe gentrified neighborhoods as "diverse" and "cosmopolitan" enclaves.

References

Albert-Blanco V (2019) Valoriser le quartier par la diversité religieuse: regards croisés entre la Goutte d'Or (Paris) et le Raval (Barcelone). Cahiers De Géographie Du Québec 63(178):21–35

Albert-Blanco V (2022a) Diversidad religiosa, políticas públicas y gentrificación en París y Barcelona. Papeles Del CEIC 2022(1):1–20

Albert-Blanco V (2022b) Encadrer le religieux: une politique de gentrification? Le cas de l'Institut de cultures d'islam de Paris, Métropoles, n° 31, online

Becci I, Burchardt M, Casanova, José (dir.) (2013) Topographies of faith: Religion in urban spaces. Leiden, Éditions Brill

Becci I, Burchardt M, Giorda M (2017) Religious super-diversity and spatial strategies in two European cities. Curr Sociol 65(1):71–90

Bidou-Zachariasen C, Poltorak J-F (2008) Le «travail» de gentrification: les transformations sociologiques d'un quarter parisien populaire. Espaces Et Sociétés 132–133(1–2):107–124

Bourdieu P (2002 [1984]) Questions de Sociologie. Paris, Editions de Minuit

Bourdieu P (2021 [1979]) La distinción. Criterios y bases sociales del gusto. Madrid, Tauros

Bridge G (2001) Estate agents as interpreters of economic and cultural capital: the gentrification premium in the Sydney housing market. Int J Urban Reg Res 25(1):87–101

Casanova J (1994) Public religions in the modern world. University of Chicago Press, Chicago

Chabrol M, Collet A, Giroud M, Launay L, Rousseau M, Ter Minassian H (2016) Gentrifications. Éditions Amsterdam, Paris

Chabrol M, Fleury A, Van Criekingen M, (2014) Commerce et gentrification. Le commerce comme marqueur, vecteur ou frein de la gentrification. Regards croisés à Berlin, Bruxelles et Paris, In: Gasnier A, Lemarchand N. (dir.), Le commerce dans tous ses états. Espaces marchands et enjeux de société, Rennes, PUR, 277–291

Charmes E, Bacqué M-H, (dir.) (2016) Mixité sociale, et après ? Presse universitaire de France Paris

Clot-Garrell A, Albert-Blanco V, Martínez-Cuadros R, Esteso, C (2022) Religious tastes in a gentrified neighbourhood, J Relig Eur, 15(1–4): 291–318
Griera M (2012) Public policies, interfaith associations and religious minorities: A new policy paradigm? Evidence from the case of Barcelona. Soc Compass 59(4):570–587
Kong L, Woods O (2016) Religion and space. Competition, conflict and violence i the contemporary world. London-New York, Bloomsbury
Ley D (2003) Artists, aestheticisation and the field of gentrification. Urban Stud 40(12):2527–2544
Ley D, Martin B (1993) Gentrification as secularization: The status of religious belief in the post-industrial city. Soc Compass 40(2): 217–232
Martínez-Ariño J (2021) Urban secularism. Negociating religious diversity in Europe. Routledge
Martínez-Ariño J, Griera M (2020) Adapter la religion: négocier les limites de la religion minoritaire dans les espaces urbans. Soc Compass 67(2):221–237
Saint-Blancat C (2019). Introduction. L'inscription religieuse dans l'espace urbain. État de l'art. Soc Compass 66(1):3–23
Smith N (1996) The new urban frontier: gentrification and the revanchist city. Routledge, London-New York
Summers BT (2021) Race, authenticity, and the gentrified aesthetics of belonging in Washington, D.C. In: Lindner C, Sandoval GF (dir.) Aesthetics of gentrification: Seductive spaces and exclusive communities in the Neoliberal City. Amsterdam University Press, 115–136
Tissot S (2011) De bons voisins. Enquête dans un quartier de la bourgeoisie progressiste. Paris, Raisons d'Agir
Tissot S (2012) Les centres-villes: modèles, luttes et pratiques. Actes De La Recherche En Sciences Sociales 195(5):4–11
Zukin S (2008) Consuming Authenticity. From outposts of difference to means of exclusion. Cult Stud 22(5):724–748

Chapter 20
Metropolitan Spatial Reconfiguration and the Mobility Transition: Sustainability Challenges in the Fragmented City

Fernando Gil-Alonso and Cristina López-Villanueva

Abstract Within the context of the European Commission's Sustainable and Smart Mobility Strategy for a transition to a green, smart, and affordable transport system, local governments of large cities have implemented private vehicle restriction policies. However, do these policies come into conflict with current metropolitan suburbanisation spatial trends? (i.e. the fact that a growing share of the urban population is residing in increasingly large and fragmented metropolitan peripheries). First, this text reflects on the reasons for the spatial reconfiguration of urban and metropolitan areas; the consequences of these changes on daily mobility; and the design of European and local policies for the transition to sustainable mobility, which—this is our hypothesis—can collide with the present population and residential mobility trends in urban cores and their peripheries. This hypothesis is verified in the second part of the chapter, taking the region of Madrid as a case study. Results show that population suburbanisation trends in the last decades have led to an increase in daily mobility, and particularly, in the use of private vehicles, despite policies hampering their use and promoting public transport.

Keywords Suburbanisation · Daily mobility · Metropolitan areas · Madrid · Barcelona

---

F. Gil-Alonso (✉)
Department of Geography, University of Barcelona, Barcelona, Spain
e-mail: fgil@ub.edu

C. López-Villanueva
Sociology Department, University of Barcelona, Barcelona, Spain
e-mail: clopez@ub.edu

E. Navarro-Jurado et al. (eds.), *Urban Dynamics in the Post-pandemic Period*,
The Urban Book Series,
https://doi.org/10.1007/978-3-031-36017-6_20

20.1 Introduction

The aim of this chapter is to analyse the apparent contradiction between, on the one hand, the political interventions that local governments have planned and implemented seeking a transition towards more sustainable mobility—for instance, encouraging the use of public transport and of the so-called 'active mobility' (walking, cycling, etc.)—and, on the other hand, population and migration trends observed in urban/metropolitan areas over the last decades. The initial idea is that the spatial reconfiguration of large metropolitan areas is currently determined by the diversification of mobility and migration flows, leading to an increasingly fragmented city from the social, economic, and demographic perspective. Focusing on population, residential mobility towards urban cores, and other phenomena such as gentrification, touristification, and the replacement of the native population by foreign immigrants, have increased in the last decades. However, the magnitude of these centripetal flows is still smaller than that of those towards the suburban peripheries, where medium- and low-density urban sprawl grows. These centrifugal flows, known as decentralisation or suburbanisation, are mainly residential in nature and driven by families (adults and their children) who seek housing that meets their economic, environmental, or status expectations. The initial hypothesis is that suburbanisation trends—which potentially imply an increase in the number of movements, the time spent in them, and the distance travelled, and also a greater use of cars—counterbalance policies for a transition to a more sustainable mobility. In fact, this may lead to their failure, at least in suburban peripheries. Thus, we believe that urban population change and residential mobility trends should be taken into account when elaborating local mobility plans. This would enable to predict how metropolitan spatial reconfiguration will condition daily mobility and the use of different means of transport in the near future. This chapter shows the challenge of implementing the transition to sustainable mobility in a specific metropolitan reality, that of post-Fordist neoliberal capitalism, which is increasingly spatially diversified and fragmented. This is done by using the metropolitan region of Madrid as a case study, and by analysing its demographic (population growth in the urban core and the peripheral areas) and daily mobility data.

20.2 Urban Spatial Reconfiguration and Its Impacts on Mobility in a European Framework to Combat Climate Change

20.2.1 Mobility as a Determinant of Urban Growth and Its Spatial Differentiation

Studies on the spatial structure of urban areas from the point of view of the population's demographic and social characteristics have a long tradition in urban geography and sociology. Whilst the sociological tradition originates in the Chicago School (Burgess, Park, McKenzie, etc.), geodemographic studies have been influenced by the stages of urban development model developed by Hall, Berry, Champion, Fielding, Klaassen, Van den Berg, and others. In short, the model intends to explain the population growth of urban cores and peripheries in four stages—urbanisation, suburbanisation, counterurbanisation (or deurbanisation), and reurbanisation, that repeat in a cyclical way. This mechanistic model, still being used as a classificatory framework for present urban areas (Wolff 2018), has been criticised by several authors, as it often does not fit into empirical evidence. For instance, as the model does not incorporate contextual economic factors—which would for instance explain the existence of shrinking cities (Bagchi-Sen et al. 2020; Haase et al. 2021)—the model cannot shed light on why diverse urban areas of the same country can be at different stages of the cycle. In addition, the fact that several urban areas have suburbanised whilst their urban cores have continued to grow also shows the limits of a model that treats cities as closed systems, and does not consider the increasing role of external migration in the growth of urban cores and peripheries.

However, the role of mobility in its different forms—not only international migration but also internal migration, intra-metropolitan residential mobility, and daily trips—is increasingly relevant. Indeed, so is its impact on the growth, and the spatial differentiation of urban areas (Gil-Alonso et al. 2021), and on the demographic and the socio-economic characteristics of their population (Rees and Sander 2019). Thus, a proper understanding of the changes in the spatial structure of urban areas and of their social and economic consequences, requires an analysis of spatial mobility, distinguishing the different flows that compose it. For example, foreign immigrants tend to settle in urban cores alongside the 'creative class' (Florida 2017). By contrast, young families prefer to move to suburban peripheries (Rérat 2020), forming large low-density neighbourhoods (Garcia-Coll and López-Villanueva 2018) with distinctive demographic, sociological, or political features (Dioni 2021).

20.2.2 Structural and Temporal Factors of Mobilities in Urban Areas

Mobility patterns—and thus population growth and decline in (and within) urban areas—cannot be explained by a closed cyclical urbanisation model, but by a series of internal and external factors acting on them and making sense of who moves and who does not. Some of these factors are temporary whilst others are structural, such as a population's composition by age and sex and the type of households they form. These characteristics influence residential moves related to biographical events, that is, life-course transitions (Clark 2016; McCollum et al. 2020). Household and family transformations associated with the Second Demographic Transition model also entail an increase in residential moves deriving from changes in living arrangements (Mikolai and Kulu 2020). This could lead to the hypothesis that residential mobility has increased worldwide, in comparison to former decades. However, this is not true for all types of mobility or for all geographical contexts (Shuttleworth et al. 2021). In some developed countries, mobility rates are currently lower than in previous decades (Champion and Gordon 2021). In Spain—as in southern and eastern Europe—residential mobility levels are lower than in northern and western Europe (Rowe et al. 2019) and continued to decline further during the recession years (Bayona and Pujadas 2020). This was the consequence of fewer foreign immigrants—who move more frequently than Spanish people—arriving in Spain, and of fewer young natives leaving their parental homes and starting new families. Even though the first cause has been reversed in the post-crisis period, the second has not, because smaller cohorts have reached the age of leaving parents' households.

Amongst the temporary factors, phases in the economic cycle influence urban growth and decline in multiple ways: directly, by affecting both migration and natural growth—through fertility changes—and indirectly, by their impact on the labour and housing markets.

The labour market and, more specifically, the impact of economic cycles on labour insertion levels of both Spanish workers and foreign immigrants has influenced the spatial distribution of urban population through diverse types of mobility. International migration was the main cause of urban population growth during the economic expansion years (since the mid-1990s to 2007) and was one of the main factors in population redistribution within urban areas (Otero-Enríquez et al. 2019). Some of the immigrants left during the economic crisis (2008–2014), but inflows have resumed in the post-crisis period (2015–2019). Changes introduced by the COVID-19 pandemic and the extent to which previous international flows will recover remain to be seen. This will depend on Spain's capacity to generate employment and attract population (Esteve et al. 2021). In labour markets described as 'dual'—using Piore's (1979) words—foreigner flows will very probably recover, especially in the case of Spain, where most foreign immigrants work in occupational niches within low added value economic sectors (Moreno-Colom and López-Roldán 2018).

The relationship between economic cycles, the labour market, and mobility within urban areas—intra-metropolitan flows—and its spatial consequences, seems to be even more complex, as it is influenced by local factors. In general terms, suburbanisation is the dominant process during economic expansion phases, whilst recentralisation was relatively more common during the last economic crisis period. This latter phenomenon could be explained by the fact that peripheries had higher unemployment levels than metropolitan cores, as industrial employment and construction were more important in the former (Rodríguez and Pozo 2019). Additionally, the stock of rental housing is larger in urban cores and these areas also have denser public transport networks. As suburbanisation reduced during the economic crisis, differences in growth between urban cores and peripheries also did so (López-Villanueva and Pujadas 2015). The post-crisis has also influenced migration flows, though in the opposite direction, by leading intra-metropolitan mobility to increase and, as consequence, to an upsurge in suburbanisation. However, given the extent of population ageing and fertility decline, this new periphery growth period is not expected to be as strong as that before 2008. In fact, urban cores have continued to attract certain types of population during the post-crisis period (Gil-Alonso et al. 2021).

The housing market is also influenced by the moment in the economic cycle, and by other institutional mechanisms such as access to mortgages and public policies regarding housing (López-Colás et al. 2021). In Spain, some specific groups, such as young people and immigrants, find it particularly difficult to access housing. Their problems have worsened in the post-crisis period, because housing purchase and rental prices have increased, especially in large cities, because of financial globalisation. Instead of satisfying a basic need, housing has become an attractive investment for small investors and especially for large funds (Méndez 2021), intensifying gentrification in some neighbourhoods of large cities (Alexandri and Janoschka 2020). Tourist pressure or 'touristification' is another factor to be considered (Cócola-Gant and López-Gay 2020): it is the reason why part of the housing stock is no longer used for residential purposes but for tourism demands, consequently reducing the housing supply and increasing its prices. All these processes related to housing affordability have increased residential mobility and displaced residents with less purchasing power from urban cores, particularly from those neighbourhoods most affected by gentrification, to the periphery. Therefore, residential mobility has been acting as a way to relocate people by social class, income, or educational level (López-Gay 2018).

20.2.3 Diversified Daily Mobility in Increasingly Fragmented Cities

The suburbanisation of low-income groups—which some authors call the 'suburbanisation of poverty' (Hochstenbach and Musterd 2021; Porcel and Antón, 2020; Bailey and Minton 2018)—partly explains the renewed immigrant and native suburbanisation flows in the post-crisis period. Simultaneously, another form of precarious housing, that of sublet rooms, has gained force in urban cores and the first metropolitan rings (Orozco et al. 2022). As these authors claim, it is the only way for the most precarious groups, particularly foreign immigrants, to access housing. This confirms that rather than a suburbanisation of poverty or a simplistic dichotomy between affluent urban cores and deprived peripheries, there would be a 'fragmentation of poverty'. In other words, vulnerable areas and also those occupied by high-income groups are growing both in urban cores and peripheries, in contiguous and even overlapping areas.

These complex socio-residential patterns, within this growing inequality context (Piketty 2019), respond to what authors such as Florida (2017) call a 'new urban crisis'. Indeed, residential mobility plays an essential but extremely complex role in the social reconfiguration of metropolitan spatial interactions, generating new 'fragmented' areas (Borsdorf and Hidalgo 2009). However, socio-spatial fragmentation—with growing segregation, polarisation, and the creation of new vulnerable areas—is not the only form of urban fragmentation in the twenty-first century. It is also occurring for environmental reasons, as in the case of the 'return to the countryside', observed during the COVID-19 pandemic. In 2020, urban areas lost population and some rural ones gained it. More specifically, those municipalities close to (or well connected to) urban areas, and those that had many secondary homes having become the main place of residence, were those that benefited the most (González-Leonardo et al. 2022). Leaving the temporary nature of this counterurbanisation caused by the COVID-19 pandemic aside, a prediction can be made. Due to teleworking and the improvement of transport infrastructures, metropolitan areas will extend increasingly further away from the core city, so that the transitional spaces between what is considered urban or rural will be increasingly blurred and fragmented.

Simultaneously, demographic fragmentation, because of the diversification of population and household structures, is also occurring and generating spatial impacts. For instance, gentrified urban cores are being able to attract youth and young adults. In the coming years, working-class neighbourhoods built in the 1960s and 1970s will be rejuvenated, as its population, which is becoming increasingly elderly, will gradually be replaced by a younger one, mainly international immigrants. Similarly, many low-density suburban housing developments, where young families settled a few decades ago, will also be affected by ageing in the future.

Finally, an economic fragmentation is also taking place and is introducing spatial changes in production, distribution, and consumption, which the COVID-19 pandemic has intensified. Yet, the impact of the skyrocketing growth of e-commerce (that has increased the areas devoted to logistics) and of the expansion of teleworking on urban spaces is difficult to predict. The tendency of shops in the centre of large cities to specialise may intensify, ensuring their survival, but, at the same time, the crisis of retail shops in non-central neighbourhoods and of suburban shopping malls will deepen (Guimarães 2019; Townsend et al. 2017). These spatial consequences could affect both commuting and daily trips to consumption and leisure centres. In sum, urban reconfiguration for environmental, demographic, social or economic reasons will change daily mobility and the way modes of transport are used, although these impacts will vary by sex, age, place of residence (in the core city or in any of the peripheral rings), and social class.

20.2.4 Local Policies Promoting the Transition to an Urban Sustainable Mobility

Within the global climate change context, large cities have been promoting policies to protect the environment and reduce greenhouse gas emissions, for several decades. Policies are mainly based on reducing the number and the use of cars, as urban transport accounts for 40% of total road transport in the EU, and their emissions are now higher than in 1990. Moreover, the pollution it generates is believed to be responsible for around 400,000 premature deaths in Europe per year (Bernardo et al. 2020). These local and metropolitan policies accelerated when the Sustainable and Intelligent Mobility Strategy was launched by the European Commission (2020), with the aim to make modes of transport more sustainable, whilst improving the health of the population. This strategy is part of the European Green Deal that aims to achieve an ecologically and socially just transformation towards a climate-neutral development model (net-zero target) by 2050 (European Commission 2019).

Within this policy framework, and the context set by COVID-19 pandemic, which has reduced mobility and introduced major changes in the use of urban transport, European cities have accelerated policy actions to reduce air pollution and congestion levels. This has been done along three axes:

a) the promotion of the use of public transport, car sharing, and micro-mobility (bicycles, motorised, or electric scooters), better adapted to short-distance daily trips (Gragera 2021);
b) 'tactical urbanism' or the intervention on the physical urban space by the reduction of private vehicle lanes, the extension of bus lanes and those for the exclusive use of bicycles and skates, the enlargement of pavements or the creation of pedestrian streets, thus implementing the so-called 'low traffic neighbourhoods' (Leach 2021). These are areas that reduce car traffic by restricting circulation (through speed limits, changes in the direction of traffic, or road closure to private vehicle, amongst other measures); and
c) the implementation of traffic restrictions by the prohibition of polluting vehicles in urban cores. In this sense, Madrid is one of the more than 300 European cities that has opted to ban the circulation of the oldest and most polluting vehicles by implementing the Madrid Central Low Emission Zone (LEZ) in 2018. The new municipal government has renamed this LEZ as *Zona de Bajas Emisiones de Especial Protección* (ZBEDEP according to its acronym in Spanish) *Distrito Centro* (Ayuntamiento de Madrid 2022) or, Low Emission Zone of Special Protection, Central District, in English. This LEZ is like the one implemented in Barcelona on 1 January 2020, called ZBE-*Rondes*. Though LEZs have shown to be effective in combating pollution, only tolls on vehicles entering the city—a mechanism implemented in several European cities such as London, Stockholm, or Milan—reduce congestion (Simeonova et al. 2019; Gibson et al. 2015). In fact, whilst LEZs reduce traffic in urban cores, they increase congestion in the surrounding areas.

This text aims to be a useful exercise to develop public policies such as those mentioned above, intending to diminish pollution in cities by reducing the use of cars. However, we believe that they may miss their target if they do not consider population and spatial trends in urban reconfiguration, as we underline in the following section using the case study of the metropolitan region of Madrid.

20.3 The Case Study of the Region of Madrid

20.3.1 Definition of the Study Area and Data Sources Used

This chapter uses the metropolitan region of Madrid as a case study. There are several ways of defining its boundaries. One is the 'Functional Urban Area' (FUA) delimitation used in the *Indicadores urbanos* (urban indicators) published by the Spanish National Statistics Institute (INE). However, it does not provide up-to-date data on journeys and modes of transport. Therefore, we chose to use the population and transport data published by the Community of Madrid's Household Mobility Survey (*Encuesta Domiciliaria de Movilidad*, EDM) in its 1996, 2004, and 2018 editions. The survey covers the entire Autonomous Community of Madrid (CAM), made up of 179 municipalities. It is very similar, from the spatial and demographic point of view, to the Madrid FUA—though the FUA includes some municipalities in the bordering provinces of Toledo, Avila, and Guadalajara, and excludes others in the northern, eastern, and western parts of the CAM—for this reason, the CAM is used here as the equivalent of the metropolitan region of Madrid.

The main advantage of this source is that it publishes population and transport data for four transport rings. Population data come from the 1996, 2004, and 2017 *Padrón continuo* (continuous local register), which provides official annual population data for each municipality. As for those on movements and modes of transport, they proceed from EDM surveys, which in the case of the 2018 one was done to 10,000 households or 50,000 people. All this information allows us to analyse changes in both variables for the central city (the municipality of Madrid) and its periphery (178 municipalities). In fact, the city of Madrid is divided into two rings: *Madrid Almendra* (Madrid core districts) and *Madrid Periferia* (rest of the city districts). The first, with 981,000 inhabitants in 2017, covers Madrid's most central districts, inside the M-30 ring road. As for *Madrid Periferia*, it includes the districts within the municipality of Madrid but outside the M-30, so it is located around the city core, and had 2,201,937 dwellers in 2017. Those municipalities belonging to the periphery of Madrid are also divided into two rings: the *Corona Metropolitana* (inner ring) and the *Corona Regional* (outer ring). The first, which is made up of the 49 most densely populated municipalities located closest to the city of Madrid, forming its inner ring, had 2,847,633 residents in 2017. Concerning the *Corona Regional*, which consists of the remaining 129 municipalities of the CAM, creates a less densely populated

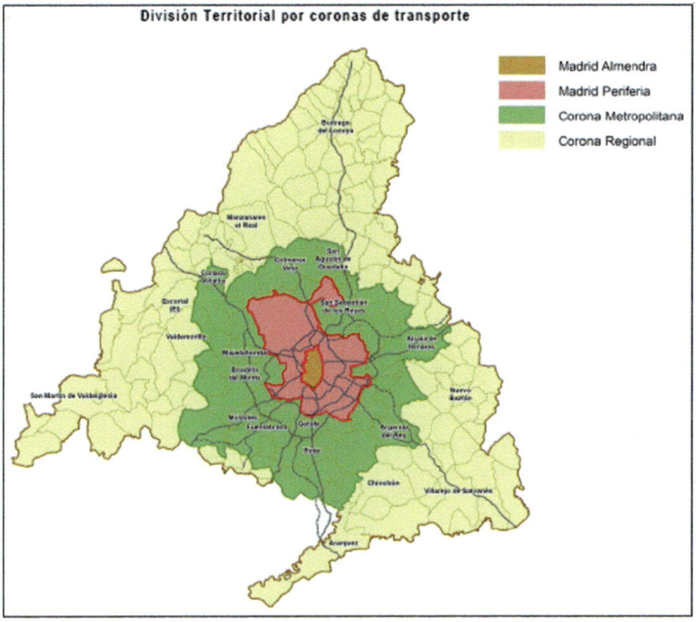

Fig. 20.1 Map showing the four metropolitan rings in the Community of Madrid, used by the EMD surveys in 1996, 2004, and 2018 (*Source* CRTM (2019), p. 3)

outer ring of municipalities situated further away from the core city, with 476,570 inhabitants in 2017 (Fig. 20.1).

20.3.2 Population Changes: The Growing Suburbanisation of Madrid's Population

Table 20.1 shows that suburbanisation grew throughout all the periods analysed. The population of periphery rings increased, and particularly that of the municipalities situated furthest away from the city, which almost doubled their residents between 1996 and 2017. In fact, the further out a municipality was situated, the more it grew.

Similarly, inside the city of Madrid, the population of the districts situated outside the M-30 grew more than core city ones, which lost population between 2004 and 2017 (population data published in the 2018 EDM correspond to the *Padrón* as of 1 January 2017). Therefore, in 2017, 7.3% of the population lived in the *Corona Regional* or outer ring, which means a significant increase with respect 1996, when less than 5% of the population resided there. In 2017, the highest share corresponded to the *Corona Metropolitana* or inner ring, as almost 44% of the population of the Community of Madrid lived in it, compared to 38% in 1996. As a result, the population of the city of Madrid represented less than half of the total of that of

Table 20.1 Absolute and relative (%) population by residential rings in the Community of Madrid, 1996, 2004, and 2017

Ring of Residence	Population 2017	% Population 2017	Population 2004	% Population 2004	Population 1996	% Population 1996	Population change (%) 1996–2004	Population change (%) 2004–2018
Madrid core districts	981.044	15,1	1.018.732	17,5	915.318	18,2	11,3	-3,7
Rest of the city districts	2.201.937	33,8	2.116.513	36,3	1.951.532	38,9	8,5	4,0
Inner ring	2.847.633	43,8	2.348.884	40,3	1.913.804	38,1	22,7	21,2
Outer ring	476.570	7,3	342.143	5,9	241.635	4,8	41,6	39,3
Community of Madrid	**6.507.184**	**100**	**5.826.272**	**100**	**5.022.289**	**100**	**16,0**	**11,7**

Source CRTM (2019, Table 3 and Charts 1 and 2, p. 10) and CRTM (2005, Table 11, p. 15)

the CAM (48.9% in 2017), whilst in 2016 it constituted well above half of it, i.e. 57.1%. In fact, this was so despite the arrival of a significant number of international immigrants to the urban core and the municipalities closest to Madrid. Yet, they were not able to compensate for the outflow of Spanish-born inhabitants who moved from the central areas to the peripheral municipalities (Gil-Alonso et al. 2021). The impact of this increasing suburbanisation on the daily mobility of the Community of Madrid's inhabitants, and their use of different modes of transport, will be analysed in the following section.

20.3.3 Daily Mobility Trends: The Increasingly Dominant Use of Private Vehicle

According to the 2018 EDM, the inhabitants of Madrid have made an average of 2.44 trips per person per day, diminishing by one-tenth those of 2004. In fact, there are hardly any differences between residential rings (Table 20.2). As for the number of journeys per household, a growing centre-periphery gradient can be observed, as households are larger in the periphery (where families with children have settled) than in the core city, where there are more one-person and single-parent households.

Concerning the mode of transport, active mobility users (those travelling on foot or by bike, the latter being included in the category 'other') stand out in the city of Madrid, and especially in the core districts. There, 40% of all trips were made on foot, and almost 5% by other means of transport (Table 20.3). By contrast, cars were dominant in the inner and outer ring. In the city of Madrid, more people used public transport than cars though in the outer districts of the city (outside the M-30) both

Table 20.2 Population, daily trips, and number of daily trips per person and per household, by residential rings, in the Community of Madrid, 2018

Ring of Residence	Population	Number of journeys	Number of journeys / person	Number of journeys / household
Madrid core districts	981.044	2.402.684	2,45	5,22
Rest of the city districts	2.201.937	5.325.948	2,42	5,84
Inner ring	2.847.633	6.957.688	2,44	6,75
Outer ring	476.570	1.160.947	2,44	6,83
Community of Madrid 2018	**6.507.184**	**15.847.266**	**2,44**	**6,16**
Community of Madrid 2004	**5.782.072**	**15.173.407**	**2,55**	**7,18**

Source CRTM (2019, Table 3 and Chart 19, p. 20 and 22) and CRTM (2005, Table 18, p. 20)

Table 20.3 Distribution of daily trips by residential rings and by mode of transport in the Community of Madrid in 2018, comparing the total regional shares to those of 1996 and 2004

Ring of Residence	On foot	Public transport	Private vehicle	Others	Total
Madrid core districts	40,0	34,8	20,3	4,9	100,0
Rest of the city districts	32,2	32,8	32,4	2,6	100,0
Inner ring	34,0	16,4	47,7	1,9	100,0
Outer ring	29,6	10,8	56,2	3,4	100,0
Community of Madrid 2018	**34,0**	**24,3**	**39,0**	**2,7**	100,0
Community of Madrid 2004	**31,0**	**32,0**	**34,0**	**3,0**	100,0
Community of Madrid 1996	**37,0**	**32,0**	**28,0**	**3,0**	100,0

Source CRTM (2019, Table 4 and Chart 26, p. 30)

modes of transport were used similarly. From a time perspective, the proportion of trips made by public transport has decreased since 1996, whilst the use of cars and walking has increased significantly since 2004.

Comparing the Community of Madrid and the Metropolitan Area of Barcelona (Gil-Alonso et al. 2022), it is worth noting that active mobility was more frequently used in Barcelona (representing around 50% of all trips). In Madrid, it only accounted for just over a third of all moves. Daily trips by public transport represented around a quarter of the total in both metropolitan areas, though they became slightly less important compared to previous years. As for those by car, they grew more than those by sustainable means of transport, accounting for almost 40% of the trips in the Madrid region, but only a quarter in the Catalan metropolis.

20.3.4 *Characterisation of the Different Types of Mobility: Seeking Why the Use of Public Transport Diminishes*

In Madrid, walking was especially popular amongst retired people, women, children, and teenagers (i.e. students). The latter three have also been the main users of public transport, whilst adult men (aged 30–59 years), particularly workers, are those that mainly use private vehicles (Table 20.4).

As a consequence of an increase in commuting because of the post-crisis economic recovery, the percentage of car users has risen. In fact, despite the diminution of the ratio of jobs per inhabitant from the centre to the periphery (0.96 in the core and only 0.27 in the outer ring), employments have tended to centrifuge, in the last decades. Whereas the core districts concentrated almost 40% of jobs in 1996, this percentage has dropped to less than 30% in 2018. By contrast, it has risen from 29 to 35%, in the inner ring, in the same dates (Table 20.5). Therefore, suburbanisation and gradual population ageing (fewer young people and concentration of baby boomers at adult ages), together with economic recovery (more commuting), and a growth of jobs in

Table 20.4 Daily trips by sex, age, and occupation, and their distribution by mode of transport in the Community of Madrid, 2018

Ring of Residence	On foot	Public transport	Private vehicle	Others	Total
Male	31	21	44	4	100
Female	36	27	35	2	100
4–13	51	8	38	3	100
14–17	44	29	22	5	100
18–21	24	57	17	2	100
22–29	20	43	34	3	100
30–39	26	23	48	3	100
40–49	25	19	53	3	100
50–59	28	24	45	3	100
60–64	40	22	36	2	100
65–74	49	25	25	1	100
> 74	58	25	14	3	100
Worker	20	26	51	3	100
Unemployed	47	18	33	2	100
Student	42	26	29	3	100
Retired	51	23	24	2	100
Ohers	52	15	32	1	100
Community of Madrid 2018	**34**	**24**	**39**	**3**	100

Source CRTM (2019, Charts 29, 30, and 31, p. 32 and 33)

peripheral municipalities, have led to an increase in the use of private vehicles and a diminution in that of public transport in the Community of Madrid.

Why does the use of public transport diminish because of population and economic suburbanisation? Peripheral areas are generally less densely populated (Garcia-Coll and López-Villanueva 2018), so the distance travelled between the place of residence and work/study grows as the distance to the central city also does. Indeed, the average distance of each trip, or of the set of trips made by an average person, triples between those living in core districts and those doing it in the outer ring (Table 20.6). In fact, because of the lower density, the accessibility and efficiency of public transport diminishes as distance from the core districts increases. Thus, this should also result in an increase in the average time travelled. However, this is not so (Table 20.7). The average daily travel time was 29 min in 2018, only two more minutes than in 1996. When all the trips made by one person per day are added (on average, about 2.4 trips per day) the figure rises to 69 min. Surprisingly, these times are similar regardless of the area of the Community of Madrid where people live, and this is also the case for walking and car trips. The only visible difference is their duration by public transport, which increases as the distance does when the trip to be done is short. Therefore, in the municipality of Madrid, people choose to walk when the

Table 20.5 Population and job distribution by residential rings in the Community of Madrid, 1996, 2004, and 2018

Ring	% Population 2018	Distribution (%) of jobs 2018	Jobs / inhabitant ratio 2018	Distribution (%) of jobs 2004	Jobs / inhabitant ratio 2004	Distribution (%) of jobs 1996	Jobs / inhabitant ratio 1996
Madrid core districts	15	29	0,96	34,7	0,94	38,7	0,75
Rest of the city districts	34	30	0,47	29,5	0,39	28,7	0,26
Inner ring	44	35	0,43	32,4	0,38	29,1	0,27
Outer ring	7	4	0,27	3,4	0,27	3,5	0,26
Community of Madrid	**100**	**98***	**0,52**	**100**	**0,47**	**100**	**0,35**

Source CRTM (2019, Charts 1, 6, 9, 10, and 11, p. 10, 13 and 15) and CRTM (2005, Chart 12 and Table 16, p. 19)
* *Note* 2% of employees working outside the CAM are not counted

trip to be done is short, whilst in the two peripheral rings, they opt for the car when it takes longer.

In sum, the growing number of people (many of whom are working-age adults who move for work-related reasons, i.e. commuters) living in peripheral suburban municipalities—where public transport is no longer competitive, as it involves longer journeys—helps to explain the growth in the use of private vehicles in the Community of Madrid in the analysed period. This translates into an increase in the number of vehicles per capita and per household, both in the inner and outer rings, whilst in peripheral districts of the city of Madrid they have increased more slowly. Finally,

Table 20.6 Mean distance travelled by ring and mode of transport in the Community of Madrid, 2004 and 2018

	Mean distance travelled (in km) 2018					2004
Ring of Residence	Daily distance travelled by persons travelling	On foot	Public transport	Private vehicle	Mean distance of each journey	Mean distance of each journey
Madrid core districts	11,1	0,8	4,7	8,7	4,6	5,6
Rest of the city districts	13,8	0,7	5,9	7,8	5,2	6,6
Inner ring	19,6	0,7	12,9	9,4	8,0	10,2
Outer ring	30,2	0,6	24,6	13,2	13,7	18,5
Community of Madrid	**17,1**	**0,7**	**8,2**	**9,3**	**6,7**	**8,8**

Source CRTM (2019, Table 9 and Chart 44, p. 56 and 57)

Table 20.7 Mean duration of the trip by ring and mode of transport in the Community of Madrid, 1996, 2004, and 2018

	Average time (in minutes) 2018					2004	1996
Ring of Residence	Daily time travelled by persons travelling	On foot	Public transport	Private vehicle	Mean duration of each journey	Mean duration of each journey	Mean duration of each journey
Madrid core districts	69	19	31	26	28	28	26
Rest of the city districts	72	18	36	23	29	30	29
Inner ring	67	17	46	22	28	27	26
Outer ring	70	15	57	24	29	28	23
Community of Madrid	**69**	**17**	**38**	**23**	**29**	**28**	**27**

Source CRTM (2019, Table 10, p. 58)

Table 20.8 Motorisation rate (vehicles per 1,000 inhabitants) and Motorisation index (number of vehicles per household) by residential rings in the Community of Madrid, 1996, 2004 and 2018

Ring of Residence	Motorisation rate 2018	% Households without private vehicle 2018	Motorisation index / household 2018	Motorisation index / household 2004	Motorisation index / household 1996
Madrid core districts	356	37,6	0,76	0,72	0,75
Rest of the city districts	388	30,0	0,94	0,90	0,87
Inner ring	472	16,9	1,28	1,13	1,10
Outer ring	506	12,0	1,41	1,09	0,97
Community of Madrid	**429**	**24,8**	**1,08**	**0,97**	**0,93**

Source CRTM (2019, Charts 13 and 14 and Table 2, p. 16 and 17) and CRTM (2005, Chart 10, p. 18)

in central core ones, where public transport is more efficient and more widely used, the motorisation rate remains stagnant in the period analysed (Table 20.8).

In other words, policies to reduce the use of private vehicles have had some success in the city of Madrid, where there is a significant use of public transport and active mobility, and where the rate of motorisation has not increased. However, this has not been the case in metropolitan periphery municipalities, where the use of private vehicles has increased significantly. As it has been argued, in the inner ring, cars are used for almost half of all daily trips, whilst in the outer ring this percentage rises to more than half of all of them (Table 20.3). Thus, the more population lives in the periphery, the higher the use of private vehicles and the lower that of public transport and active mobility—though their use has also increased in absolute numbers since 2004, that of cars has increased even more.

20.4 Discussion and Conclusion

The case study research, which analyses the relationship between recent spatial population trends in the metropolitan region of Madrid and their effects on daily mobility, confirms the initial hypothesis. Indeed, population suburbanisation trends in the last decades have led to an increase in daily mobility, and particularly, in the use of private vehicles. This same trend has been observed in other Spanish cities, both in the post-economic crisis and the post-COVID periods—these last years, the fear of contagion has delayed the recovery in the use of public transport (Greenpeace 2021). Although the European Commission's Sustainable and Smart Mobility Strategy stimulates policies that discourage the use of private vehicles and promote public transport, there would be several reasons for this increased use of cars.

The first is changes in population structure, in which there is a decreasing number of young people—the main users of public transport—and a concentration of baby boomers at adult ages, where the use of private vehicles reaches its maximum. The second is the growing relevance of commuting because of the post-crisis economic recovery as mobility by labour reasons involves a greater use of cars than daily trips for other reasons. This trend was later accentuated by the COVID-19 pandemic, when mobility restrictions reduced congestion and therefore shortened commuting times, encouraging a greater use of private transport instead of the metro, buses or trains. The third reason, and we consider it as the most important, is the increasing percentage of population living in Madrid's peripheries—most of them, working-age adults and their descendants. A growing suburban population implies an increase in both the distance travelled and in car use, to reduce the time that travelling by public transport would involve.

Obviously, there are differences between the city of Madrid—where there is a very significant use of public transport and active mobility—and the suburban municipalities, where car use predominates. The slow transition to sustainable forms of mobility in both the Madrid's core and the rest of the municipality must be considered a success. However, this tendency is counterbalanced by what occurs in the inner and outer rings, where an increasing proportion of the population of the Community of Madrid lives. This result, applicable to Barcelona (IERMB 2020; Gil-Alonso et al. 2022) and other large cities, shows that interventions to reduce car use in large cities can be affected by the socio-demographic and residential trends observed at the metropolitan level, which seem to favour the use of cars over that of public transport or active mobility. Therefore, policies must be planned beyond the strict municipal boundary of the core city and be extended to the metropolitan level, in which environmental and transport planning policies develop by definition.

The measures that have been implemented so far to reduce road traffic and reclaim public space for citizens have strengthened the transition to low-carbon mobility in many European cities. However, they have also widened the so-called 'mobility divide' or the mobility gap between social groups (García Lamarca et al. 2020). For instance, the implementation of the LEZ in Madrid and Barcelona—and of entry tolls in other European cities—has negatively affected the mobility of the most vulnerable population. Whilst middle- and high-income commuters, who own newer and less polluting cars, have been able to continue commuting by private car after the low emission zone came into force, low-income commuters find it more difficult to do so, unless they renew their vehicle.

The relationship between the demographic and the socio-economic characteristics of the population, mobility and urban inequality has even more angles. Indeed, the latter not only involves an unequal access to housing but a growing uneven accessibility to teleworking and an increasing diversity of modes of transport. Green gentrification (Triguero-Mas 2021; Frago and Graziano 2021) makes it possible for a significant proportion of gentrifiers to commute on foot, by bike, or scooter. By contrast, the ban on polluting cars from the city is increasing the use of public transport—and travel time—by lower-income groups. This mobility divide widened even more during the first COVID-19 lockdown. Workers in essential sectors—in jobs that are not compatible with teleworking—continued to commute, with the risk of catching the coronavirus both at the workplace and on public transport (Checa et al. 2020). At the same time, teleworking in higher value-added jobs—such as those in ICT-intensive sectors—increased dramatically (Florida et al. 2021). In sum, the results have shown that it is impossible to separate mobility, environment, and equity when designing policies for the transition to sustainable mobility. Therefore, their implementation should not only be based on environmental criteria, but also on social ones.

References

Alexandri G, Janoschka M (2020) Post-pandemic transnational gentrifications: A critical outlook. Urban Stud 57(15):3202–3214

Ayuntamiento de Madrid (2022) Zona de Bajas Emisiones. Ayuntamiento de Madrid, Madrid

Bagchi-Sen S, Franklin RS, Rogerson P, Seymour E (2020) Urban inequality and the demographic transformation of shrinking cities: The role of the foreign born. Appl Geogr 116:102–168

Bailey N, Minton J (2018) The suburbanisation of poverty in British cities, 2004–16. Urban Geogr 23:1–24

Bayona J, Pujadas I (2020) Las grandes áreas metropolitanas en España: del crecimiento y la expansión residencial al estancamiento poblacional. Documents D'anàlisi Geogràfica 66(1):27–55

Bernardo V, Fageda X, Flores-Fillol R (2020) Mitigating congestion in urban areas: Urban tolls versus low emission zones. Papeles De Energía. 10:67–92

Borsdorf A, Hidalgo R (2009) The fragmented city. Changing patterns in Latin American cities. Urban Reinventors Online J 3:1–18

Champion T, Gordon I (2021) Linking spatial and social mobility: Is London's "escalator" as strong as it was? Popul, Space & Place, 27(7)

Checa J, Martín J, López J, Nel.lo, O. (2020) Los que no pueden quedarse en casa: movilidad urbana en el área metropolita de Barcelona durante la pandemia COVID-19. Boletín De La Asociación Española De Geografía 87:1–35

Clark WAV (2016) Life events and moves under duress. Longitud Life Course Stud, 7(3)

Cocola-Gant A, López-Gay A (2020) Transnational gentrification, tourism and the formation of 'foreign only' enclaves. Urban Stud 57(15):3025–3043

CRTM (2005) Encuesta Domiciliaria de Movilidad de 2004 en la Comunidad de Madrid—Documento de Síntesis. Consorcio Regional de Transportes de Madrid, Madrid

CRTM (2019) Encuesta de Movilidad de la Comunidad de Madrid 2018—Documento Síntesis. Consorcio Regional de Transportes de Madrid, Madrid

Dioni J (2021) La España de las piscinas. Arpa, Barcelona

Esteve A, Blanes A, Domingo A (2021) Consecuencias demográficas de la COVID-19 en España: entre la novedad excepcional y la reincidencia estructural. Panor Soc 33:9–23

European Commission (2019) The European Green Deal. Communication COM 640 Final, 11 December 2019. Brussels: European Comission

European Commission (2020) Sustainable and Smart Mobility Strategy—putting European transport on track for the future. Brussels: European Comission, 9.12.2020.COM 2020 789 final

Florida R (2017) The new urban crisis. Basic Books, New York

Florida R, Rodríguez-Pose A, Storper M (2021) Cities in a post-COVID world. Urban Stud, 00420980211018072

Ll, Frago, Graziano T (2021) Public space and the green city: Conflictual narratives of the superblock programme in Poblenou, Barcelona. J Urban Regen and Renew 15(1):113–128
Garcia-Coll A, López-Villanueva C (2018) The impact of economic crisis in areas of Sprawl. Urban 2(4):113
García-Lamarca M, Connolly J, Anguelovski I (2020) Green gentrification and displacement in Barcelona. Housing Displacement. Routledge, New York, pp 156–170
Gibson M, Carnovale M (2015) The effects of road pricing on driver behavior and air pollution. J Urban Econ 89:62–73
Gil-Alonso F, López-Villanueva C, Bayona-i-Carrasco J, Pujadas I (2021) Towards an even more spatially diversified city? new metropolitan population trends in the post-economic crisis period. Urban 5(2):41
Gil-Alonso F, López-Villanueva C, Thiers-Quintana J (2022) Transition towards a sustainable mobility in a suburbanising urban area: the case of Barcelona". Sustainability 14(5):2560
González-Leonardo M, López-Gay, A.; Recaño-Valverde, J. and Rowe, F. (2022). Cambios de residencia en tiempos de COVID-19: un poco de oxígeno para el despoblamiento rural. Perspectives Demogràfiques, 26
Gragera A (2021) How can shared mobility contribute to the post-pandemic urban mobility transition? In Urban Mobility after COVID-19. In: Abdullah H, Serrano E (eds) Urban mobility after COVID-19. Long-term strategies for the sustainable mobility transition in European cities. Barcelona: CIDOB, Ajuntament Barcelona, pp 63–696
Greenpeace (2021) Análisis del estado de la movilidad poscovid. Recovered from: https://es.greenpeace.org/es/wp-content/uploads/sites/3/2021/09/Analisis-de-la-movilidad-poscovid_compressed.pdf
Guimarães PPC (2019) Shopping centres in decline: Analysis of demalling in Lisbon. Cities 87:21–29
Haase A, Bontje M, Couch C, Marcinczak S, Rink D, Rumpel P, Wolff M (2021) Factors driving the regrowth of European cities and the role of local and contextual impacts: a contrasting analysis of regrowing and shrinking cities. Cities 108:102942
Hochstenbach C, Musterd S (2021) A regional geography of gentrification, displacement, and suburbanisation of poverty. Area 53:481–491
IERMB (2020) Enquesta de mobilitat en dia feiner 2019 (EMEF 2019). La mobilitat a l'àmbit de l'Àrea Metropolitana de Barcelona. Bellaterra: Institut d'Estudis Regionals i Metropolitans de Barcelona
Leach J (2021) Challenging the car's dominance to bring life back to residential streets and support high streets and town centres. In: Abdullah H, Serrano E (eds) Urban Mobility after COVID-19. Long-term strategies for the sustainable mobility transition in European cities. Barcelona: CIDOB, Ajuntament de Barcelona, pp 45–48
López-Colás J, Azevedo AB, Módenes JA (2021) Insécurité résidentielle en période d'incertitude. Populations et crises en Méditerranée. Franco Angeli, Milan
López Gay A (2018) Cambio en la composición social y gentrificación en Barcelona: una mirada a través de los flujos migratorios y residenciales. Papers. Regió Metropolitana De Barcelona 60:80–93
López-Villanueva C, Pujadas I (2015) Transformaciones espaciales y demográficas en las regiones metropolitanas de Madrid y Barcelona. In: Domínguez M, López-Villanueva C (eds) Barcelona y Madrid. Procesos Urbanos y dinámicas sociales. Madrid: Síntesis, pp 71–106
McCollum D, Keenan K, Findlay A (2020) The case for a lifecourse perspective on mobility and migration research. In: Handbook on demographic change and the lifecourse. Elgar
Méndez R (2021) Financiarización urbana y burbuja del alquiler en España: tendencias y contrastes en perspectiva multiescalar. Documents D'anàlisi Geogràfica 67(3):441–463
Mikolai J, Kulu H, (2020) Family changes, housing transition, and residential mobility. In: Handbook on emographic change and the lifecourse. Elgar
Moreno-Colom S, López-Roldán P (2018) El impacto de la crisis en las trayectorias laborales de las mujeres inmigrantes. Cuadernos De Relaciones Laborales 36(1):65

Orozco-Martínez C, Bayona-i-Carrasco J, Gil-Alonso F (2022) Inmigración y vivienda durante el confinamiento domiciliario: El caso de las habitaciones subarrendadas. Migraciones 54:1–21

Otero-Enríquez R, García-Abad J, Domínguez-Mujica J, Pérez-Caramés A (2019) Inmigración y dinámicas territoriales en España: crisis y recuperación. In: Anuario CIDOB Inmigración. Barcelona. CIDOB, pp 190–217

Piketty T (2019) La crisis del capital en el siglo XXI: crónicas de los años en que el capitalismo se volvió loco. Anagrama/Siglo XXI, Madrid

Piore MJ (1979) Birds of passage. Cambridge University Press

Porcel S, Antón F (2020) Ciudad postindustrial y dinámicas socio-residenciales en España Scripta Nova, 24(649)

Rees P, Sander N (2019) The role of internal migration as a driver of population change in Europe. Comp Popul Stud 44:533–544

Rérat P (2020) Residential mobility. Handbook of Urban Mobilities. Routledge, New York, pp 224–233

Rodríguez-Moya JM, Pozo-Rivera E (2019) Análisis territorial del desempleo por género en la Comunidad de Madrid durante la crisis. Cuadernos Geográficos 58(1):277–299

Rowe F, Bell M, Bernard A, Charles-Edwards E, Ueffing P (2019) Impact of internal migration on population redistribution in Europe: urbanisation, counterurbanisation or spatial equilibrium? Comp Popul Stud 44:201–234

Shuttleworth I, Foley B, Champion T (2021) Internal migration in Northern Ireland: Are people becoming more stuck in place? Popul Space Place 27(7):e2338

Simeonova E, Currie J, Nilsson P, Walker R (2021) Congestion pricing, air pollution, and children's health. J Hum Resour 56(4):971–996

Townsend M, Surane J, Orr E, Cannon C (2017) America's 'Retail Apocalypse' Is Really Just Beginning. Bloomberg, New York

Triguero-Mas M (2021) Transitioning to sustainable urban mobility in a just and quitable manner: how to prevent environmental gentrification and enhance social equity?. In: Abdullah H, Serrano Roble, E (eds) Urban mobility after COVID-19. Long-term strategies for the sustainable mobility transition in European cities. CIDOB, Ajuntament de Barcelona. pp 49–53

Wolff M (2018) Understanding the role of centralization processes for cities. Evidence from a spatial perspective of urban Europe 1990–2010. Cities 75:20–29

Chapter 21
Sustainability and Urban Development Strategies: Ciudad Real

Maria Carmen Cañizares Ruiz and Maria Angeles Rodriguez-Domenech

Abstract In many medium-sized cities, the process towards achieving sustainability involves applying effective urban policies. Strategies for making cities sustainable first emerged in the 1980s with environmental policies, but it was not until the signing of the Pact for Urban Sustainability and the approval of Local Agenda 21 projects that significant progress was made. The Integrated Sustainable Development Strategy (ISUDS) is currently the most transformative territorial policy; this is mainly due to its participatory methodology and to the fact that it involves an integrated approach. Here, Ciudad Real is analysed as a typical example of a city in which the path towards sustainability has been complex and the implementation of these policies has been accompanied by both successes and failures, with civic participation being essential, but only relatively recent.

Keywords Sustainable development · Sustainability · Urban Agenda · ISUDS

21.1 The Long Road to Urban Sustainability

Temporal analysis suggests that urban sustainability has survived because of the elasticity of the term. Sustainability is certainly a concept that has been characterised by its longevity: after several decades, it is still popular with public organisations, and also within more general public discourse (Adolfsson et al. 2021).

When we talk about sustainability today, we refer to a process of improvement, of moderation in the use of resources, of respect for the environment and for ourselves; in short, to a "social virtue" (Panadero 2010, p. 79). This is undoubtedly a reference to the Sustainable Development Goals (SDGs), which were approved in 2015, within the 2030 Agenda for Sustainable Development, and to the need to address the 17

M. C. Cañizares Ruiz · M. A. Rodriguez-Domenech (✉)
Geography Department, University of Castilla La Mancha, Ciudad Real, Spain
e-mail: Mangeles.rodriguez@uclm.es

M. C. Cañizares Ruiz
e-mail: MCarmen.Canizares@uclm.es

E. Navarro-Jurado et al. (eds.), *Urban Dynamics in the Post-pandemic Period*,
The Urban Book Series,
https://doi.org/10.1007/978-3-031-36017-6_21

challenges that form part of an ongoing debate about how to tackle the main problems currently affecting our planet. This debate recognises that several key issues, including poverty, hunger, health, education, gender equality, and environmental degradation are interlinked and should therefore only be addressed together. The main challenge is that of how to develop a more integrated approach to sustainable development and one that embraces new governance frameworks and helps us to facilitate and manage systemic transformations (Weiland et al. 2021).

However, the road to get here has been long and complicated. It has involved raising awareness at all levels (scientific-academic, media, and society) and has been built on a series of summits, events, and documents. It was the Stockholm Conference (1972), under the slogan "Man and his environment: the basis for a better life", that first brought together world leaders and scientists to discuss the proliferation of international environmental problems, such as transboundary air and water pollution. This was the first of the "mega" conferences on the environment (Seyfang 2003) to be sponsored by the United Nations. Since then, a huge amount of documentation has been dedicated to environmental issues. Growing media attention has also led to greater social awareness and, above all, to the development of specific environmental policies. In 1987, the Brundtland Report launched the concept of "sustainable development"; this provided a framework within which it was possible to combine and integrate the promotion of environmental protection and economic development (Jacobs 1999, p. 21). The Rio Summit of 1992 was then of paramount importance for the consolidation and international dissemination of environmental impact assessment, which was officially recognised as a tool for informed decision-making in the pursuit of sustainable development and the protection of biodiversity (Sánchez and Croal 2012). Today, its initial "spirit" seems rather distant, as progress towards these objectives has been slower and less effective than initially expected. This has been a particularly arduous process in countries like Spain, where territorial awareness and environmental sensitivity still require a great deal of attention.

The documentation that underpins all the theory behind environmental protection and sustainability has not always been accompanied by practices that have helped to minimise the impact of human activity and population on the environment. In the specific field of cities, and urban sustainability, there have been numerous campaigns, charters, and agreements since the Aalborg Charter on Healthy Cities (1994). These have included various declarations issued within the frameworks of conferences and initiatives such as the European Healthy Cities Campaign, Urban Charters, and Habitat Conferences, and which culminated in the UN's SDGs. Within this process, the focus has gradually shifted from discussing the urban environment to urban sustainability.

Today, at least at the theoretical and operational level, urban sustainability is embodied in Sustainable Development Goal 11 (SDG 11), relating to "Sustainable Cities and Communities", which aims to "make cities and human settlements inclusive, safe, resilient, and sustainable" by pursuing 10 targets and taking 713 specific measures. SDG 11 seeks to address the challenge of urban sustainability and its scope includes the following cross-cutting issues: affordable housing; sustainable transport; the planning and management of human settlement; green and public spaces;

and, helping to establish positive economic, social, and environmental links between urban, peri-urban, and rural areas. The indicators proposed for measuring the achievement of these targets include, but are not limited to, monitoring: the proportion of population living in slums; the ratio of land consumption to population growth; total and per capita expenditure on the conservation of natural and cultural heritage; solid waste collection; and, the proportion of local government organisations that adopt disaster risk reduction strategies (Verma and Raghubanshi 2018, p. 283).

The evolution and importance of the urban dimension of EU policies and, in particular, of EU Cohesion Policy has grown in relevance over the last three decades. It is important for promoting and supporting an Urban Agenda for the EU, and for fostering the gradual adoption and implementation of Integrated Sustainable Urban Development Strategies (ISUDS). Within this framework, EU cohesion policy has taken up a number of urban issues through its URBAN I (1994–1999) and II (2000–2006) initiatives and the URBAN Initiative (2007–2013). More recently, since 2014, it has also adopted the ISUDS programme.

Despite their relatively limited impact, these EU-funded urban programmes (URBAN, POLIS, ISUDS) have made a positive contribution to promoting a more integrated policy approach and have helped to enhance the physical and social environments of many deprived urban neighbourhoods. This holistic approach offers numerous important advantages over previous, predominantly sectoral strategies (Medeiros and Van Der Zwet 2020).

This objective is related to the New Urban Agenda, which was adopted at the Habitat III Conference (2016), and to its transposition to each country. It is also linked to the implementation of effective planning tools, which is an area in which Local Agenda 21 projects have met with unequal treatment and success. As a result, in countries like Spain, these initiatives have been giving way to Sustainable Development Strategies (EDUSI)[1] or ISDS.

Ciudad Real provides an example of a typical Spanish city that has made this journey, with a mixture of failures and successes. As in many other urban areas in Spain, it has faced environmental problems, although none of any great severity. The journey has been slow and has called for institutional and social awareness. It has also required the support of the media and civic participation, both of which have been traditionally scarce. Here, we address the keys to urban sustainability in Ciudad Real as we analyse one of the planning tools: the ISUDS 2022: Ciudad Real Eco-Integrator[2] and its most important actions.

---

[1] Integrated Sustainable Urban Development Strategies (ISUDS). The acronym in Spanish is EDUSI (*Estrategias de Desarrollo Urbano Sostenible Integrado*).

[2] ISUDS 2022: Ciudad Real Eco-Integrator. The original title in Spanish is *Estrategia Ciudad Real 2022 Eco-Integradora.*

21.2 Methodology and Case Study: Ciudad Real

The methodology used was mixed. It started with a descriptive analysis of urban sustainability and a presentation about the urban nucleus of Ciudad Real. This was largely based on a bibliographical review of both the most relevant concepts and milestones and the most significant publications about the city. This was combined with the qualitative approach of interviewing agents involved in the ISUDS, as well as an applied analysis of the main actions carried out, in which the collaboration of Ciudad Real's city council was of key importance. More quantitative methods were also used, such as a statistical treatment of data relating to actions undertaken in the city. The results focused on the importance of action and proposals for improving the urban environment.

Ciudad Real, which had 75,104 inhabitants in 2021, is the capital of the province of the same name. It is located in the interior of the Iberian Peninsula and is in the autonomous community of Castilla-La Mancha. The city has undergone the most important transformation in its history during the last forty years: since the restoring parliamentary democracy. During this period, the city has established its position as the leading centre in its province, both in terms of its demographic weight and functions. This has been reflected by both its urban growth and the functions that it provides to the territory that it administers and has been the result of two major driving forces. Both of these emerged in 1992, but they were conceived several years earlier. One was the arrival of the AVE high-speed train line (Madrid-Seville connection, initially) and the other was the creation of the University of Castilla-La Mancha. Both of these developments created outstanding synergies that were further reinforced and intensified by the very favourable economic climate at the end of the twentieth century. This had a significant impact on the city's urban development (Rodríguez Domenech 2012, p. 665), as a clear tertiary-sector vocation emerged. In the following years, the University General Hospital of Ciudad Real and other significant pieces of infrastructure and strategic projects continued to improve the city.

The recent 600th anniversary of the upgrading of Ciudad Real from a town to a city provided a significant opportunity to focus on its urban evolution (Fig. 21.1). The city has continued its "oil slick"-like development, which has produced a relatively concentrated urban area; this contrasts with the dispersed, or diffuse, growth observed in the majority of today's Spanish cities (Pillet and Rodriguez 2021). Although enjoying quite a privileged setting, with a relative absence of industrial facilities in its vicinity, Ciudad Real has had to overcome a number of historical problems, especially related to its water supply; these were particularly evident from the last third of the eighteenth century until the 1980s (Pillet 1984).

Ciudad Real is located in a territory: Castilla-La Mancha, in which urban policies related to promoting sustainability began with a certain delay with respect to those implemented in other autonomous communities (Cañizares 2008; Marchán et al. 2018). As a result, many measures have taken time to emerge and it is only relatively recently that there has been an awakening of awareness amongst local citizens.

Fig. 21.1 Plaza Mayor of Ciudad Real, 2022. *Source* Own elaboration

21.3 Results and Discussion

Any analysis of urban sustainability applied to Ciudad Real must be framed within the context of action undertaken in Spain in recent decades. In fact, there was no national urban policy prior to the relatively recent Spanish Urban Agenda (2019), although several municipal urban policies had previously focused on urban development (Bellet 2020). Within this framework, the urban policy of the European Union (EU), and in particular the two rounds of the URBAN Community Initiative: its most specific instrument, has made urban regeneration the most relevant practice in Spain, from the perspective of Collaborative Planning. The European Union's cohesion policy initially focused on a number of urban issues through its URBAN I (1994–1999) and URBAN II (2000–2006) initiatives, and then through the URBAN Initiative (2007–2013). More recently, since 2014, this work has been continued through the ISUDS programme. This last project, which we will analyse below, seeks to ensure the continuity of an integrated, strategic, and more participatory approach to urban development (De Gregorio and González 2017, pp. 70–71). This type of approach had already been included in the Europe 2020 Strategy and the ISUDS seeks to implement four of its 11 thematic objectives for promoting a smarter, more sustainable, and more inclusive economy. The novelty, in this last case, lay in "the assignment of 5% of European Regional Development Funds for integrated actions to: promote sustainable urban development; strengthen the Local Entities (urban authorities) that select which operations to carry out within the framework of their respective strategies; and give a greater role to the Urban Initiatives Network" (Nasarre et al. 2017, p. 804), all of which are essential for coordination.

21.3.1 Urban Sustainability in Ciudad Real: Background and Local Agenda 21

The most recent study of urban sustainability in Spain (Observatorio de la Sostenibilidad 2018), which evaluated the progress made by cities towards achieving more sustainable development, based on the UN Sustainable Development Goals (SDGs) and the New Urban Agenda (NUA), offered an x-ray of the country's 52 provincial capitals. It was conducted using 59 indicators, relating to 26 different themes, which were divided into four large blocks: economy, social issues, the environment, and transparency and cooperation. As shown in Fig. 21.2, on average, Ciudad Real ranked 38th within the country. Its best position was obtained in the social area (poverty, hunger, health, etc.) where it occupied 17th place, whereas in the environmental area (water and sanitation, climate action, and ecosystem life) it dropped down to 43rd position. In the economic area, it was in 31st place, and it came 36th in Transparency and Governance (Fig. 21.2).

In general, and as has occurred in other medium-sized cities, planning has tended to show very little concern for environmental issues. In fact, "the first general urban development plans, of 1963 and 1978, favoured land speculation, setbacks, and high-rise construction. The 1988 Plan sought to reduce the height of buildings, to protect and rehabilitate the built heritage, and to consolidate the introduction of green areas" (Pillet 1995, p. 529). Existing problems were exacerbated in the early 1990s by a high degree of speculation in the building sector. In Ciudad Real, this saw the construction of 31,355 new dwellings, the AVE effect, and the arrival of the university. This real estate boom doubled the amount of built space in a period of 15 years. Noise pollution and traffic then became the most important environmental problems in the opinion of local citizens (Rodriguez-Domenech 2012, pp. 505, 625). They also highlighted

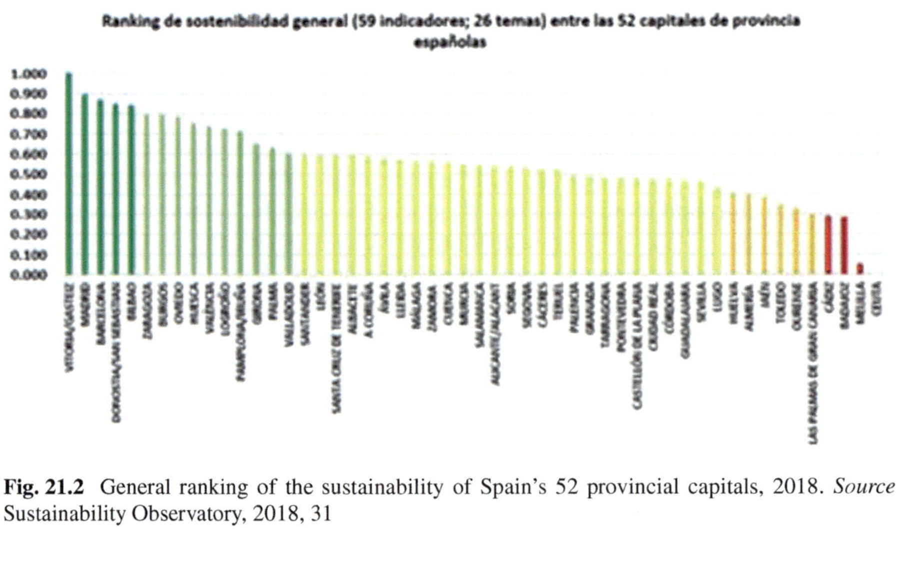

Fig. 21.2 General ranking of the sustainability of Spain's 52 provincial capitals, 2018. *Source* Sustainability Observatory, 2018, 31

the need for urban recycling, including the use of specific containers, and for the pedestrianisation of city centre streets in order to favour mobility on foot.

It should be noted that although various valuable management tools—including the Local Agenda 21, which was an essential antecedent—already existed, they largely remained unused, or proved ineffective. It is well-known that this strategy originated from Rio, in 1992, with chapter 28 of Local Agenda 21 specifically urging local authorities to initiate a process of dialogue with all sectors of their respective municipalities in order to improve the environment. In the European context, this was also linked to the European Healthy Cities Campaign (now Sustainable Cities and Towns) and to the Aalborg Charter (1994), which commit cities to initiating the transition towards sustainability.

Ciudad Real signed the Charter in 2004 and immediately began work on a Technical and Qualitative Diagnosis of the city's current situation, applying the SWOT (Strengths, Weaknesses, Opportunities, Threats) matrix in four main areas: (1) environmental and territorial considerations; (2) socio-economic policy; (3) structuring; (4) and the natural environment. This was a very positive document, given the aim of creating a more sustainable city. However, although it sought to ensure coherence between its objectives and proposals, its list of targets was very long and did not lay out a hierarchy of priority problems (Rodriguez-Domenech 2012: 695 et seq.). A more workable list of objectives and proposals was subsequently formulated, based on five lines of strategy within an Action Plan: (1) Environmental sustainability and urban development; (2) Improving social welfare; (3) Socio-cultural revitalisation; (4) Responsible economic development and promotion; and (5) The application of sustainable criteria in municipal management. This document was presented in 2009, together with a Monitoring Plan that included a commitment that could probably have been better organised. This could, for example, have been achieved by considering territorial aspects that reached beyond the purely municipal level and by including objectives related to sustainable accessibility and mobility. In fact, the Local Agenda 21 project did not address the city model. In the case of Ciudad Real, this could have been that of a "university city" providing competitive services. This was particularly relevant as one of the main problems detected was that many local professionals expressed a lack of confidence in the city's local services and stated that this often led them to use external providers (Rodriguez-Domenech 2012, p. 696). Nevertheless, it is important to recognise that this was just a first step towards approaching the issue of the local urban environment. It involved diagnosing existing problems and presenting proposals for improvement. This, in itself, did, however, represent a significant step forward in institutional work and in raising public awareness. We can only regret that an opportunity was lost to put many of these measures; they would certainly have helped to improve Ciudad Real, and the wellbeing of its inhabitants, but their final application was quite limited in scope.

Other tools which were associated with the ISUDS, albeit indirectly, and which we will only mention in passing, included: the Municipal Management Plan of 1997, which is currently pending an update; the Strategic Plan for Socioeconomic Development and Innovation of Ciudad Real (2021), which is aligned with the UN's 17 SDGs; and the 2025 Modernisation Plan for Ciudad Real, which is an institutional strategic plan (Pillet and Rodriguez 2021) and the fruit of an agreement between the regional and provincial government and the city council. The latter has four pillars of action: (1) Freeing the city of traffic and transferring protagonism to pedestrians; (2) Recovering, renovating, and regenerating buildings and urban spaces; (3) Improving connections and mobility; and (4) Carrying out interventions on existing buildings. For the moment, all of this is down on paper, but it has yet to be put into action.

21.3.2 Integrated Sustainable Urban Development Strategies (ISUDS): Ciudad Real 2022 Eco-Integrator

For the first time ever, more than 140 Spanish municipalities suddenly faced the challenge of managing a significant amount of European funding which they had to use, in a very short period of time, to define lines of action, draw up projects, and contract works. This was in addition to the complexity of the procedures and controls established by the Thematic Objectives (TO), Specific Objectives (SO), and Action Plan (AP) imposed by the ERDF (European Regional Development Fund).

The urban dimension of the European Union (EU) cohesion policy had a major influence on Spanish urban policy through the initiatives mentioned above. This continued during the period 2014–2020, when Spain's Ministry of Finance launched 173 ISUDS (De Gregorio et al. 2021, p. 3).

Five projects in Castilla-La Mancha were selected in the first phase of this initiative: Albacete, Almansa, Ciudad Real, Tomelloso, and Cuenca, with a total value of €44.1 million. In the second phase, three more were added: the Diputación de Albacete, which covered the municipalities of Campo de Hellín (Hellín, Albatana, Tobarra, Pozohondo, Liétor, Férez, and Socovos); Guadalajara; and Toledo, with total funding of €18.9 million. Finally, in the third phase, seven more cities joined the initiative: Valdepeñas; Puertollano; Alcázar de San Juan; Azuqueca de Henares; Seseña; Talavera de la Reina; and Illescas, with funding of €45 million. In total, there have been 15 ISUDS, linked to 14 Local Agenda 21 projects, 13 Municipal Management Plans, and five Strategic Plans, with total funding of €108 million euros: 9% of the national total (De Gregorio et al. 2021, pp. 37, 51). With the exception of the Diputación de Albacete, these have all been essentially local projects, despite recommendations that they should contemplate territories that extend beyond urban centres. This has led to criticism that the ISUDS model has tended to conceive such projects in isolation from their most immediate contexts and that they have not significantly contributed to the articulation of local territories. This is all the more relevant when the distribution criterion is the number of local inhabitants (Santos 2019, pp. 2–3), with this being a particularly important factor in Castilla-La Mancha.

Ciudad Real was included in the first phase of the programme, which was aimed at municipalities, or functional areas, with populations of at least 20,000 inhabitants. It then began to address the challenges associated with the different methodological phases: identifying urban problems and challenges; analysing the city from an integrated perspective; diagnosing the existing situation (SWOT analysis); implementing a strategic plan; describing mechanisms for civic participation; guaranteeing administrative capacity; and, establishing horizontal principles and cross-cutting objectives. The thematic objectives (TO) established were:

- TO2. Improving access to, and the use and quality of, information and communications technologies.
- TO4. Favouring the transition to a low carbon economy.
- TO6. Conserving and protecting the environment and promoting greater resource efficiency.
- TO9. Promoting social inclusion and the fight against poverty.

According to the city council, the ISUDS: Ciudad Real 2022 Eco-Integrator strategy is a project that aims to make Ciudad Real a more inclusive and modern city and one which will be environmentally sustainable. It is also hoped that it will help to increase the quality of life of its inhabitants. Following a strategic vision, and based on an updated diagnosis, the city council seeks to develop its existing capacities and to take advantage of all the opportunities available to it. This approach implies creating a city of the future and one which will: be efficient; offer services that are accessible to the whole of its population; be attractive for the creation of opportunities; be walkable; be integrated, socially balanced, and modern; and provide transparent management and efficient municipal services. Its priority focus is on social inclusion, although the environment and energy efficiency are also important issues. The initial proposal was approved by the Secretary of State for Budgets and Expenditure, in 2016, and it was co-financed through the ERDF Operational Programme for Sustainable Growth 2014–2020. However, it is only in recent years that some of the actions initially proposed have actually begun to materialise.

European Union funding for the project was €10 million and the city council was obliged to contribute a further €2.5 million. The total budget allocation was therefore €12.5 million. The Implementation Plan was laid out in a set of Thematic Objectives (TO) and also in the Specific Objectives (SO) and Action Plan (AP) presented in Table 21.1.

The plan takes into account the principles of guaranteeing equal opportunities for women and men and not discriminating against people on the grounds of their age, religion, ethnic origin, sexual orientation, or disability. It also looks to disaggregate data by sex for relevant indicators, such as demographic and employment data. The plan also seeks to promote the employment of unemployed women and to give priority to training programmes aimed at female entrepreneurs. With regard to the inclusion of sectors of the population at risk of exclusion, it gives special attention to problems currently faced by women (premature pregnancies, literacy, etc.).

The diagnosis covers environmental, economic and social, and demographic issues. These include: the characterisation of the city and its natural spaces; land use; building stock; green areas; economic sectors; business creation and sectors; different occupations; unemployment; and demographic evolution. It does this relating all of these questions to sustainable development.

Table 21.1 Thematic objectives and action plan

Thematic Objectives (TO) and Specific Objectives (SO)	Action Plan (AP)
TO2 SO233. Promoting TICs within DUSI strategies through actions involving electronic administration and *Smart cities*	AP1 Civic Platform of Open Government and Electronic Administration
TO4 SO 451. Promoting sustainable urban mobility	AP2 Plan for Sustainable Urban Mobility and Accessibility of Ciudad Real
TO4 SO 453. Improving energy efficiency and increasing the use of renewable energy in urban areas	AP3 Buildings and municipal installations: large-scale rehabilitation
TO6 SO 652. Integrated action to revitalise cities and improve the urban environment and its neighbouring natural environment	AP4 Ciudad Real without noise
	AP5 Project for the complete regeneration of the *Barrio del Pilar* neighbourhood
TO9 SO 982. The physical, economic, and social regeneration of the urban setting in disadvantaged urban areas through DUSI strategies	AP6 Plan to strengthen the business fabric and to promote commerce in Ciudad Real
	AP7 Programme for social inclusion (*San Martín de Porres, Barrio del Pilar* and *Barriada de la Esperanza*)
	AP8 Management
	AP9 Communications

Source https://edusi.ciudadreal.es/edusi-ciudad-real-2022-eco-integrador/

The challenges and goals have been determined by integrating the three dimensions in each area. For example, it has been proposed to improve the capacity for efficiency in those areas with the greatest environmental quality and to foster some economic development there. Another idea, in areas where there is a focus on economic development, is to develop new activities which have a low environmental impact. Another target is to increase and improve accessibility to housing, the urban environment, and public services via various means including digitalisation. One of the main goals involving digital accessibility involves catering for the needs of different age groups (providing accessibility to digital technology and mobility, and preventing exclusion). There are also proposals to prevent the ageing of the population by trying to generate new job opportunities for university students and entrepreneurs.

With regard to climate change (mitigation and adaptation), the diagnosis has examined factors such as: temperature and precipitation; the main sources of GHG emissions, most of it is carbon dioxide from the burning of fossil fuels: coal, oil and natural gas (building and transport); and initiatives implemented to reduce emissions (SUMP and renewable energies). It also includes a risk analysis. In addition, it highlights the types of operations that most contribute to reducing GHG emissions and even outlines a number of targets that have been set to help reduce emissions (including: citizen cards, e-government, improved transport, cycling, and energy efficiency).

21.3.3 The Main Actions of the ISUDS: Ciudad Real 2022 Eco-Integrator

As in other cities, the ISUDS implies taking action to address urban economic, environmental, and social challenges, with city councils being responsible for promoting such initiatives. Its most ambitious objective is that of improving the post-crisis socio-economic situation and fostering urban-social regeneration, particularly through reducing unemployment, promoting entrepreneurship, increasing training for employment, and strengthening the local business fabric (Marchán et al., 2018, pp. 739, 748).

The actions of the ISUDS: Ciudad Real 2022 Eco-Integrator project will come to an end in 2023. At present, however, barely 24% of the budget has actually been executed. This highlights a somewhat paradoxical contrast between the measures that were initially proposed as priorities for action and those that have so far been put into action (Table 21.2). The Action Plan (AP) proposals that have been most actively pursued to date are: the Plan for Sustainable Urban Mobility and Accessibility of Ciudad Real (AP2), the Programme for social inclusion (San Martín de Porres, Barrio del Pilar, and Barriada de la Esperanza) (AP7), and the Civic Platform of Open Government and Electronic Administration (AP1). Other proposals that have been executed by more than 50% (up to March 2022) include Communications (LA9), with an executed budget of 75%, the Large-scale Rehabilitation of Buildings and Municipal Installations to improve their energy efficiency (AP3), with 70.2%, and the Plan for Sustainable Urban Mobility and Accessibility of Ciudad Real (AP2) with 57.9%.

Of all the actions already underway, the one that perhaps most stands out is the design, management, and maintenance of the web portal (€14,399) to provide transparent management (AP9). Another relevant area of action is the rehabilitation of public buildings, such as the Town Hall (€11,9783.51), to improve their energy efficiency. These interventions have included action on: the Town Hall's thermal envelope (€242,613.29); the Vertical Garden (Fig. 21.3) on one of its facades (€91,902.11); and the natural gas connection in the Plaza Mayor (€5,043.28) (LA3). The purchase of six buses that run on compressed natural gas (Fig. 21.4) (€1,836,780) forms part of the Mobility Plan (AP2). Also worthy of mention are some of the projects that form part of the social inclusion programme (San Martín de Porres and Barriada de la Esperanza) with funding of €383,048.37 (AP7) and the reorganisation of the public space in the Barrio del Pilar neighbourhood (AP5), with funding of €161,627.34.

Table 21.2 Budget distribution and the value of the work carried out, broken down by parts of the Action Plan (AP)

Action Plan	Value of Actions (Act)	% Value of Actions (Act)	% Completed 2022	Value completed
AP1. Civic Platform of Open Government and Electronic Administration	€2,019,445	16.16%	21.43%	€432,767.06
AP2. Plan for Sustainable Urban Mobility and Accessibility of Ciudad Real	€2,925,422	23.40%	57.92%	€1,694,404.42
AP3. Buildings and municipal installations: large-scale rehabilitation	€496,800	3.97%	70.02%	€347,859.36
AP4. Ciudad Real without noise	€333,333	2.67%	22.39%	€74,633.26
AP5. Project for the complete regeneration of the Barrio del Pilar neighbourhood	€2,633,333	21.07%	2.83%	€74,523.32
AP6. Plan to strengthen the business fabric and to promote Commerce in Ciudad Real	€1,538,889	12.31%	0	€0.00
AP7. Programme for social inclusion (San Martín de Porres, Barrio del Pilar, and Barriada de la Esperanza)	€2,052,778	16.42%	18.66%	€383,048.37
AP8. Management	€470,000	3.76%	15.99%	€75,153.00
AP9. Communications	€30,000	0.24%	75.13%	€22,539.00
Total	€12,500,000	100.00%	24.9%	€3,106,250.00

Source ISUDS Ciudad Real 2022 Eco-Integrator, 2022

21.4 Conclusion

ISUDS have provided paths via which to improve many Spanish cities and municipalities. They have been applied under the umbrella of sustainability and supported by the ERDF, which has provided 80% of their funding. Based on a case study of Ciudad Real, this analysis argues that despite their limited impact, EU-financed urban programmes (such as URBAN, POLIS, and ISUDS) have made a positive contribution to promoting an integrated approach to policy which has enhanced the physical and social environments of many deprived urban neighbourhoods. Other studies conducted in Portugal (Medeiros and Van Der Zwet 2020), based on the recent implementation of 103 ISUDS, have revealed the gradual adoption of more holistic

Fig. 21.3 Vertical garden and gas-driven buses in Ciudad Real, 2022. *Source* Own elaboration

Fig. 21.4 Vertical garden and gas-driven buses in Ciudad Real, 2022. *Source* Own elaboration

and integrated approaches to urban development by urban and local authorities and the increasing effectiveness and efficiency of the policies applied.

This initiative offers a great opportunity to rethink existing city models. Cities must not forget their citizens, nor the essential functions that they already perform. They must continue to concentrate on meeting the needs of their citizens, promote greater social inclusion, and fight poverty by re-qualifying their most disadvantaged neighbourhoods and integrating urban development undertaken on their peripheries. In the case of Ciudad Real, the city must articulate its university campus and make a firm commitment to establishing a well-communicated "University City" with adequate access to information technologies.

The ISUDS horizon has been extended to 2023 and has made possible the adaptation, redevelopment, and updating of the city. It is currently enabling Ciudad Real to meet the new technological, environmental, and accessibility and inclusion-related needs demanded by modern standards of quality and social welfare.

Acknowledgements This work has been carried out within the framework of the research project (PID2021124511NB C21-C22), funded by the Spanish Ministry of Science and Innovation.

References

Adolfsson P, Lindblad J, Peacock S (2021) Translations of sustainability in urban planning documents. A longitudinal study of comprehensive plans in three European cities. Cities 119. https://doi.org/10.1016/j.cities.2021.103360

Bellet C (2020) Las políticas urbanísticas municipales en España: 40 años de ayuntamientos democráticos (1979–2019). Boletín de la Asociación de Geógrafos Españoles 85:1–38. https://doi.org/10.21138/bage.2877

Cañizares MC (2008) Claves de la sostenibilidad urbana a comienzos del siglo XXI. In: F Cebrián (Coord.), Turismo rural y desarrollo local. Ediciones de la Universidad de Castilla-La Mancha, Cuenca, pp 29–37

De Gregorio S, González M (2017) Las EDUSI en el contexto de las políticas de regeneración urbana en España (1994–2018). WPS Rev Int Sustain Hous Urban Renew (RI-SHUR) 6(2):54–80. Accessed https://dialnet.unirioja.es/servlet/articulo?codigo=6368214

De Gregorio S, Do Santos V, Batti A (2021) La Europeización de la política urbana en España en el periodo 2014–2020: análisis de las Estrategias de Desarrollo Urbano Sostenible Integrado (EDUSI). Cuadernos de Investigación Urbanística 134:1–99. https://doi.org/10.20868/ciur.2020.134.529

Jacobs, M (1999) Sustainable development as a contested concept. In: Fairness and futurity: essays on environmental sustainability and social justice, pp 21–46

Marchán CP, Tabasco JP, Martínez HS (2018) La movilidad urbana sostenible como estrategia de integración urbanística y social en las nuevas agendas urbanas de las ciudades medias de Castilla-La Mancha. In Cebrián F (Coord.), Ciudades medias y áreas metropolitanas: de la dispersión a la regeneración. Ediciones de la Universidad de Castilla-La Mancha, Cuenca, pp 733–751

Medeiros E, Van Der Zwet A (2020) Evaluating integrated sustainable urban development strategies: a methodological framework applied in Portugal. Eur Plan Stud 28(3):563–582. https://doi.org/10.1080/09654313.2019.1606898

Nasarre y de Goicoechea F, Baiget Llompart M, Fernández Hernando M, Ángeles, Rodríguez Arce LM, González Fernández-Mellado S (2017) Las Estrategias de Desarrollo Urbano Sostenible e

Integrado (estrategias DUSI). La apuesta por un crecimiento inteligente, sostenible e integrador en las ciudades españolas. Ciudad y Territorio Estudios Territoriales 49(194):801–819. Accessed https://recyt.fecyt.es/index.php/CyTET/article/view/76597
Observatorio de la Sostenibilidad. (2018). Informe sobre las Ciudades Sostenibles en España. Accessed https://www.observatoriosostenibilidad.com/informes/
Panadero M (2010) Geografía Humana y sostenibilidad. Hitos y perspectivas. In: Pillet F, Cañizares MC, Ruíz Pulpón AR (eds) Territorio, Paisaje y Sostenibilidad. Un mundo cambiante. Ediciones del Serbal, Barcelona, pp 49–86
Pillet F (1984) Geografía urbana de Ciudad Real (1255–1980). Akal Universitaria, Madrid
Pillet F (1995) El medio ambiente urbano de Ciudad Real. Anales de Geografía de la Universidad Complutense de Madrid (Monográfico Ciudad y Medio Ambiente) 15:527–537. Accessed https://dialnet.unirioja.es/servlet/articulo?codigo=86445
Pillet C, Félix y Rodríguez-Domenech MA. (2021). El urbanismo de Ciudad Real: desde mediados del siglo XVIII hasta la actualidad. En el Vía Cívitas una ciudad en la historia. Ciudad Real Juan II - 1420-2020. Ciudad Real. pp. 123–144. ISBN. 978-84-87248-63-4
Rodriguez-Domenech MA (2012) Nueva realidad urbana y Territorial de Ciudad Real (1980–2010). Instituto de Estudios Manchegos (CSIC), Ciudad Real
Sánchez LE, Croal P (2012) Environmental impact assessment, from Rio-92 to Rio+ 20 and beyond. Ambiente & Sociedade 15:41–54
Santos A (2019) Sobre el modelo de ciudad. La transición de la ciudad competitiva a la ciudad integradora. In: XIII Congreso Español de Sociología. Actas del Congreso. València, 3–8 de julio, pp 1–6. Accessed https://ruidera.uclm.es/xmlui/bitstream/handle/10578/28282/AntonioSantos_13CongresoFES.pdf?sequence=1
Seyfang G (2003) Environmental mega-conferences from Stockholm to Johannesburg and beyond. Glob Environ Chang 13(3):223–228. https://doi.org/10.1016/S0959-3780(03)00006-2
Verma P, Raghubanshi AS (2018) Urban sustainability indicators: Challenges and opportunities. Ecol Ind 93:282–291
Weiland S, Hickmann T, Lederer M, Marquardt J, Schwindenhammer S (2021) The 2030 agenda for sustainable development: transformative change through the sustainable development goals? Polit and Governance 9(1):90–95

Websites

Ayuntamiento de Ciudad Real (2022) EDUSI Ciudad Real 2022 Eco-Integrador. Accessed https://edusi.ciudadreal.es/edusi-ciudad-real-2022-eco-integrador/
Grupo de Geografía Urbana (2021) Un poco de historia. En Grupo de Geografía Urbana de la Asociación Española de Geografía. Accessed https://www.ggu2015.com/proyectos-de-investigacin
Naciones Unidas (2017) Nueva Agenda Urbana. Accessed https://www.agenda2030.gob.es/recursos/docs/HABITAT.pdf
Naciones Unidas (2021) Objetivo 11. Accessed https://sdgs.un.org/es/goals/goal11

Part IV
Postscript

Chapter 22
Intra-Urban Borders in Border Cities: The Nationally Interchangeable Dynamics of Urban Centrality

Fernando Carrión Mena

Abstract *Methodological nationalism* denied the possibility of the existence of cities under the influence of different national States. Globalization has revealed this incoherence, because many capitals are home to several States, as well as several private global corporations. A case that illustrates this discordance is that of border cities, which configure *multinational urban regions*, thanks to the fact that *complementary asymmetry* made borders enter into a logic of integration. Hence, interstate limits behave as lines of integration of neighboring states and borders become regions of universal articulation. In this context, not only are border cities a unified multi-state territory, but they also operate as pivots of articulation of trans- and inter-border logics, as well as of *global border systems.*

Keywords Border cities · Urban regions · Urban centrality · Urban systems · Borders · Urbanization process.

> Centrality is a form, in itself empty, but which requires a content: objects, natural or artificial beings, things, products, works, signs and symbols, persons, actions, situations.
>
> Lefebvre Henri (2013)

22.1 Introduction

Borders or Frontiers can be defined through some of the following meanings: from anthropology, originating in the looks around the symbolic and belonging (otherness); from *sociology*, about the socio-economic differences and inequalities coming from racism, xenophobia and classism; from *urbanism*, through the logic of urban segregation that is expressed in the space of the city (fragmentation). Also from the etymology, since it is a compound word, originating in Latin, where *front* comes

F. Carrión Mena (✉)
Universidad Central Ecuador, Quito, Ecuador
e-mail: fcarrion@flacso.edu.ec

E. Navarro-Jurado et al. (eds.), *Urban Dynamics in the Post-pandemic Period*, The Urban Book Series,
https://doi.org/10.1007/978-3-031-36017-6_22

from *frontis*, frontier or front, and *era* (arius) from place or spot; however, it can also be interpreted as a phase or historical moment, that is, from a site in front of a barrier, a place or a historical moment. Additionally, there is the definition of the *Dictionary of the Real Academia Española (DRAE)*, which considers it as the *border*, *limit* or *in front of*. This set of meanings of the word border or frontier operates simultaneously within the borderline cities, because they are *complex urbs* where the intra-, trans- and inter-urban—at various levels—are present concurrently.

The condition of borders will be approached in this article through two structural themes of the contemporary city: centrality and its areas of influence located in differentiated but articulated spaces. What is relevant is that both demarcate their own limits or borders, under the dynamics of what is known as urban segregation, which is a traditional and classic theme in Latin American urban studies (Sabatini 2006). In the case of *border cities*, to the general processes of social division of space according to population and activities, the boundaries with other national States must be added.

In Latin America, *border cities* undergo acute transformation processes, superior and different from those occurring in other cities, which lead to changes in the way they were conceived within the national States. Nowadays, the previous logic of Latin American urbanization composed of *nuclear cities* is being overcome in order to move toward *urban regions* (Carrión 2017b). In the case of the urbs in border areas, an urban agglomeration is formed that originates from the confluence of several states simultaneously and not only from the influence of a single state, as it used to be.

Interpretations of the existence of several cities on the borders are falling away, leading to a new pattern of urbanization characterized as a *multinational urban region*. That is, there is a transition from a conception that assumes several nuclear cities inscribed in each adjoining national State, to what actually occurs at the present time: an integrated urban territory, regardless of whether its location is in several national States. In other words, the border condition erroneously led people to think that there were several cities, one on one side and another on the other side, just because the territory was flanked by the interstate limit. A conception of this type ignores that the essence of the city is heterogeneity, born of the interaction produced by the fragmentation of activities, populations and infrastructures (urban segregation); to which, in this particular case, national boundaries are added.

Undoubtedly, the determining element of this situation is based on an interpretation sustained on visions circumscribed in the domestic spheres of each country, which impedes the understanding of the new characteristics of Latin American urbanization, emerging since the end of the last century. It is an optic inscribed in power relations in the exclusive national sphere. For this reason, when analyzing urbanization processes, these cities are characterized as small or, at most, intermediate, defined within the urban rank-size hierarchy of each country.

The frontiers in border cities are based on an urban structure with important inequalities and asymmetries, stemming from at least two determinations. The first is the traditional socio-spatial inequalities originated in the dynamics of urban segregation (centralities-peripheries). Its most extreme expression nowadays are the cities of gated neighborhoods or of walls, according to Teresa Caldeira (2008). The second comes from power asymmetric of interstate relations, which mark each of the national States (sovereignty), like Ipiales, Melilla and Juarez. Unfortunately, it is an entry that has been little worked on in academic terms, which has impeded comparisons with non-border cities and with those located, for example, within the European Union, where the border topic is not so gravitating. In any case, the two inputs producing intra-urban borders—local and international—are interrelated and inscribed in the same multinational urban-regional unit.

In this context of change, there is a significant growth of the economy and, consequently, of urbanization in the border zones (population), events that occur above the averages of each of their countries, with which these regions achieve an important national and international protagonism.

Nowadays, border societies are represented through the strengthening of local governments, thanks to the processes of decentralization, democratization and economic accumulation, which has allowed them to have an unprecedented local and international presence. As a result, these territories are claiming greater autonomy from state centralism, represented mainly by the capital cities.

In addition, these cities assume a new role, given the weight of the functional attributions they acquire in the national and global context. Therefore, border cities operate as *pivots* of integration of border regions, as well as of articulation with world economies, legal or illegal. This change leads to another not minor one: interstate relations in border zones begin to be constituted on the basis of trans- and inter-urban links; that is, of an integration that is projected from the local to the national and global. Not as it used to happen before, when they were confined exclusively to the local sphere and the national integration passed hegemonically through the multilateral entities (OEA, CELALC, CAN MERCOSUR).

A few facts that should be highlighted in this process. First, the interstate limit behaves as an urban centrality, since this line integrates the urban territories of one side with those of the other—not separates, as it was formerly conceived—through the complex dynamics of *complementary asymmetry* (Carrión 2013). This asymmetry is even more complex because the design of urban policies in border areas is marked by the logic of defending sovereignty, customs control and migratory restrictions, contrary to the reality of local integration. Unfortunately, public policies are not designed with an emphasis on transboundary issues, in accordance with the processes occurring in the zone, but rather with those that have a strong nationalist content, increasing the sense of the interstate border.

Secondly, COVID-19 illustrates the disruptive impact of national health policies in border zones, because they are homogeneous in heterogeneous realities, which leads to greater distancing from the borderline (Fig. 22.1). As well as the effects it produces on the border side where there is more population and more poverty (social vulnerability). The greatest contagion and lethality occurred on the borders with a larger population (due to social interaction), with greater insertion in globalization (due to importation of the virus) and with a higher index of poverty (pre-existing diseases) (Lara-Valencia and García-Pérez 2021).

But it should also be noted the impact produced by the multi-scale closure of territories, in the areas of housing (#Stayinhome), the neighborhood, the city and the

Fig. 22.1 COVID-19 in a border zone in Latin America. *Source* Galería fotográfica PCM. Licensed under CC BY-NC-SA 2.0

country, through policies of physical distancing, restriction of mobility and border closure. Especially this last one showed—by absence—that the border is the real umbilical cord of the transformation of the urban region, generating a dangerous process of state recentralization that weakened the transboundary region at local and global levels. It was a very strong recessive process of the previous interaction obtained.

In order to develop this general thesis, which is inscribed in this text, there will be an expository logic with the following sections. It begins with the necessary historical vision of the borders, since they are regions with a high level of transformation, which allows to situate and understand the dynamics of urbanization and the cities contained therein. Then follows the central theme of the article: to show, on the one hand, how the border city operates under the form of an urban region (city of cities), and on the other hand, the way in which border cities construct intra-urban boundaries. It will conclude with a set of recommendations that seek to highlight some characteristics of these border cities, of the urban border region.

22.2 Historical Borders, with Bordering Urban Regions

The fundamental starting point for the analysis of borders is their historical sense, due to their highly changing dynamics. In this perspective, Augé (1998) states that a typical characteristic of these zones is that: "*borders are never erased, but are redrawn.*" It is a reality of strong mutation—but not linear—because they are permanently reconfigured. The limit, the border and the transboundary act together, although with differential relative weights according to place and time.

The borders have undergone a profound metamorphosis since the end of the last century, when they acquired their own political and economic weight. Its territory gained autonomy thanks to its economic development and its important political role, with which it tended to reconfigure its sense of periphery and redefine its relationship with the center (State). But it also overcomes its local-national quality to become an interstate integration zone, since its structural relations are conformed by other borders, some contiguous and others discontinuous-distant.

The processes of globalization and decentralization, which occur simultaneously, have much to do with this mutation. The mundialization of the economy, culture, technology and society is taking place, with its counterpart, the concentration of its effects at the local level (Borja 1994). In other words, this global phenomenon needs strategic places, such as cities, to project itself ubiquitously throughout the planetary territory. This is what Robertson (2003) describes as *glocalization*, which is expressed clearly and with greater force in border cities. As a result, the structure of the border region reaches a *cosmopolitan* dynamic, because today *borders are universal.*

In the process of transformation of the borders, the State Reform had much to do with it, bringing some unprecedented consequences, among which the following should be mentioned: (i) *Decentralization*, which allowed the promotion of the autonomy and power of the territories, with representative local governments,[1] with more administrative competencies and greater economic capacity (budget); (ii) the sign of *international openness*, which closed the cycle of the import substitution model and placed the borders in a condition of regional and global pivot; and (iii) *Privatization*, which strengthened the market logic and weakened the State, increasing inequalities and reducing its regulatory capacity.

Undoubtedly, under these new conditions, the borders acquire unprecedented power, because they acquire greater political legitimacy thanks to the popular election of local authorities, to the significant improvement of management, administration and government mechanisms (management models). But also because of the increase in accumulation processes, thanks to the capture of substantial economic resources, both legal and illegal. With this new condition, more direct relations are established with the other side to form a unified multinational transboundary territory, which tends to question the centralism of the national States. As well as to look for new forms of linkage to more distant borders, where the logic of illegal trafficking contributes a lot.

[1] While in 1985 only seven countries in the region elected local authorities, by 1997 all did so.

Perhaps this expression of illegality is one of its most complex factors. Because it undoubtedly represents a strong disregard for the national state regulatory order, which is initially expressed in smuggling and then in trafficking. Clearly, the national membership of public, private and formal community institutions is bypassed by the local illegal activities, remaining outside the control and regulation of the State. In this context, autonomy comes into play and debate, with a totally different stamp than in the rest of the national territories.

The accelerated and significant urbanization of the borders throughout the region is contributing to this process. First, because the borders are undergoing a major population growth process, which takes place in the cities as its nodal point; and second, because they are the starting point for multinational regional integration.

22.2.1 From Limit to Border

The historical process of borders in Latin America is related to three sequential moments, which show the transition: the *limit* is a line that separates (succession of points), the *border* is a region-plane that relates its parts (inter-border), and the *transboundary* is a space of global integration.

- The first one originates with the formation of national States, which are born from the processes of independence and decolonization, in general, in Latin America. In this context, the sense of *limit*[2] takes form, under the traditional definition of the *boundary*; that is, of the territorial demarcation of the States or, in other words, of how far they reach and from where they start. It is, in short, a real or imaginary line of encounter-separation with the other (alterity), through a strange tension between cleavage (fracture) and otherness (integration), between different adjoining national States.

  The limit is a concept initially conceived as separation, which even led to military disputes for the defense of the so-called territorial integrity (sovereignty). But it also led to the splitting of the native communities located in these created regions, because the constitution of the new national States caused several of these ancestral populations to fragment, sometimes even ending up as enemies, regardless of the fact that their origin and kinship were exactly the same.
- The second one was established from the second post-world war, when the so-called *inward or import substitution model* began to develop (Rodríguez 1980; Fitzgerald 1993). This mutation allowed the transition from the limit-line to the border region, establishing the sense of *inter-border*, because there are several borders that interact with each other.

  Each of the States began to think of its development inward, whereby the limit becomes a border, related to that of the other side under inter-border dynamics.

[2] According to the DRAE, the word *limit* is: "Real or imaginary line that separates two lands, two countries, two territories".

That is, a relationship of two or more border regions, different and distinct, that share the same territorial demarcation line.

Borders are nothing other than the space where interstate relations are expressed, given that the State on the one side is different from that on the other. This situation involves a relationship that operates similarly to the condition of a magnet: it is an area that has a particular magnetism that makes opposite poles attract each other, that is, borders of distinct States that need each other because of their differences.[3]

- The third moment originates in the context of the globalization process, when *borders are redrawn* under the *transboundary*[4] dynamic. With this, the previous meanings of limit-line and border-plane are overcome, to give way to the unified transboundary-plurinational region. That is, an integrated territory based on a limit that demarcates several countries.

  This process should be interpreted in terms of *multinational transboundary integration*, through two interrelated modalities of border regions: on the one hand, that end up establishing a single regional entity, forming a plurinational assembly; and, on the other hand, that articulate with other border regions located in distant places, to structure a *global border system*, where cities behave as strategic points or interconnection links, forming an urban complex.

  Nowadays, the dynamics of borders operate as a hub or router, since this set of relationships leads to the confluence of multiple flows of products, capital, raw materials, services, symbols and people immersed in the processes of import and export (legal and illegal), with which transboundary regions, according to Castells (1999), experience "*the transit from the space of places to the space of flows*."

Undoubtedly, this new reality of the borders shows us the character of a complex reality, where politically the actors and institutions of the region question the central power located in distant places (capitals), which becomes an additional incentive to act jointly in this pluri-state space.

22.2.2 Urbanization Patterns

In Latin American urbanization, there are two major moments: one originated in the second post-world war period, in which a nuclear or central city is configured, inscribed in a specific territory (whether countryside or metropolitan region), but within a single national State. In other words, it is a traditional and classic urbanization pattern with a *nuclear city, inscribed in a territory and within a national State*. This characteristic leads to the formation of the so-called *urban hierarchy*, constituted by

---

[3] Perhaps smuggling is the clearest expression of this statement: What is not available or is more expensive here is obtained on the other side, but under informal or illegal modalities. This logic originates in the difference and is resolved in the same relationship of the existing market.

[4] According to the DRAE, the prefix *trans* (of Latin origin) means: "behind," "on the other side of" or "through".

attributes that arise from the rank size of the population, within the same nation-State (Carrión 2017a, 2017b).

Subsequently, a new one was foreshadowed from the 1980s of the last century, when the pattern of urbanization was modified: from the nuclear or central city to an *urban region*, where cities are not urbanized but rather urban regions, under the quality of a *city of cities*. This urban region is no longer inscribed in a surrounding territory (rural or metropolitan), but within the framework of other cities, in order to build the *global urban system.*

This phenomenon introduces a new characteristic: the transboundary region is not inscribed within a single State, but within several, in addition to countless private corporations of a global order that have local effects and induce a public-private capitality of a global nature. This consolidates new forms of autonomy, especially in capital cities (capitality) and in border cities (nationalism), which are different from those of other cities as a whole.

New urban forms are taking shape in the border zones, in line with existing urbanization patterns in Latin America, among which we can mention the ones described in Table 22.1.

In the first moment of Latin American urbanization (Graphic No. 1), cities are configured in immediate and contiguous spaces (nuclear city), with administrations structured on the basis of local or municipal governments. In the second moment, the inter-urban logic, proper of urban systems, prevails, producing a complex problem to be solved: there is no harmony between the emerging territorial organization and the current institutional framework, producing a multilevel government that is exercised in a multi-scale territory with an institutionally with juxtaposed autonomies.

It is even more complex, if we take into account that the urbanization pattern is not homogeneous but diversified and heterogeneous, and at least the following analytical typologies can be found:

- *Clusters*, which are the result of the sum of cities within the framework of a region that is a single producer of services or goods, as in the case of the Salmon Cluster in Chile, with 45 integrated municipalities, or the Bajío in Mexico, which produces vehicles of North American brands under advantageous conditions for the US market. The government concretizes with the implementation of free trade zones, exercised by various public administrations, but explicitly submitted to the market.
- *Urban regions,* which are heirs to metropolitan logic (Soja 2008), operate under the sense of *urban agglomeration*, which is nothing more than the concentration of population and activities in the same *multi-scale* territory. The most evident case is Mexico City, which has a population of less than 9 million 200 inhabitants, but its metropolitan area can have more than 22 million inhabitants. And the most complex thing is that it has 60 municipalities and 3 states, which makes very complicated the management of different autonomies, coming from different levels of government (multilevel).
- *Imagined cities*, which are constituted thanks to international urban-urban migratory processes, after the closing of the rural-urban migration cycle. These are

Table 22.1 Patterns of urbanization and institutional frameworks for city governance in Latin America

	Urban types	Spatial model	Relations	Government
Cities urbanization	Nuclear city	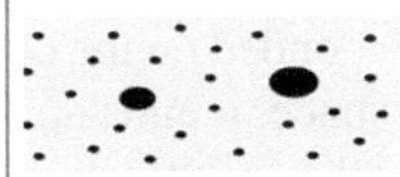	City—Countryside	Municipal isolated
	Metropolitan city	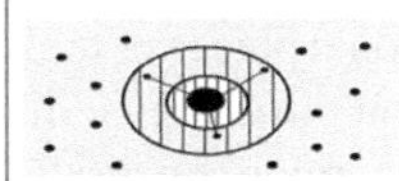	Polarized city	Metropolitan
Inter-urbanization	Urban system	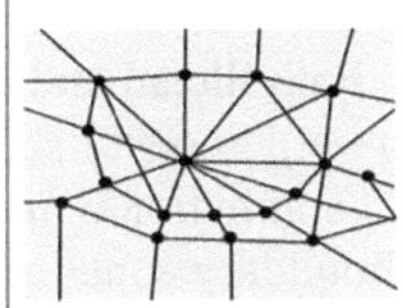	Articulated city	Municipal horizontal
Transurbanization	Clusters	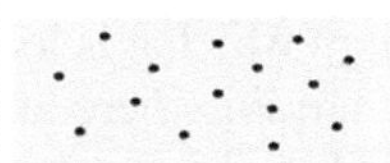	Factory city	Public/private multilevel
	Borders cities	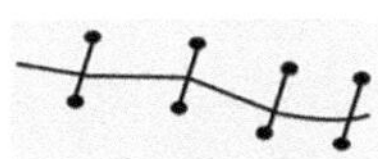	Cities couples	Multistate
	Imagine cities		Multisited cities	Multistate
	Regional urbanization	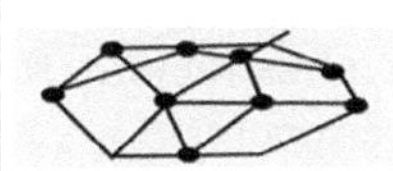	Cities of cities	Multilevel

Source Own elaboration

cities articulated in distant and discontinuous spaces thanks to the development of communication technologies, which make it possible to diversify remittances (economic, cultural) and integrate spaces. They are known as multisite urban regions. This is the case of the second and third cities of our countries that are located outside the national territories. Examples: Cuba's second city is in Miami, El Salvador's in New York, Mexico's in San Antonio or Houston, Ecuador's in Murcia and Venezuela's in Bogota, among many other cases.

- *Border cities* historically originated in the boundary line that demarcates the confines of a State with respect to others. The interstate demarcation line defines the separation of cities by States, that is, the starting place of a State and of a city, with respect to those of the other State and of the other city, regardless of whether they are geographically contiguous. But this consideration is modified when multinational borders are inscribed in global circuits. In this case, the border city reproduces the condition of a multisite city within several neighboring states. One of the most relevant cases is the urban area formed by Ciudad Juarez (Mexico) and El Paso (USA), where only a bordering river separates them. If in the previous phase, the existence of one city depended on the other, nowadays, due to the interstate dynamics occurring in this space, it ends up unifying them, to the point that they are beginning to be conceived as multinational metropolitan areas.

In this sense, border cities are traditional places of trade, exchange and services, which are positioned as nodes of integration through two dimensions: the one coming from the multinational transboundary dynamics (integrated region) and the other one of a global borders network or system. This is because border cities facilitate flows and movements of people and products, both illegal and informal (Vera da Silva 2015), even more nowadays, when there is a perceived shift from *offshore* to *nearshore* logics, which give rise to a stronger supranational integration with an urban base.

In general, the dynamics in transboundary zones are given by the sense of enclave economies arising from different States, where accumulation develops without integrating with the border environment in which it is located. This is the case of border cities, which operate as a strategic terrain where an economy, legal and illegal, is located under an import-export market. To make this work, there are routes and circuits, operating within a global urban system.

22.2.3 Borders Are Urbanizing: Regional Urban Complex

The border region in Latin America is experiencing an accelerated population growth. In the mid-twentieth century, the border population was very small because it was inhospitable, dangerous and lacked infrastructure (Reboratti 1992). At that time, the borders were the periphery of the periphery.

Subsequently, since 1950, with the acceleration of the urbanization process throughout the continent, as well as the need to control borders and reaffirm national sovereignty, States sought to redistribute population and resources for geopolitical purposes (Rodríguez 2002). The public policy proposal was based on the slogans of "*to govern is to populate*" or to build "*living borders*," accentuating the protectionist function of border territories, inscribed in nationalist logics and, therefore, to promote separation.

From the 1980s of the last century onwards, the borders began to experience a significant population growth and economic diversification (legal and illegal), which resulted in increasing trade and industrial production (maquila) flows. In this perspective, national interstate integration programs were promoted (Rodríguez 2002), as well as those coming from multilateral organizations (CAF, WB, BID, OEA).[5]

---

[5] For example, one CAF program is the Border Development and Integration Support Program (PADIF) for regional integration purposes.

Borders become *universal spaces*, because they are no longer only the link between neighboring states, but also form part of the world economic system. From this moment on, new forms of integration of the local with the supranational urban region began, which became a central and gravitating element.

In these border regions, positive externalities, both legal and illegal, are beginning to be generated, operating as a gravitational mass of population attraction. The evident result is the significant demographic changes. For this reason, the traditional migratory flows that have historically operated in the continent are being modified: rural-urban migration, rural-urban migration or internal mobility within the national States show traits of being in their terminal cycle.

First, because the rural origin of emigration no longer has the population flow it used to have. In Latin America, according to UN-HABITAT, in 1950 the urban population was 41%, whereas now it is over 83%. This means, if the migratory process is finite, that only 17% of the population is potentially left to migrate from rural areas, which ends up being impossible. Moreover, when there are inverse processes, with two clear characteristics: the countryside cannot continue to expel population, no matter how much mechanization is introduced, because someone must sustain agricultural and livestock production. And second, because an inverse population movement is beginning to take place: urban-rural, although still embryonic, sustained by the search for a better quality of life and remote work, based on high technology in work and services.

This trend is strengthened by the presence of the COVID-19 epidemic. Thus, for example, in Peru, several cities experienced processes of population return (Zolezzi 2020), where Lima stands out, a place that produced an emigration of about 300 thousand people, because of the need to reduce contagion factors and to capture products of basic consumer goods. But several cities in the region have also seen a growth in nearby areas and cities, due to the closure of businesses and activities, especially in areas of urban centrality, which have relocated their activities (Buenos Aires, Quito, Mexico City).

And secondly, in terms of the migratory destination, because there is a redirection of the places to which the population used to go, as were generally the largest cities, capitals or metropolitan areas, to give way to a strong urban-urban migration with two final places: one, inscribed within the national territories in the border regions and cities, and the other, of international character, in cities located in other countries (Lattes 2001).

In these new migratory processes, border cities have a strategic function: to be the natural filter assigned to them by national States for internal-external contact. Hence, they become highly conflictive spaces, because that is where the illegal organizations dedicated to these purposes operate, most of which are international.

In other words, a new cycle of population mobility based on urban-urban migration, but of a border and international nature. That is why the second and third cities of Latin American countries are located outside the national territories, as stated by García Canclini (1997), forming the urban pattern of *imagined or multi-situated cities* (Carrión 2017a, 2017b). For example, in the USA live about 40 million people of Mexican origin and the absolute majority of them in cities. More than 7 million people have left Venezuela in this century, but also from Haiti, Colombia, Ecuador and Guatemala, among others. In general, this international migration has opened the new cycle, predominantly urban-urban.

What is evident is the accelerated population growth in border areas, a phenomenon that can be seen in Table 22.2. The average percentage of urbanization in the countries is quite high, but even higher in the border areas. Guatemala, which starts with the lowest percentage of urbanization, is the country with the highest growth, both at the national level and at its borders. In contrast, Brazil and Argentina, which have the highest national urbanization rates, are the countries with the highest percentages at their borders.

If the demographic behavior of the two intercensal periods is compared, it can be affirmed that the average population growth rate of the border areas in Latin America is faster than the average for the national States. The intercensal growth of the urbanization rate of the countries is 4.95%, while that of the border areas is 6.26%. These data comparatively mean that in this period, the average population rate of the border areas grew 26% more than that of the countries. And this is a relatively recent phenomenon, typical of the turn of this century.

It should be noted that the growth of the border population is sustained in border cities, which leads us to the conclusion that there is an accelerated process of urbanization of the borders in Latin America. For example, Ciudad Juarez in Mexico has a high population growth from 1980, when it had 544,000 inhabitants, to 2020, when it reached 1,501,000; that is, an increase of almost three times in the period. Cúcuta in Colombia, on the border with Venezuela, had 379,000 in 1985 and reached 777,000 in 2020, a two-fold increase. Between 1971 and 2001, the population of the Triple Border (Brazil, Argentina and Paraguay) grew by an average of 30% per year, reaching 664,000 inhabitants in 2012 (Bello 2013).

Table 22.2 Urbanization of the borders in relation to that of countries

	%	%	%	%	%	%
Country	National Urbanization Previous Census	Border Provinces Urbanization Census	National Urbanization Last Census	Border Provinces Urbanization Last Census	National Intercensal Growth	Intercensal Growth Border Provinces
Argentina	89.4	**83.03**	91.02	**85.16**	**(+) 1.62**	**(+) 2.13**
Ecuador	61.13	**43.48**	62.76	**48.01**	**(+) 1.63**	**(+) 4.52**
Bolivia	62.42	*56.11*	67.50	*61.01*	*(+) 5.08*	*(+) 4.90*
Colombia	70.90	**40.17**	76.00	**50.72**	**(+) 5.10**	**(+) 10.55**
Brazil	81.23	**76.56**	84.36	**80.15**	**(+) 3.13**	**(+) 3.59**
Guatemala	34.98	*20.43*	48.48	*31.78*	*(+) 13.05*	*(+) 11.35*
Peru	70.11	**58.62**	76.70	**68.89**	**(+) 6.59**	**(+) 10.27**
Mexico	74.60	*76.10*	78.00	*78.89*	*(+) 3.40*	*(+) 2.79*

Source National Census 2000 and 2010 and own elaboration

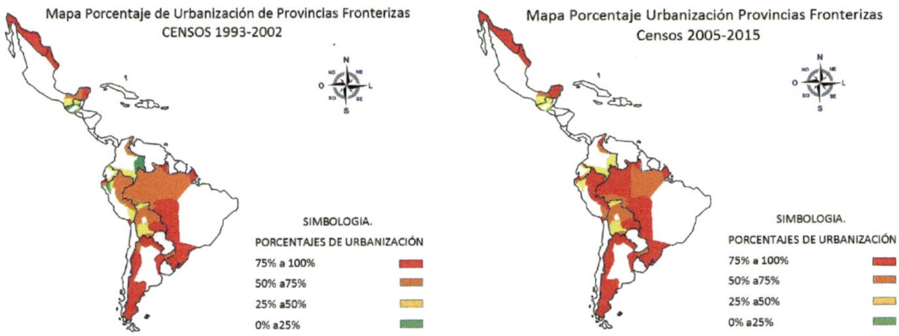

Fig. 22.2 The urbanization of the borders modifies the pattern of urbanization in Latin America. *Source* Table 22.2 of this text and own elaboration

Another fundamental characteristic of demographic behavior in this period has to do with the substantial change in the profiles adopted by urbanization in Latin America. As the borders are located in the interior of the continent and there is population growth in cities or the birth of new ones located there, the logic of urbanization is redirected toward unprecedented zones.

In the past, cities were preferentially located on the coastal profiles, in strategic places for trade (seas, rivers) or in areas with a vocation for agricultural or livestock production. However, in general, there is a redirection that is expressed, for example, in the change from the capital of Brazil, located on the Atlantic coast (Rio de Janeiro), to the interior (Brasilia). But the most evident process has to do with what can be seen in Fig. 22.2, especially in South America where the phenomenon of mutation is evident: urbanization tends to move toward the interior of the continent. And this trend stems from the growth of the border economy and, as a consequence, the settlement of transboundary regions.

There is an additional demographic connotation to the border population (Picech 2017). The highest population concentration is on the "weaker" side of the multinational transboundary link. For example, in Mexico's relationship with Guatemala, the highest proportional concentration of population is on the Guatemalan side, while the situation is reversed in Mexico's relationship with the USA. Similar cases occur in the Dominican Republic and Haiti, where in the first one 4.7% of the total population lives on the border while in the second one it is 25.6%. In the border between Argentina and Bolivia, the population is 4.7 and 13.9%, respectively (Canales et al. 2010). The explanation for this inequity is in line with the need of the weaker borders to take advantage of the more developed ones, anchored according to the physical law of communicating vessels.

Hence, social and economic demands tend to be redirected to the places where there is more population. For this reason, important infrastructure and service works are promoted and built to satisfy the social demands and functions of the borders, in many cases different depending on the side and the needs of integration to the multinational territories. Roads, highways, airports, bridges, ports, as well as hotels, pharmacies, restaurants, schools, universities, industries and shopping centers are built to meet the needs of their population, their activities and multinational interaction.

22.2.4 The Initial Binary Logic of the Urban Border Region

The creation of border cities has several origins, the first and painful one are the settlements of the native peoples, which were split when the national States were formed. The boundary lines drawn by the emerging States use geographic accidents as coordinates and do not take into account social criteria. In this process, the societies settled there are disregarded, so much so that they end up cutting them off for reasons of national sovereignty.[6] There are, for example, the cases of the transboundary indigenous peoples and nationalities of Colombia, Ecuador and Peru: Awá, Inga, Quillacinga, Pastos, A'i Kofán, Eperara Siapidaara, Kamentsá, Murui, Kichwa, Siona, Secoya, Coreguaje, Nasa, Embera Chamí, Shuar and Achuar.

Subsequently, the cities that existed prior to the delimitation of the national States began to strengthen their inter-urban relations and to develop along the boundary line. In this context, they maintain their specificities, but under a structural determination: the limit acts under the metaphor of the magnet: *different poles attract each other*. But this attraction is fickle, depending on the structural conditions of the poles and their interrelations. Hence, at one moment one border city may have a greater development over the other,[7] while at another it may be the other way around, structuring substantial momentary differences.

It is precisely the border logic that gives this dynamism to its cities, allowing a new form of integration, subject to asymmetrical relations. In this context, inter-urban links are established; city-city, regardless of their proximity or distance, because it is the complementary functionality that determines the formation of the urban region.

In the past they were conceived as different cities, because national limits divided them, being known under the names of: (i) *mirror cities*, as a metaphor to say that one urbe is reflected in the other, when in reality they are absolutely different in size and characteristics; (ii) *twin cities*, as an allegory that serves to affirm that they have the same origin, where the limit operates as a matrix, but which in no way leads to

---

[6] According to Tohono O'odham on the border between Mexico and the United States since 1853 the people have moved freely over their ancestral lands, but the recent militarization of the border has not only made their travel difficult but has also divided the people. https://news.un.org/es/story/2018/08/1439422.

[7] For example, a devaluation of the Colombian peso causes the centrality of the city of Ipiales (Colombia) to attract the consumption of the population of Ecuador, depressing the Ecuadorian city of Tulcán. The price of Venezuelan oil derivatives in Cúcuta (Colombia) generates unthinkable accumulation processes.

their being similar; (iii) and the *peer cities*, simply it is a figure to reaffirm that they are two similar urbs, which does not occur at all.

In terms of the use of these notions, it can be affirmed that in Brazil the concept of the twin city prevails, while in Mexico it is the concept of the mirror city and in the world of academia that of the peer city. These three conceptual denominations refer to cities that have supposedly similar patterns, when they are very different, to the point that there is a relationship of determination of one over the other. In other words, in reality they are not two, neither are they equal cities.

The city on the one side is interconnected with the one on the other, thanks to the confluence of three processes: those of globalization (expansion of markets and technologies), national ones (economic regulation, integration agreements) and the local ones (trade, industry). This confluence imprints a key quality: cities structure transboundary regions, linked to other border regions and, in this context, operate as *hinges of integration* (Boisier 1987), in the manner of a broad-spectrum *zipper*.

The *zipper* behaves as a device that has a set of gears that serve to join the poles of the equation (border). It consists of two parallel ribbons (limits) that have a set of devices that interact with the purpose of integrating what is separated.

A good example of this process can be seen in Fig. 22.3, in which the *zipper* logic is formed with the cities of the US and Mexico border region, which operate as a gear. Undoubtedly, this is an emblematic case of the urbanization process in border zones. Along the demarcation line of these two countries, there are 11 urban gears that make up the urban complex within the transboundary region. It should be noted that this characteristic is reproduced in all the border zones of Latin America, although in some cases it is more explicit than in others.

The urbanization process of this border has been accelerated: if in 1960 it had 1,179,910 inhabitants, by 2010 it reached 7.2 million and now exceeds 17 million residents. The interesting thing about this population is that it is mostly urban, because of internal and international migration, attracted by the labor supply located in the

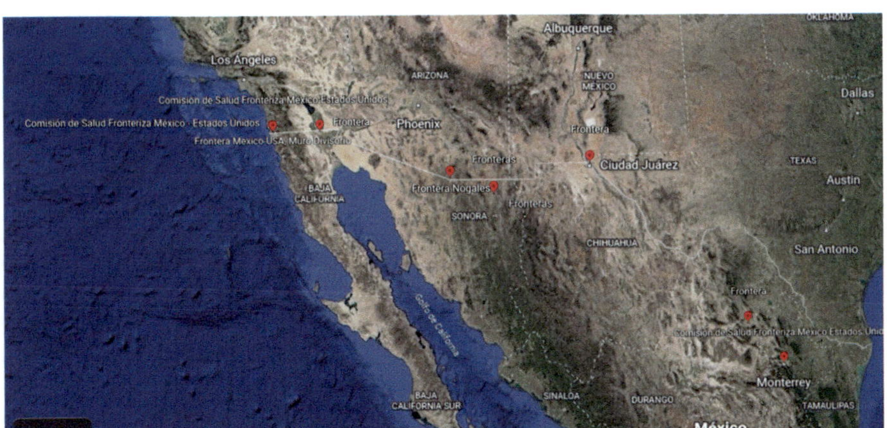

Fig. 22.3 The US/Mexico border urban complex operates as a zipper. *Source* Google Earth

cities, converted into the centers of gravity of the new global economy, sustained by digitalization and specialized services, among which there is the maquiladora industry (Fuentes and Peña 2018).

In Latin America's borders, two types of urban complexes operate, determined by the countries that conform them:

The first one, *Binational Urban*, is formed between border cities. Here we can point out the cases of Brazil, which has borders with all the countries of South America, with the exception of Ecuador and Chile. There are also those of Colombia (Ipiales) with Ecuador (Tulcán), where the Guáitara River separates and integrates them. Or, in the case of Rivera and Livramento, it is a street that fulfills this dual function to form a single city, as can be seen in Fig. 22.4.

Another significant example of a river that integrates-separates the cities on the one side with those on the other is the Río Grande, with the cities of Ciudad Juarez in Mexico and El Paso in the USA.

Interesting cases of binational border cities are those that have the same name on one side of the border as on the other. There are, for example, Nogales in the U.S. and Mexico; Paso Canoas in Costa Rica and Panama. But there is also the city of Mexicali, whose name is a fusion of Mexico and California, and its peer on the other side, Calexico, which comes from California and Mexico.

Fig. 22.4 Rivera (Uruguay) and Livramento (Brazil). *Source* Miguel Chaves. Licensed under CC BY-SA 3.0

This is the case of the city of Desaguadero in Bolivia and Peru that share the same name, which comes from the toponymy of the Desaguadero River, which is the geographical boundary between the two countries (Fig. 22.5).

Another significant example where a river integrates-separates the cities of one side with the other is the Río Grande, with Ciudad Juarez in Mexico and El Paso in the USA (Fig. 22.6).

In the case of the triple border between Brazil, Argentina and Paraguay, there is an integrated urban region formed around the Paraná and Iguazú rivers. The paradox lies

Fig. 22.5 Desaguadero (Bolivia) and Desaguadero (Peru). *Source* Google Earth

Fig. 22.6 Ciudad Juárez (Mexico) and El Paso (USA). *Source* C. Sigi. Licensed under CC BY-NC-SA 2.0

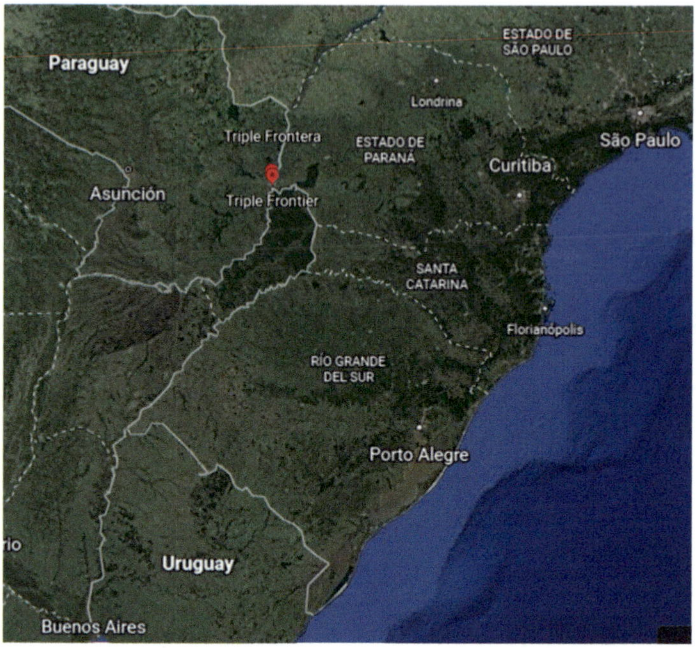

Fig. 22.7 Triple border an urban region: Foz de Iguazú (Brazil), Ciudad del Este (Argentina) and Puerto de Iguazú (Paraguay). *Source* Google Earth

in the fact that the rivers operate as national boundaries that separate, but also—given their natural condition—as meeting points, as can be seen in Fig. 22.7.

Additionally, there are those cities where spatial discontinuity is present, but which have complementary logics: these are the cases of Tacna (Peru) and Arica (Chile) with 53 km of distance[8] and those of Cucuta (Colombia) and San Antonio del Táchira (Venezuela) with 11 km, among many other cases of unitary integration, where physical distance is not an obstacle.

A second, *Urban Multinational*, composed of some emblematic cases, such as the Triple Border, where Ciudad del Este has 308,983 inhabitants (INE, 2022, Argentina), Foz de Iguazú with 272,656 (DGEEC, 2022, Brazil) and Puerto de Iguazú with 82,849 (OMA, 2022, Paraguay). These nuclei form a tri-nucleus city around the Paraná and Iguazú rivers. The urban region has 664,488 inhabitants and its key economic sectors are tourism, electricity and commerce, with markets oriented to the transboundary region and the entire world, both legally and illegally (Rabossi 2013).

On the other hand, there is the Tetra Frontera, with four distant cities but articulated through twinning agreements, which work with specific functions in an integrated

[8] However, the greatest distance is not spatial but temporal, because Santiago-Chile determines its hour according to the conditions of its metropolitan area, which leads to the existence of an hour difference in the two cities and, therefore, the public administration does not temporally match the other side.

manner. There are four countries in this border region located in the Amazon basin that includes the cities of Manaus (Brazil), Iquitos (Peru), Leticia (Colombia) and Sucumbios (Ecuador). In general, they are developed along the Putumayo River (affluent of the Amazon), which has become the material basis for integration, with the intention that Brazil joins the Pacific Basin, through the articulation of the cities of Manaus (Brazil) and Manta (Ecuador), a port on the Pacific Ocean. Another interesting case is the Trifinio between: Honduras, El Salvador and Guatemala in Central America, where in a multinuclear or multifocal way 45 border municipalities are integrated.

In all cases of urban complexes, there is a sui generis conurbation dynamic, as an historical expression of an undeniable fact: the integration process in a double sense: urban-urban and urban-regional, in a context of high national diversity.

22.3 The Multinational Transurban Region

The current border city is framed within the general patterns of urbanization in Latin America, but with its own specificities. There, the modality of a *multinational urban region* is configured, because it is inscribed in several national States that build the sense of *continuum or conurbation*, to be configured as a *city of cities*, which has several types of intra-urban borders.

The transboundary logic, in times of globalization, introduces transurban dynamics that go beyond the sense of the urban stain, to form this new pattern of urbanization, arising from the territorial confluence of the old city units, one for each State, which acted in an independent manner.

Its logic is contextualized in a unified territory, where there is spatial segregation by location of activities, population, infrastructure, in addition to that coming from interstate limits. The particularity of this urbanization pattern has to do with the type of urban structure that is configured from the *urban centrality* and its various *areas of influence*, to which are added those coming from the weight of the interstate determination (limit).

22.3.1 The Urban Structure of the Urban Border Region

Border cities develop along the interstate boundary line, which leads to two divergent positions regarding urban patterns: on the one hand, that there would exist one city and, on the other, that there would be several. The majority thesis is the one that points out the presence of urban territories, one on each side of the interstate demarcation line. This type of interpretation demands a new explanatory paradigm, given the mutations that the globalization process introduces into the nationalist-based dynamics that sustain them.

The strongest argument for denying the existence of several cities in border zones is based on the principle of *methodological nationalism* (Faist 2019; Beck 2005). The interesting thing about this conception is that it operates not only in the borders but also in the general process of urbanization in Latin America. Its thesis is based on the existence of a nuclear city inscribed in a single national State.

It is about an erroneous interpretation that gains more strength at the borders, because supposedly on each side of the interstate limit there would be a different city. This is because national sovereignty is the element that builds this plurality of cities, denying, for example, the sense of the *global city*, as sustained by Saskia Sassen (1999); that is, of the real processes.

Methodological nationalism introduces distortions in institutional frameworks, in the design of public policies,[9] in social organization and in social imaginaries. It is a principle that overemphasizes the weight of the national, behaving as the essential unit of analysis and the organizing principle of institutional, social and territorial[10] structures. Moreover, it is the one that builds identity, belonging and representation systems within its territorial jurisdiction. All of this is related to the sense of cities.

However, this vision tends to be eroded historically by the process of globalization, as transnationalization liquefies the weight of the national.[11] This is the basis for the thesis that in transboundary regions there are not several cities, but only one, under the modality of a transboundary urban complex or *urban region*, deployed in a supranational territory, which must be thought and practiced as such.

The urban region behaves as a matrix of localization of urban activities, under the logic of *complementary asymmetry* (Carrión 2022), originated in national policies such as protectionism, monetary parity, production costs and employment offer, among others. The paradox: these asymmetries tend to become complementary thanks to smuggling and trafficking; because the product on the other side is not available here, due to the fact that the price of the good is lower or that it is of better quality than here; with this, the illegal markets complement them, when it should be the opposite. Its result is the construction of a strategic space that becomes a HUB, as a place of entry and exit of information, services, capital, products and people that come and go from the glocal.

In this case, the different parts that constitute the *urban region* conform an *urban structure* composed of multi-scalar borders (multi-scalar space), of different origin, which come from the social and economic division of space, expressed in the varied nature, where the demarcation line of the adjoining States is one of them.

The location of urban activities in the space forms what is known as zoning, which is nothing more than the land uses of the main activities of each city; in this

[9] In 2008, a round table discussion was held on urban planning in the cities of Tulcán (Ecuador) and Ipiales (Colombia). The interesting and paradoxical thing: the planning secretaries of the two cities had never met and did not even know each other. Strange, because the determination of the development of one city depends on the other.

[10] Even as a basis for multilateral cooperation, which is of a national nature.

[11] The nation state ceases to be the starting point for the organization of society, because the principle of identity between nation state and society, considered as a unit of analysis closed in itself, is broken (Smith 1979).

case coming from the complementary asymmetry. Thus, for example, on the US-Mexico border, there is a set of armories all along the border, because in the USA the commercialization is legal while in Mexico it is prohibited. In contrast, on the Mexican side, there are pharmacies on a recurrent basis because in the USA the purchase of a medicine has to be made with a prescription, which makes the product more expensive. On the Brazil/Uruguay border, Uruguay has legalized the consumption of marijuana and abortion while Brazil does not, which generates population flows from Brazil for these purposes.

On the other hand, the location of the population in the cities follows the same conditions of residential segregation, to which are added those of nationality. In this way, to the paradigm of intra-urban segregation, the multinational perspective is added, shaping the sense of foreignness, which also exists in Latin American cities,[12] but with more evidence in this case, due to nationality and estrangement.

Additionally, infrastructures, considered the material basis of cities, are in tune with the demands of the transboundary urban region.[13] In this case depends on the meaning and content of the different levels of government (multilevel): at the multinational level, mobility and accessibility (bridges, roads, transportation); at the national level, the state apparatuses with their policies (customs, migration, sovereignty, Armed Forces); and at the local level, with the weight of the furniture sector and services (commerce, banks).

22.3.2 Urban Centrality: Nodal Point of Integration and Projection

Walter Christaller (1966), in his book on central places, argues that the centrality of a place comes from the hierarchical supply of services and from the attraction of the population, located in its areas of influence or in its market zone (demand), under the costs of distance and transportation. Although Christaller was referring to the city as a whole, there has always been an interpolation to intra-urban central places.

Beyond the criticisms of this definition, it is important to highlight the weight of the relationships that define it. Thus, urban centrality cannot be understood within itself or only under certain attributes, but from the relationships that configure it, in this case, from the offer (centrality) and toward the demand (peripheries). In sum, it is a construction of a relationship of plural content, which comes from the concentration of powers and capitals in the territory that produce a gravitational attraction.

---

[12] http://www.arquitecturatropical.org/EDITORIAL/documents/CARRION%20ESPACIO%20PUBLICO.pdf.

[13] Thus, for example, if it is a tourist city, it must have hotel infrastructures, restaurants, cultural centers and beaches. If it is a capital city, it must be dominated by those linked to the condition of headquarters of the state apparatus. If it is an industrial city, it must have a density of general production conditions. And so on.

Urban centrality is the expression of urban segregation, that is, of inequality expressed territorially, in the following sense: it is the hierarchical and polarized concentration of central functions in a particular part of the city (center).

In this perspective, it is important to distinguish between the *center*, as a specific space or central place (site), with that of *urban centrality*, which has the property of attraction born of the unequal concentration of functions in the territory. Thus, urban centrality generates an attraction of users and consumers to the center, since it is the place from which this attraction is exerted (gravitational mass).[14] This supposes an articulation of the center (nodes) with its edges (peripheries) through an accessibility (mobility) that ends up shaping all the roads that lead to the center. A key element that constructs and processes inequality in the city space is the price of land. It is the one that segregates and expels activities and people.

In order to understand urban centralities in border cities, it must be recognized that the limit that separates the states behaves as a gravitational center of attraction of the parties, which is possible because its central function is the multinational complementarity of the diverse.

But it is also essential to understand centrality under a dynamic that has a historical trajectory, in which at least it can manifest itself in the following three sequential moments:

- Initially, the cities created their *foundational centrality* when their border condition was not yet foreshadowed as a determining factor, since this interstate functionality did not exist. Their areas of influence were the intra-urban borders and the surrounding rural region.
- Subsequently, with the formation of national States and the delimitation of their territorial boundaries, this centrality added the components of *business centrality* (financial and commercial capital). In this context, a second type of centrality was born, that of a longitudinal nature, located along the border cordon.
- Nowadays, with the globalization process, a third moment is being experienced, in which the centralities incorporate the function of the *upper tertiary*, adopting the sense of *longitudinal centralities of global articulation* (Carrión and Cepeda 2021), in this case, along the boundary zone.

In the border city these three centralities (foundational, business and global) are expressed simultaneously, but under two explicit logics: the *longitudinal* centrality, originated in the border limit, acquires the functional quality of a *flow centrality*, allowing the integration of multinational, adjacent and distant territories. While the second, of *zonal* character (foundational and/or business), is structured on the basis of the privileged location of capital to valorize it and of power to exercise it, inscribed within the national State, forming a *centrality of place*. That is, inscribed in the relationship of flows and places that define urban centralities, according to Borja and Castells (1997). And the two centralities can be cataloged as *historical foundational*

[14] COVID-19 helps to understand this gravitational condition of urban centrality, since the accounting of cases of infected people according to territorial units, those with a significant expression are their areas of influence.

centralities, because they embody a double birth: that of the city cloistered in the national sphere and that of the multinational urban region.

In the two centralities, power and business are concentrated, prioritizing exchange relations promoted by multinational economic agents, both legal and illegal. And, in addition, they combine the central functions of the national and multinational order, which locally manifest transnational inequalities. There are the customs, migratory, territorial sovereignty and symbolic expressions of identity, belonging and nationality functions. These functions are so strong that they generate a level of attraction that promotes very high accessibility, with highways, roads, ports and bridges for multipurpose vehicles (trucks, ships, automobiles), as well as state-of-the-art technology.

The central functions are linked to trade (wholesale, retail), to administration (public, private), to certain services (health, education) and to some productive activities, among which is the *maquiladora industry*. This industry benefits from the complementary asymmetry, since foreign capital is attracted to the other side of the border because from there it can import inputs and export products with tariff benefits, in addition to taking advantage of low salaries and benefiting from the scarce social benefits.

This plural condition of the centralities in the border cities leads to the formation of a *system of centralities*, where one and others are articulated under the modality of a multicentric pattern. In this perspective, an important infrastructure has been developed, especially in transportation, which allows important accessibility with its peripheries and integration of the different centralities.

However, the complexity of this reality has led certain key actors to turn their backs on them, to the extreme of denying their existence.[15] Historically, societies ignore the origin of cities—that is, their history—which could be defined as *urban patricide* (Carrión 2010), which can lead to a decentralization or relative loss of centrality (Carrión 2010).

The longitudinal centrality—that of flows—is not recognized as such, because it is conceived as a space of separation and not of integration. And the centrality of places is undergoing a process of abandonment and stigmatization by the elites, due to the concentration of low standard commerce, the residential location of popular sectors and the promotion of informality in every sense. In both centralities, mobility is a key issue, which has turned these centralities into zones of high migration with global contact; as well as a plurinational form of integration of the urban region.

[15] Just as the existence of a multinational urban region was denied, because of the emphasis on the sense of the national, in this case the historical origin is denied, because capital is interested in the speed of its rotation and the elites in its comfort.

22.3.3 Areas of Influence: The Plurinational Sense

Centrality and its areas of influence are part of the urban structure and should be understood as two unequal forms of specialization of activities and people in space. Between one and the other there is a close relationship, where centrality creates a network of interaction, under a gravitational logic. This relationship has changed throughout history, giving rise to the presence of old and new centralities and peripheries, at least in the following three periods:

- The first, in the context of a small city with low mobility demands, the areas of influence of its centralities are located, on the one hand, within the city itself, in what could be defined as its urban edges or peripheries (suburbs). And, on the other hand, the surrounding rural structure or hinterland (countryside-city).
- The second, when national States are formed, the area of influence of centrality expands and changes, extending to the other side of the border, to build an inter-urban logic based on a multicentric condition. In other words, to the intra-urban peripheries and those inscribed in the rural sphere, this inter-border periphery is added, with which the logic of expansion and the urban stain are questioned.
- And third, when the multi-scale and multilevel transboundary logic penetrates, the centralities and their peripheries are configured within the global/local and flows/places referents (Borja and Castells 1977). This makes the peripheries, in a way, ex-central, because they acquire a relative autonomy from the centralities. The dormitory city is one of its evident expressions, because it is defined by a meaninglessness: there are no monofunctional cities. In this perspective, a centrality located on the other side can become the periphery of the other and vice versa.

Centrality for labor, services, administrative activities or commercial reasons can be developed within one or several states, depending on their gravitational mass. In general, the attraction of the centrality of the other side has to do with issues related to complementary asymmetries, while the one that develops in the national space is related to daily and less specialized issues. In other words, the inhabitant chooses the centrality to which to go, while its location has a more nationalistic sense and more typical of residential segregation.

What does happen, following Wacquant (2007), is that territorial stigmas are built, where one of the key elements is xenophobia, which leads to the dilution of the sense of citizenship within cities and strengthens the sense of foreignness.

22.4 Conclusions

Nowadays, the borders are part of a global border system, where their cities are the core poles of multinational articulation, because they have been able to generate a solid economy (legal and illegal), an important urbanization process, a strengthening of local powers and close transboundary links.

For this reason and in this context, mirror cities, pairs or twins, as cities deduced from methodological nationalism, fall into disuse to give way to the creation of *multinational urban regions*. This new territorial order questions the institutional frameworks and traditional government policies, in order to begin to rethink them from an integrated local-border-global perspective.

The urban region becomes the center of gravity of the global border system, because it acts as the pole or pivot of the equation, becoming the articulating element of the economy and life on the border, as well as the fundamental link that articulates national economies with international ones. In this perspective, the transurban is the definitive element, thanks to the fact that the centralities and their areas of influence acquire a multinational condition, behaving, one and the other, alternately.

In this way, the linkages between cities occur over and above the relations between nations, sustained by new migratory flows, communication technologies and legal and illegal markets. Thus, cities located in distant territories are integrated by economy, culture, society and technology.

These structural mutations result in the unprecedented protagonism achieved by the border regions with respect to the national States, thanks to the greater political autonomy deduced from decentralization, economic growth, population increase and the new form of national integration, originating from the pluri-state urban regions.

In terms of the innovation of the governments of these territories, some iconic and instructive examples should be highlighted. There are the cases of Ciudad Juarez (Mexico) and El Paso (USA) that think of a dual management metropolitan area. There are also the agreements for the formation of commonwealths between Tulcán (Ecuador) and Ipiales (Colombia) for solid waste management.[16] Something more ambitious are the cases of the Trifinio (Honduras, El Salvador, Guatemala), in which international cooperation, national governments and 45 municipalities of the three countries are involved. Likewise, on the border between Peru and Ecuador, there is a group of municipalities working in the same direction, especially after the signing of the Peace Accords in 1998, thanks to international and national cooperation.

Internally, these urban regions have borders deduced from the unequal location of activities, inhabitants and services, as well as from the demarcation line of the states (multinational). This is how intra-urban borders have historically been constructed in border cities, a conceptual statement that makes reference to the title of this article.

[16] Something complex because the United Nations prohibits the final deposit of solid waste produced in a different State. In this case, it has been surpassed by the proposal to capture a larger scale of garbage for recycling.

Something that should be emphasized: the centralities and their areas of influence have a unique dynamic, which makes them behave nationally interchangeable, because in some cases they can behave as an area of influence and in others as a center, which makes the borders of border cities to be itinerant, therefore, as well as their intra-urban limits.

Something that should also be emphasized in a particular way: the centralities and their areas of influence have a unique dynamic, which makes them behave in a nationally interchangeable way, because at some moments they can behave as an area of influence and at others as a centrality, which makes the centralities of border cities itinerant, just like their intra-urban limits.

However, these links do not develop without conflicts, because they represent new relationships of power, even more if the borders begin to have a greater political weight. But also of violence and illegalities (smuggling, trafficking), because of the characteristics of the asymmetry they provoke, which leads these regions to have homicide rates significantly higher than national averages (Carrión and Gotsbacheer 2021).

These negative processes increased with COVID-19 and the policies promoted to counteract it. Borders were closed which increased inequalities due to the strong emphasis placed on national government. As a result, local functions were redefined to the benefit of national power or, in other words, generation shifted to the center, which ended up by increasing illegality and violence, as well as harming the health of border inhabitants, especially the most vulnerable (pre-existing diseases).

Fig. 22.8 The Wall U.S. and Mexico. *Source* Jonathan Mcintosh. Licensed under CC BY-NC-SA 2.0

In structural terms, COVID-19 introduced a process of re-frontierization, at a time when there had been much progress toward inter-border integration. But it also accelerated, with the return of criticism of globalization and the repositioning of nationalist visions, which once again positioned the concept of the border as a protective shield against external threats (Lara-Valencia and García-Pérez 2021).

And the paradox: with it, it seeks to go against the historical processes of integration, building not only ideological walls, but also physical ones, both within cities and between neighboring countries (Fig. 22.8). Undoubtedly, methodological nationalism is a real danger and a bet against history.

Bibliography

Augé M (1998) *Los no lugares, espacios del anonimato*. Ed. GEDISA, Barcelona

Beck U (2005) La mirada cosmopolita o la guerra es la paz. Paidós, Barcelona

Bello D (2013) La Triple Frontera como polo de atracción de actividades ilícitas: Condiciones endógenas institucionales e ilegalidad. En: Atenea No 508, Concepción

Boisier S (1987) Notas en torno al desarrollo de regiones fronterizas en América Latina. Estudios Internacionales (78). Revista del Instituto de Estudios Internacionales de la Universidad de Chile

Borja J (1994) Notas sobre ciudades, gobiernos locales y movimientos populares. Revista EURE-Revista de Estudios Urbano Regionales 20(59)

Borja J, Castells M (1997) Local y global: la gestión de las ciudades en la era de la información. Ed. Taurus. Barcelona

Caldeira T (2008) Ciudad de muros. Editorial Gedisa, Barcelona

Canales A, Martínez J, Reboiras L, Rivera F (2010) Migración y salud en zonas fronterizas: informe comparativo sobre cinco fronteras seleccionadas. CEPAL—Serie Población y desarrollo, (95): Santiago de Chile, Naciones Unidas

Carrión F (2010) Ciudad, memoria y proyecto. Ed. IMQ-FLACSO. Quito

Carrión F (2013) Asimetrías en la frontera Ecuador-Colombia: entre la complementariedad y el sistema. Ed. FLACSO-IDRC, Quito

Carrión F (2017a) Derecho a la ciudad y gobierno multinivel en América Latina. Monografías CIDOB, Barcelona

Carrión F (2017b) Nuevos Patrones de urbanización en América Latina, En: Borja, Carrión y Corti (Editores). Ciudades resistentes, ciudades posibles, Barcelona

Carrión F, Gotsbacher M (2021) Border violence in Lain American: an expression of complementary asymmetries. In: The Oxford handbook of the sociology of Latin America. New York

Carrión F, Cepeda Pico P (2021) Corredores urbanos: centralidades longitudinales de articulación global. Revesta INVI 36(102):183–207

Carrión F, Zepeda B, Enriquez F (2022) Latin America's Global Border Systema, an introduction. Ed Routledge, New York

Castells M (1999) La era de la información. Ed. Siglo XXI, Barcelona

Chaves M (s/f) Obelisco—Plaza Internacional—Frontera de la Paz—Livramento—Rivera. https://wordpress.org/openverse/image/b700c946-ab0b-40de-a190-8d71234c3339

Christaller W (1966) Central places in Southern Germany. Prentice Hall, Englewood Cliffs, NJ

Faist T (2019) The transnationalized social question: migration and the politics of social inequalities in the 21st century. Oxford University Press, Oxford

Fitzgerald (1993) ECLAC and the formation of Latin American economic doctrine, Los Angeles

Fuente C, Peña S (2018) Las fronteras de México: nodos del sistema global de las drogas prohibidas: Ciudad Juárez. FLACSO Ecuador, IDRC-CRDI, El Colef

Galería fotográfica PCM, P (s/f) Retorno De Migrantes a Sus Regiones. https://wordpress.org/openverse/image/15952e68-1a3b-4666-8025-2675a34f92c3

Garcia Canclini N (1997) Culturas Híbridas. Ed. Eudeba, Buenos Aires

Google Earth (2022a) Desaguadero. https://earth.google.com/web/search/Desaguadero/@-16.5641376,-69.0410471,3830.09571799a,5427.04322143d,35y,0h,45t,0r/data=CnYaTBJGCiUweDkxNWMxODlkYmRlMDk4ZmI6MHhkMzgxNjY4MjE1YTE2NTRlGYYOcPcokDDAIQH76NSVQlHAKgtEZXNhZ3VhZGVybxgCIAEiJgokCaQHevEW7jxAEdRkv0YZwSHAGbRYw0J6t0jAIXIXIYuPJ2TAKAI

Google Earth (2022b) Fronteras México y Estados Unidos. https://earth.google.com/web/search/Fronteras+M%c3%a9xico+y+Estados+Unidos/@33.45315694,-105.34315198,2933.57391318a,4136812.48179138d,35y,0h,0t,0r/data=CigiJgokCY2Q9TIq-UFAETepzgJb5TNAGYyDsXt1qVXAIYC_uBtXdWDA

Google Earth (2022c) Triple Frontera. https://earth.google.com/web/search/Triple+Frontera/@15.82562399,-105.3348455,106.3410378a,23040320.95009824d,35y,0h,0t,0r/data=CigiJgokCbos2ph5R0BAEeN2BaD3AhHAGTyHNvwiBEfAIdFLxFxllGTA

Lattes A (2001) Población urbana y urbanización en América Latina. En: Carrión F. La ciudad construida, urbanismo en América Latina. Ed. FLACSO-JA, Quito

Lefebvre H (2013) La producción del espacio la production. Ed. Capitan Swing, Madrid

Lara-Valencia F, García-Pérez H (2021) Las fronteras de la pandemia: lecciones para la gobernanza y la cooperación en las ciudades de la frontera México-Estados Unidos. En Revista: Estudios Fronterizos, Mexicali

Martins H (1974) Time and theory in sociology. en: Rex J (ed) Approaches to sociology. An introduction to major trends in British sociology. Routledge & Kegan, London, pp 246–294

Mcintosh J (s/f) Double Wall Near Tijuana. https://wordpress.org/openverse/image/d618feac-2c36-4853-960c-fa936c53f06d

Picech M (2017) Ciudades de fronteras: pivotes locales de procesos globales. En: Revista Cuestiones Urbanas, Ed. IC-IMQ, Quito

Rabossi F (2013) "Dinámicas económicas de la Triple Frontera (Brasil, Paraguay y Argentina)" Seguridad, planificación y desarrollo en las regiones transfronterizas, Fernando Carrión (Comp.). Ed. FLACSO Ecuador, IDRC-CRDI. Quito

Reboratti C (1992) Fronteras agrarias y población en América Latina. In: International Union for the Scientific Study of Population, El poblamiento de las Américas, Actas, Lieja, vol 4

Robertson R (1992) Globalization: social theory and global culture. Sage, Londres

Robertson R (2003) Glocalización: tiempo-espacio y homogeneidad heterogeneidad. Cansancio del Leviatán: problemas políticos de la mundialización. Madrid

Rodríguez O (1980) Teorías del subdesarrollo de la CEPAL. Ed. Siglo XXI. México, D.F

Rodríguez J (2002) Vignoli, Distribución territorial de la población de América Latina y el Caribe: tendencias, interpretaciones y desafíos para las políticas públicas: Santiago de Chile. CELADE División de Población, Naciones Unidas

Sabatini F (2006) La segregación social del espacio en las ciudades de América Latina. Ed. BID. Washington, DC

Sassen S (1999) La ciudad global. EUDEBA, Buenos Aires

Sigi C (s/f) Grenze zwischen USA und Mexiko (El Paso/Ciudad Juarez). https://wordpress.org/openverse/image/82b81622-2c67-48bb-88b0-4c872ca9a02f/

Soja E (2008) Postmetrópolis, estudios críticos sobre las ciudades y las regions. Ed. Mapas, Madrid

Vera da Silva T (2015) Los ilegalismos y la ciudad de São Paulo. La globalización desde abajo. La otra economía mundial, Carlos Alba Vega, Gustavo Lins Ribeiro y Gordon Mathews (coord.), Fondo de Cultura Económica, Ciudad de México

Wacquant L (2007) Los condenados de la ciudad: Gueto, periferias y Estado, siglo XXI, Buenos Aires

Zolezzi M (2020) La ciudad. La Covid-19 y "el desborde inverso". En: Revista de Sociología No 30. Lima

GPSR Compliance

The European Union's (EU) General Product Safety Regulation (GPSR) is a set of rules that requires consumer products to be safe and our obligations to ensure this.

If you have any concerns about our products, you can contact us on ProductSafety@springernature.com

In case Publisher is established outside the EU, the EU authorized representative is:

Springer Nature Customer Service Center GmbH
Europaplatz 3
69115 Heidelberg, Germany

Batch number: 09151336

Printed by Printforce, the Netherlands